TOPICS IN MODERN MATHEMATICS
FOR TEACHERS

TOPICS IN MODERN MATHEMATICS
FOR TEACHERS

Anthony L. Peressini
Donald R. Sherbert

University of Illinois

HOLT, RINEHART AND WINSTON, INC.
New York   Chicago   San Francisco
Atlanta   Dallas   Montreal
Toronto   London   Sydney

Copyright © 1971 by Holt, Rinehart and Winston, Inc.
All rights reserved
Library of Congress Catalog Card Number: 71-130861
SBN: 03-084912-8
Printed in the United States of America
1 2 3 4   22   9 8 7 6 5 4 3 2 1

To Joan and Carolyn

PREFACE

This book has evolved over the past several years as a result of the authors' experience in teaching mathematics courses primarily intended for prospective or practicing secondary school mathematics teachers. This is not a teaching methods text but rather a mathematics book dealing with a variety of topics that seem particularly relevant to the needs of the secondary school teacher. Although most colleges offer separate courses on some of the topics considered in this book, the crowded nature of the teacher training program often makes it difficult for students to take these courses as electives, desirable as they may be. As a result, many colleges now offer a course in their teacher training programs dealing with several special topics in mathematics to ensure greater breadth in the subject matter preparation of their students. We have used the material in this book in such topics courses at the University of Illinois.

The various chapters of this book are largely independent of one another so that the instructor may select the topics he wishes to cover in accordance with the background and needs of his students. Each chapter concludes with a section entitled "Remarks and References" in which we relate the subject matter of the chapter to the content of the secondary school program and provide references for further reading. Our selection of topics was motivated by the following objectives: (a) to enrich the reader's background in several areas of mathematics that are closely related to the secondary school mathematics program, but that are not ordinarily discussed in current secondary school texts; (b) to deepen the reader's understanding of some of the more difficult topics that he will be teaching in a modern secondary school mathematics program. Thus, we have included such topics as number theory, graph theory, the geometry of complex numbers, and the application of Boolean algebras to computer design with objective (a) in mind, while the material on set theory, mathematical induction, number systems, the binomial coefficients, and probability theory is directed primarily toward objective (b). Needless to say, there is considerable overlap between these two objectives.

This book contains more than enough material for a two-semester course at the sophomore-junior level meeting three days a week. However, because of the independence of the various chapters, this book can be used for a variety of one-semester courses. For example, we have used Chapters 1, 2, 3, 6 and the first section of Chapter 8 for a course stressing the algebraic foundations of mathematics, while Section 2.4 of Chapter 2 and Chapters 7, 8, 9 can be used for a topics course in geometry. We have also found the material in Chapters 4, 5, 7 together with Chapter 3 or Chapter 6 to be an appealing one-semester grouping for our own students.

We wish to express our gratitude to the many students who provided helpful comments concerning the mimeographed lecture notes that formed the basis for the final form of this book. We also wish to thank Professors John Wetzel, Franz Hohn, William Frascella, and Clinton Kennel for their useful remarks on various parts of the preliminary manuscript. The preparation of the final manuscript was greatly facilitated by Carolyn Sherbert who not only served as an expert and punctual typist but also helped to identify and eliminate a number of stylistic blunders in our exposition. Finally, we are grateful to the friendly and competent staff of Holt, Rinehart and Winston, Inc., for making the publication of this book as painless as possible. Special thanks are directed to the Mathematics Editor, Mr. Jack Murphy and to our Senior Project Editor, Mrs. Marilyn Genzer.

Urbana, Illinois   Anthony L. Peressini
November 1970   Donald R. Sherbert

CONTENTS

Preface   vii

Chapter 1   SET THEORY   1

    1.1. Basic Terminology   2
    1.2. Set Algebra   4
    1.3. Order Relations and Equivalence Relations   11
    1.4. Functions   24
    1.5. Algebraic Systems   29
    Appendix A: Infinite Sets   35
    Appendix B: A Glimpse at Axiomatic Set Theory   42
    Remarks and References   49

Chapter 2   NUMBER SYSTEMS AND THEIR PROPERTIES   53

    2.1. The Natural Number System   54
    2.2. The System of Integers and Ordered Domains   65
    2.3. The Systems of Rational and Real Numbers and Ordered Fields   80
    2.4. The Complex Number System   89
    Appendix: Decimal Representation   94
    Remarks and References   106

Chapter 3   THE THEORY OF NUMBERS   112

    3.1. Prime Numbers   114
    3.2. Greatest Common Divisors and the Euclidean Algorithm   122
    3.3. Linear Diophantine Equations   130
    3.4. Congruences   137
    3.5. The Chinese Remainder Theorem   150
    3.6. Pythagorean Triples and Fermat's Last Theorem   156
    Remarks and References   161

Chapter 4   BINOMIAL COEFFICIENTS AND COUNTING TECHNIQUES   163

    4.1. The Binomial Coefficients   164
    4.2. Binomial Coefficients and Counting Problems   172
    4.3. Multinomial Coefficients   178
    4.4. The Principle of Inclusion-Exclusion   182
    Remarks and References   187

Chapter 5   PROBABILITY THEORY   189

    5.1. Sample Spaces   192
    5.2. Probability Spaces   199
    5.3. Some Sampling Problems   212
    5.4. Conditional Probability; Bayes' Theorem   219
    5.5. Independent Events; Repeated Trials   227
    Remarks and References   236

Chapter 6   BOOLEAN ALGEBRAS AND THEIR APPLICATIONS   238

    6.1. Basic Properties of Boolean Algebras   239
    6.2. Finite Boolean Algebras   250
    6.3. Switching Networks   255
    6.4. The Design of Some Simple Computer Networks   263
    Appendix: A Simple Human Computer—A Classroom Demonstration   269
    Remarks and References   273

Chapter 7   GRAPH THEORY   275

    7.1. The Königsberg Bridge Problem—A Prelude to Graph Theory   276
    7.2. Basic Definitions; Euler Paths   281
    7.3. Planar Graphs and Euler's Formula   293
    7.4. The Four Color Problem and the Five Color Theorem   305
    Appendix: The Case of the Colored Cubes   316
    Remarks and References   321

Chapter 8   THE GEOMETRY OF COMPLEX NUMBERS   323

    8.1. The Complex Plane   324
    8.2. The Riemann Sphere   336
    8.3. Bilinear Transformations   344
    8.4. Applications to Euclidean and non-Euclidean Geometry   363
    Remarks and References   376

Chapter 9	GEOMETRIC CONSTRUCTIONS	380

    9.1. Euclidean Constructions and Constructible Numbers   384
    9.2. The Impossibility of the Classical Greek Constructions   401
    9.3. Euclidean Construction of Regular Polygons   409
   Appendix: Angle Trisection by Other Means   416
   Remarks and References   423

Index   427

Chapter 1

SET THEORY

One of the fundamental features of the so-called "new mathematics" programs in the secondary schools is the introduction and systematic use of the language of set theory. This language provides a suitable basis for a careful discussion of many mathematical concepts such as functions, variables, equations, inequalities, and so on, which were often treated in a rather casual and imprecise manner in traditional text materials. This increased precision of language was necessitated by the new emphasis in school mathematics on the conceptual and deductive aspects of the subject as opposed to the more traditional stress on computation and the classification of problems into standard groups for solution.

Set theory can be approached on several different levels. On the most rudimentary level, the basic concepts and operations of set theory are defined in an intuitive, informal fashion. No attempt is made to isolate a basic set of axioms for set theory and the meanings of such logical connectives as "and," "or," "not," and "implies" are taken for granted. Despite this rather informal basis, the resulting development of set theory is quite adequate for most purposes.

Set theory can also be approached from the formal, axiomatic point of view. Viewed at this level, set theory is a deep and important area of continuing mathematical research. Work in this area was stimulated by certain paradoxes that occurred in informal set theory; this led to the development of various axiom systems for set theory. Some of these aspects of the subject are discussed briefly in Appendix B to this chapter.

In this chapter, we shall develop some informal set theory including the algebra of sets, the notions of order relation, equivalence relation and function, and the idea of an algebraic system. In addition, an appendix on infinite sets is included. Those readers with some familiarity with set theory may choose to use this chapter primarily as a source for reference and review. It serves to highlight the more important aspects of the language of sets and to establish the basic definitions and notation.

1.1. BASIC TERMINOLOGY

The concepts of "set," "element," and the relation of "set membership" are regarded as undefined terms in set theory just as the concepts of "point," "line," and the relation of "point on a line" are undefined in plane geometry. Informally speaking, we can regard a set as any collection of objects. Concrete examples of sets occur in abundance in both mathematical and nonmathematical contexts. For example, the 1970 graduating class of the University of Illinois is a set whose objects are people, the current Library of Congress collection is a set whose objects are books, films, documents, and so on, and the natural number system is a set whose objects are numbers.

We shall refer to the objects that may or may not belong to the sets in question as *elements*. If the element a is one of the objects in the set A, we shall say that "a is a member of A" and write $a \in A$. Thus, the symbol $\in$ is used to indicate set membership. On the other hand, if the element a is not one of the objects in a given set A, we shall write $a \notin A$.

Certain sets can be specified by simply listing in braces the elements that are members of the set. For example, if A is the set consisting of the numbers 1, 7, 13, we can specify A by

$$A = \{1,7,13\}.$$

The numbers 1, 7, 13 are the members of A, that is, $1 \in A$, $7 \in A$, $13 \in A$. The fact that 8 is not a member of A is written $8 \notin A$. Observe that $\{1,7,13\} = \{13,1,7\} = \{1,7,7,13\}$; that is, the rearrangement or repetition of members of a set does not change the set.

Sometimes a set cannot be adequately specified by listing all of its elements as above. For example, the set of even numbers cannot be listed in this way since there is no last even number. In order to specify such sets adequately, it is necessary to introduce some definitions that will turn out to be also useful in other connections.

A *constant* is a name of a specific object. In general, a given object may have many names; that is, many constants may be used to denote one object. For example, 7, $6 + 1$, $15 - 8$, $\sqrt{49}$ are all names of the same object, namely, the number seven. If a and b are constants, the symbol $a = b$ means that a and b are names for the same object. Thus $3 + 1 = 4$ since $3 + 1$ and 4 are both names of the number four.

A *variable with domain D* is a symbol that can be replaced by any constant that names an element in the set D, and these constants are referred

1.1. BASIC TERMINOLOGY

to as the *values* of the variable. For example, if $D = \{1,2,3,4\}$ and if x is a variable with domain D, then x can be replaced by the constants 1, 2, 3, 4, $3 + 1$, $\sqrt{4}$, $-1 + 2$, and so on. Sometimes, we shall speak of a variable without reference to its domain D when D is either clearly understood in the context or the specific nature of D is unimportant.

A *statement* is an affirmative declarative sentence for which precisely one of the following alternatives holds: (a) the sentence is true, (b) the sentence is false. Thus, for example, the sentences "2 is a factor of 4," "$3 + 1 = 4$," "The sum of the interior angles of a plane triangle is 126 degrees," "3 is an even number or 7 is the square root of 49" are all statements; the first, second, and fourth statements are true while the third is false. On the other hand, the sentences "What is the solution of the equation?," "The function is probably a polynomial function," "This sentence is false" are not statements since the first is not a declarative sentence, the second is declarative but not affirmative, and the third sentence can be neither true nor false.

If a sentence contains a variable x with domain D and if this sentence becomes a statement (as defined above) when x is replaced by any constant naming an element of D, then the sentence is called an *open statement (with domain D)* or a *statement $S(x)$ about x*. If x is a variable with domain D and if $S(x)$ denotes a statement about x, then the set of all values of x for which the statement is true is called the *truth set* of $S(x)$ and is denoted by

$$\{x \in D: S(x)\}.$$

This notation is to be read, "the set of all x in D such that $S(x)$ is true."

With this terminology the familiar task of solving an equation can be formulated as follows. An equation in a single variable such as $x^2 + 1 = 2$ is regarded as an open statement where x is a variable whose domain is the set **R** of real numbers. When x is replaced by a constant specifying a certain real number, the resulting statement is either true or false. A solution of the equation is simply a value of x for which the resulting statement is true. Thus, we see that a value of x is a solution of the equation if and only if it is a member of the truth set of the equation. The term "solution set" is ordinarily used instead of "truth set" in the context of solving equations. Thus, for example, the solution set of $x^2 + 1 = 2$ is the set $\{1, -1\}$.

The notion of truth set can often be used to specify sets conveniently. For example, if x is a variable whose domain is the set **R** of real numbers and if $S(x)$ is the open statement "x is an even number," then

$$\{x \in \mathbf{R}: x \text{ is an even number}\}$$

specifies the set of all even numbers as the truth set of $S(x)$. Similarly, $\{x \in \mathbf{R}: x^2 + 1 = 2\}$ specifies the set $\{1, -1\}$.

4   SET THEORY

Sometimes, when there seems to be no danger of confusion concerning the intended meaning, sets will be specified by a partial list of their elements with the understanding that the remaining elements follow the same pattern as the listed elements. For example, the set **N** of natural numbers (that is, positive whole numbers) may be written as

$$\mathbf{N} = \{1,2,3,4, \cdots\}$$

while the set E of even numbers may be denoted by

$$E = \{2,4,6,8, \cdots\}.$$

Such set descriptions are simply informal presentations of sets in terms of truth sets of open statements about a variable; the appropriate open statements can be inferred from the partial lists of elements.

EXERCISES

1.1. Denote the following sets by listing their elements:
  (a) The first five positive integer multiples of 3.
  (b) The odd numbers between 8 and 16.
  (c) The solutions of the equation $x^2 - x + 6 = 0$.
1.2. Which of the following sentences are statements? open statements?
  (a) Where are you?
  (b) The number seven is a vegetable.
  (c) Golfing is a sport.
  (d) That is a game.
  (e) Eight is a multiple of two and r is a real number.
1.3. For the following open statements with domain **R**, the set of real numbers, determine the values of x for which the statements are true:
  (a) x is an integer whose square is less than 18.
  (b) $x^2 - x - 3 = 0$.
  (c) $1 < 2x - 3 < 5$.
1.4. Specify the following sets of natural numbers as truth sets of open statements:
  (a) $\{3,6,9,12, \cdots\}$.
  (b) $\{2,4,8,16, \cdots\}$.
  (c) $\{5,8,11,14, \cdots\}$.

1.2.   SET ALGEBRA

A set A is a *subset* of a set B if each member of A is a member of B. Thus, for example, the set **N** of natural numbers is a subset of the set **R** of real numbers, and the set T of all plane triangles is a subset of the set P of all plane polygons. If A is a subset of B, we write $A \subset B$; these symbols are also read as "A is contained (or included) in B." If A and B are sets, then $A = B$ if and only if $A \subset B$ and $B \subset A$. Thus, *in order to prove that two sets A and B are equal (that is, identical), it suffices to prove that each member of A is a member of B and each member of B is a member of A.*

Suppose that X is a set. The collection $P(X)$ of all subsets of X is a set called the *power set* of X. Note that the statement "A is a subset of X" is equivalent to the statement "$A \in P(X)$." Among all the subsets of X there is a largest, namely X itself, and a smallest, namely the set $\emptyset$ with no elements. $\emptyset$ is called the *empty set*. Note that $\emptyset$ can be specified by any open statement with domain X that is false for each element of X. For example,

$$\emptyset = \{x \in X : x \neq x\}.$$

Any element A of $P(X)$ satisfies $\emptyset \subset A \subset X$.

We shall now define some important operations on the power set $P(X)$ of a given set X. If $A \in P(X)$, $B \in P(X)$, we define the sets $A \cup B \in P(X)$, $A \cap B \in P(X)$, and $A' \in P(X)$ as follows:

$$A \cup B = \{x \in X : x \in A \text{ or } x \in B\},$$
$$A \cap B = \{x \in X : x \in A \text{ and } x \in B\},$$
$$A' = \{x \in X : x \notin A\}.$$

Thus, $A \cup B$ is the set of those members of X that are members of A or members of B or both, $A \cap B$ consists of those members of X that are members of both A and B, and A' is the set of members of X that are not members of A. $A \cup B$ is called the *union of A and B*, $A \cap B$ is called the *intersection of A and B*, and A' is called the *complement of A*. For example, if $X = \{1,2,4,7,8,10\}$, if $A = \{4,7,8\}$, and if $B = \{1,7,8,10\}$, then the union of A and B is $A \cup B = \{1,4,7,8,10\}$, the intersection of A and B is $A \cap B = \{7,8\}$ and the complement of A is $A' = \{1,2,10\}$.

Figure 1.1 gives a pictorial interpretation of union, intersection, and complement. In each case, the set X is represented by the set of points inside the large rectangle, the sets A and B are represented by the circular regions within the rectangle, and the shaded areas represent the sets indicated below each diagram. Diagrams of this sort are called *Venn diagrams*.

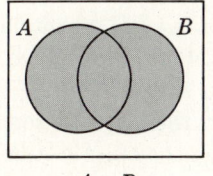
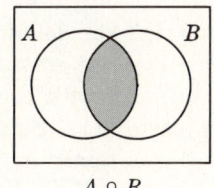
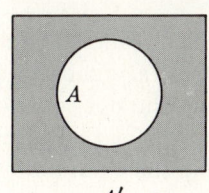

$A \cup B$ $A \cap B$ A'

Figure 1.1

6 SET THEORY

These diagrams are often helpful guides for the discovery of relationships between subsets of X. Of course, it is not essential to represent X as a rectangular region and subsets of X as circular regions. The main requirement is that subsets A, B, C, and so on of X should be put in "general position" within X; that is, the regions chosen to represent the sets in question should not imply any inclusion relations that are not assumed about the sets themselves. For example, if A and B are arbitrary members of $P(X)$, the regions representing A and B should be drawn so that they overlap and so that neither region is included in the other.

Example 1.1. Is the following equation valid for all A and B in $P(X)$?

$$A \cup B = (A \cap B') \cup (B \cap A').$$

To answer this question, we first construct Venn diagrams representing both sides of this equation. Not only do the Venn diagrams in Figure 1.2

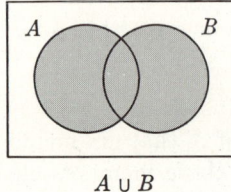
$A \cup B$

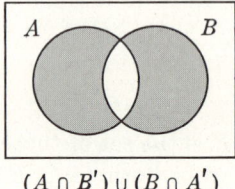
$(A \cap B') \cup (B \cap A')$

Figure 1.2

show that the above equation does not hold for arbitrary A and B in $P(X)$ but also these diagrams suggest that the following statement might be true: For all A and B in $P(X)$, the following equation holds

$$A \cup B = (A \cap B') \cup (B \cap A') \cup (A \cap B).$$

This statement is indeed true and the reader will be asked to supply the proof for himself after he has gained the experience of proving some of the basic set equations given in Proposition 1.1 below. It should be realized that while Venn diagrams are suggestive, they do not constitute formal proofs of relations between sets.

The reader should observe that the definition of $A \cup B$ given above for arbitrary subsets of X involves the statement about x obtained by joining the open statement "$x \in A$" with the open statement "$x \in B$" by means of the logical connective "or." Similarly, the open statement used in the definition of $A \cap B$ is obtained by joining the open statement "$x \in A$" with the open statement "$x \in B$" by means of the logical connective "and." Finally, the open statement "$x \notin A$" used in the definition of the complement A' of A is just the negation of the statement "$x \in A$"; that is, the statement "$x \notin A$" is logically equivalent to the statement obtained by prefixing "$x \in A$" with the logical connective "not." Thus, properties of these operations on sets depend ultimately on the properties of the logical connectives "and," "or," "not." With regard to statements in mathematics, these three connectives are assigned the following precise meanings. If P and Q are given statements, then:

(a) The statement "P or Q" is false if P and Q are *both* false; the statement "P or Q" is true otherwise.

(b) The statement "P and Q" is true if P and Q are both true; the statement "P and Q" is false otherwise.

(c) The statement "not P" is true if P is false, and "not P" is false if P is true.

For example, if P is the statement "2 is a root of the equation $x^2 + x + 1 = 0$" and Q is the statement "Man has set foot on the moon," then the statement "P or Q" is true, but the statement "P and Q" is false. The statement "not P," which would be "2 is not a root of the equation $x^2 + x + 1 = 0$," is true since P is false.

The properties of these logical connectives are summarized in the table below, where T denotes "True" and F denotes "False."

P	Q	P or Q	P and Q	not P
T	T	T	T	F
T	F	T	F	F
F	T	T	F	T
F	F	F	F	T

Observe that if $P(x)$ and $Q(x)$ are open statements with domain X, and if A is the truth set of statement $P(x)$ and B is the truth set of statement $Q(x)$, then:

(a) $A \cup B$ is the truth set of the open statement "$P(x)$ or $Q(x)$."
(b) $A \cap B$ is the truth set of the open statement "$P(x)$ and $Q(x)$."
(c) A' is the truth set of the open statement "not $P(x)$."

For example, if x is a variable with domain $X = \{2,3,6,9,12,13\}$ and if A and B are the truth sets of the open statements

$P(x)$: "$x \leq 9$,"
$Q(x)$: "x is even,"

respectively, then $A = \{2,3,6,9\}$, $B = \{2,6,12\}$, and

$$A \cup B = \{x \in X : x \leq 9 \text{ or } x \text{ is even}\} = \{2,3,6,9,12\},$$
$$A \cap B = \{x \in X : x \leq 9 \text{ and } x \text{ is even}\} = \{2,6\},$$
$$A' = \{x \in X : x > 9\} = \{12,13\}.$$

We now proceed to establish some basic properties of the union, intersection, and complementation operations on $P(X)$.

Proposition 1.1. If X is a set and if A, B, C are members of $P(X)$, then:

(1) $A \cup (B \cup C) = (A \cup B) \cup C$
    $A \cap (B \cap C) = (A \cap B) \cap C$, (Associative Properties)

(2) $A \cup B = B \cup A$
    $A \cap B = B \cap A$, (Commutative Properties)

(3) $A \cup \varnothing = A$
    $A \cap X = A$,

(4) $A \cup A' = X$
    $A \cap A' = \varnothing$,

(5) $A \cup (B \cap C) = (A \cup B) \cap (A \cup C)$
    $A \cap (B \cup C) = (A \cap B) \cup (A \cap C)$. (Distributive Properties)

PROOF. Since the proofs of the two equations in each part of (1)–(5) are quite similar, we shall only prove the first equation in each part.

(1) If x is an element of $A \cup (B \cup C)$, then $x \in A$ or $x \in (B \cup C)$. If $x \in A$, then certainly $x \in A \cup B$, so $x \in (A \cup B) \cup C$. On the other hand, if $x \in (B \cup C)$, then $x \in B$ or $x \in C$. But then it certainly follows that $x \in A \cup B$ or $x \in C$, that is, $x \in (A \cup B) \cup C$. We conclude that $A \cup (B \cup C) \subset (A \cup B) \cup C$. Similar reasoning shows that if $x \in (A \cup B) \cup C$, then $x \in A \cup (B \cup C)$; that is, $(A \cup B) \cup C \subset A \cup (B \cup C)$. Since we have also shown that $A \cup (B \cup C) \subset (A \cup B) \cup C$, it follows that $A \cup (B \cup C) = (A \cup B) \cup C$.

(2) $A \cup B = \{x \in X : x \in A \text{ or } x \in B\} = \{x \in X : x \in B \text{ or } x \in A\} = B \cup A$.

(3) By definition of union, $A \subset A \cup B$ for any subset B of X; in particular, $A \subset A \cup \varnothing$. On the other hand, the statement "$x \in A$ or $x \in \varnothing$" implies that $x \in A$ since $x \notin \varnothing$ for all $x \in X$. Therefore, $A \cup \varnothing \subset A$. Since we have already observed that $A \subset A \cup \varnothing$, it follows that $A \cup \varnothing = A$.

(4) $A \cup A' = \{x \in X : x \in A \text{ or } x \in A'\} = \{x \in X : x \in A \text{ or } x \notin A\} = X$ since each element of X either is an element of A or it is not. (In other words, the open statement "$x \in A$" with domain X must be either true or else false whenever x is replaced by a constant naming an element of X.)

(5) Suppose that x is a member of $A \cup (B \cap C)$, then $x \in A$ or $x \in (B \cap C)$. If $x \in A$, then certainly $x \in A \cup B$ and $x \in A \cup C$, so it follows that $x \in (A \cup B) \cap (A \cup C)$. On the other hand, if $x \in (B \cap C)$, then $x \in B$ and $x \in C$; consequently, it is certainly true that $x \in (A \cup B)$ and that $x \in (A \cup C)$, that is, $x \in (A \cup B) \cap (A \cup C)$. Therefore, every member of $A \cup (B \cap C)$ is also a member of $(A \cup B) \cap (A \cup C)$, that is, $A \cup (B \cap C) \subset (A \cup B) \cap (A \cup C)$.

Now suppose that x is a member of $(A \cup B) \cap (A \cup C)$. Then $x \in (A \cup B)$ and $x \in (A \cup C)$, so it follows that $x \in A$ or $x \in B$, and also that $x \in A$ or $x \in C$. Therefore, if $x \notin A$, it must be true that $x \in B$ and $x \in C$; consequently, $x \in B \cap C$, so that $x \in A \cup (B \cap C)$. On the other hand, if $x \in A$, then certainly $x \in A \cup (B \cap C)$. It follows that $(A \cup B) \cap (A \cup C) \subset A \cup (B \cap C)$. Therefore, since we have already shown that $A \cup (B \cap C) \subset (A \cup B) \cap (A \cup C)$, the first equality in (5) must hold.

EXERCISES

1.5. List all the members of $P(X)$ if $X = \{a,b,c\}$.
1.6. Let $A = \{1,2,3,4,5,6,7,8\}$, $B = \{1,3,5,7,9\}$, $C = \{4,5,6\}$, and $D = \{2,5,8\}$ be subsets of $X = \{1,2,3,4,5,6,7,8,9,10\}$.
   (a) List the elements in the sets $A \cap (B \cup C')$ and $(D' \cap B) \cup C$.
   (b) Express each of the sets $\{5\}$, $\{4,6,10\}$, $\{2,8\}$ in terms of A, B, C, D.
1.7. Demonstrate the plausibility of the equations in (5) of Proposition 1.1 by constructing the appropriate Venn diagrams.
1.8. Prove the second equality in each of the parts (1)–(5) of Proposition 1.1.
1.9. Prove that if $A \in P(X)$ and $B \in P(X)$, then $A \subset B$ if and only if $A \cap B' = \emptyset$.
1.10. Prove that for any members A and B of $P(X)$, $A \cup B = (A \cap B') \cup (B \cap A') \cup (A \cap B)$.
1.11. Two members of $P(X)$ are called *disjoint* if their intersection is the empty set. Show that for any two members A, B of $P(X)$, the sets $A \cap B'$ and $A' \cap B$ are disjoint.
1.12. The definitions of union and intersection given for pairs of sets can be extended to any collection of members of $P(X)$. Let I be a nonempty index set. If, for each $\alpha \in I$, A_α is a member of $P(X)$, define

$$\bigcup_{\alpha \in I} A_\alpha = \{x \in X : x \in A_\alpha \text{ for some } \alpha \in I\}$$

and

$$\bigcap_{\alpha \in I} A_\alpha = \{x \in X : x \in A_\alpha \text{ for all } \alpha \in I\}.$$

Show that
(a) $\left(\bigcup_{\alpha \in I} A_\alpha\right)' = \bigcap_{\alpha \in I} A'_\alpha$.
(b) For any $B \in P(X)$, $B \cap \left(\bigcup_{\alpha \in I} A_\alpha\right) = \bigcup_{\alpha \in I} (B \cap A_\alpha)$.

1.13. If **R** is the set of real numbers and if **N** is the set of natural numbers, define A_n and B_n in $P(\mathbf{R})$ for each $n \in \mathbf{N}$ by

$$A_n = \left\{x \in \mathbf{R} : \frac{1}{n+1} \leq x \leq 1 - \frac{1}{n+1}\right\},$$

$$B_n = \left\{x \in \mathbf{R} : 0 < x < \frac{1}{n}\right\}.$$

Find $\bigcap_{n \in \mathbf{N}} A_n$, $\bigcap_{n \in \mathbf{N}} B_n$, $\bigcup_{n \in \mathbf{N}} A_n$, and $\bigcup_{n \in \mathbf{N}} B_n$.

The following proposition, which contains a number of other useful set equations, can be proved by reasoning similar to that used in the proof of Proposition 1.1. The proofs of these properties are left to the reader as exercises.

Proposition 1.2. If A and B are members of $P(X)$, then

(6) $A \cup A = A$
    $A \cap A = A$,   (Idempotent Properties)
(7) $\varnothing' = X$
    $X' = \varnothing$,
(8) $A \cup X = X$
    $A \cap \varnothing = \varnothing$,
(9) $A \cup (A \cap B) = A$
    $A \cap (A \cup B) = A$,   (The Modular Identities)
(10) If for some $C \in P(X)$ it is true that $A \cap C = B \cap C$ and $A \cup C = B \cup C$, then $A = B$.
(11) If $A \cap B = \varnothing$ and if $A \cup B = X$, then $B = A'$.
(12) $(A \cup B)' = A' \cap B'$
     $(A \cap B)' = A' \cup B'$.   (The De Morgan Identities)

It is an interesting fact that the properties of intersection, union, and complementation proved in Proposition 1.1 are "basic" among all the properties of these operations in the following sense: Given any valid equation obtained by applying the operations of union, intersection, and complementation to a finite number of arbitrary members of $P(X)$ and the special sets X, $\varnothing$, it is possible to derive this equation by making use of the properties established in Proposition 1.1 only, without resort to reasoning in terms of elements as in the proof of Proposition 1.1. For example, we can give alternate proofs of the first equations in parts (6) and (7) of Proposition 1.2 in terms of the properties established in Proposition 1.1 as follows:

$$A \stackrel{(3)}{=} A \cup \varnothing \stackrel{(4)}{=} A \cup (A \cap A') \stackrel{(5)}{=} (A \cup A) \cap (A \cup A')$$
$$\stackrel{(4)}{=} (A \cup A) \cap X \stackrel{(3)}{=} A \cup A$$
$$X \stackrel{(4)}{=} \varnothing \cup \varnothing' \stackrel{(2)}{=} \varnothing' \cup \varnothing \stackrel{(3)}{=} \varnothing'.$$

(The numbers placed above the equal signs refer to the property in Proposition 1.1 that is used at that step.) For this reason, the equations in Proposition 1.1 may be regarded as "axioms" from which all other set equations involving union, intersection, and complement in $P(X)$ may be derived as "theorems." It turns out that other interesting mathematical systems have operations corresponding to union, intersection, and complement displaying the properties listed in Proposition 1.1. Such systems are referred to as *Boolean algebras* and will be studied in some detail in another chapter. We shall see that the concept of a Boolean algebra not only provides a suitable axiomatic basis for the discussion of the operations of union, intersection and complementation on $P(X)$, but also serves as a useful mathematical tool for the analysis of switching networks and computer design.

EXERCISES

1.14. Construct Venn diagrams, one for each side of the De Morgan identities, to show that these identities are plausible.

1.15. Prove the identities of Proposition 1.2 by considering elements as in the proof of Proposition 1.1.

1.16. In the discussion preceding these exercises, we used the identities of Proposition 1.1 to derive the first equations in (6) and (7). Derive the second equations in (6) and (7) as well as the equations in (8) in a similar manner.

1.3. ORDER RELATIONS AND EQUIVALENCE RELATIONS

In everyday usage, the word "relation" usually brings to mind familial connections such as the relations of father to son or uncle to niece. One can also think of many relations between mathematical objects. For example, numbers can be related in various ways; one number may be less than a second number or a number may divide another. Also, a circle is related to its center, a rectangle to its area, and so on. A feature common to all of these examples is the fact that there is a connection between an ordered pair of objects. That is, a relation could be specified by listing ordered pairs of objects where the first object in the pair is associated with the second in a particular way. Thus, the pertinent notion in formulating the mathematical definition of a relation is that of ordered pair.

An ordered pair (a,b) of elements is not the same as a set consisting of two elements; for as far as set theory is concerned the sets $\{a,b\}$ and $\{b,a\}$ are indistinguishable since they contain the same elements. The notion of ordered pair depends on being able to distinguish the "first" component from the "second" component in a pair. It is intuitively satisfactory to agree that (a,b) be read from left to right, the direction in which our eyes travel when we read. Of course, someone accustomed to a language that is read from right to left such as Hebrew would be tempted to refer to b as the "first" component of (a,b) and a as the "second" component. Although these disagreements are rare, it is worth noting that ordered pairs can be defined solely in terms of set theory and removed from any intuitive feeling of what "first" and "second" should mean. If $a = b$, then there is no need to distinguish the first element of (a,b) from the second since they are identical. If $a \neq b$, we can define (a,b) to be the set $\{\{a\},\{a,b\}\}$. The elements a and b can be distinguished because $\{a\}$ is a member of (a,b) while $\{b\}$ is not, and the first and second components of (a,b) can be defined accordingly. It can be verified from this definition that two ordered pairs (a,b) and (c,d) are equal if and only if $a = c$ and $b = d$, that is, their respective first and second components are identical.

The formal definition of relation is based on the notion of the Cartesian product of two sets. The *Cartesian product* $A \times B$ of a set A with a set B is defined to be the collection of all ordered pairs (a,b) where $a \in A$ and $b \in B$. That is,

$$A \times B = \{(a,b): a \in A, b \in B\}.$$

It is important to note that $A \times B$ is not the same set as $B \times A$. For example, if $A = \{1,2\}$ and $B = \{3,4\}$, then $A \times B$ has the four elements $(1,3)$, $(1,4)$, $(2,3)$, and $(2,4)$. However, $B \times A = \{(3,1), (3,2), (4,1), (4,2)\}$, which is different from $A \times B$.

One of the most familiar Cartesian products is a so-called Cartesian plane. A *Cartesian plane* is a plane for which a rectangular coordinate system has been specified. Each point in such a plane uniquely determines an ordered pair of real numbers, namely, its rectangular coordinates with respect to the given coordinate system. Conversely, each ordered pair (a,b) of real numbers determines a unique point in the given Cartesian plane, namely, the point P with (a,b) as its rectangular coordinates. Thus, if we identify a point P in the Cartesian plane with its ordered pair (a,b) of rectangular coordinates, we identify the Cartesian plane with **R** $\times$ **R**

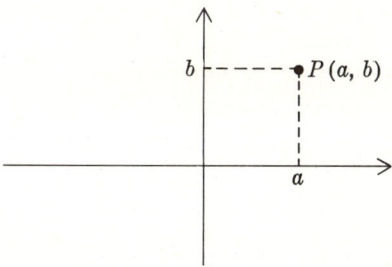

Figure 1.3

(often written as $\mathbf{R}^2$) where $\mathbf{R}$ is the set of real numbers. (See Figure 1.3.)

We are now in a position to give the formal definition of relation. A *relation* R from a set A to a set B is a subset of $A \times B$. If $A = B$, we say that R is a relation *on* A (or B). If R is a relation from A to B, then the statement $(a,b) \in R$ is often written as $a\ R\ b$ and read "a is related to b."

Here are a few examples of relations:

(1) Let A be the set of all rectangles in the plane and let B be the set of nonnegative real numbers. Define $R_1 = \{(a,b) \in A \times B : b$ is the area of $a\}$. That is, the rectangle a is related to the number b if the area of a is b.

(2) Let T be the set of all triangles and C the set of all circles in a plane. Let R_2 be the subset of $T \times C$ consisting of all ordered pairs (t,c) such that c passes through the vertices of t. Thus t and c are related if c is the circumscribing circle of t.

(3) Let L be the set of all lines and P the set of all points in a plane. Let R_3 be the relation from L to P defined as follows: $(\ell,p) \in R_3$ if and only if p lies on ℓ.

The notion of a relation is very broad, too broad to yield a significant theory without the imposition of further restrictions. After all, *any* subset of a Cartesian product $A \times B$ is a relation from A to B and it is clear that not all subsets of $A \times B$ can lead to interesting or natural relations between the elements of A and B. Most relations that are of interest in mathematics fall into one of three categories: order relations, equivalence relations, and functions. We shall now discuss order relations and equivalence relations and take up the study of functions later.

Suppose that R is a relation on a set A. We say that R is:

(a) *Reflexive* if $(x,x) \in R$ for all $x \in A$.
(b) *Symmetric* if $(x,y) \in R$ whenever $(y,x) \in R$.
(c) *Antisymmetric* if $x = y$ whenever $(x,y) \in R$ and $(y,x) \in R$.
(d) *Transitive* if $(x,z) \in R$ whenever $(x,y) \in R$ and $(y,z) \in R$.

Some intuitive feeling for the meaning of each of the preceding restrictions may be gained by considering these restrictions for relations

14   SET THEORY

on the set **R** of real numbers. If the Cartesian plane is identified with $\mathbf{R}^2$, every relation on **R** corresponds to a subset of the plane and every subset of the plane corresponds to a relation on **R**. Thus each of the sets

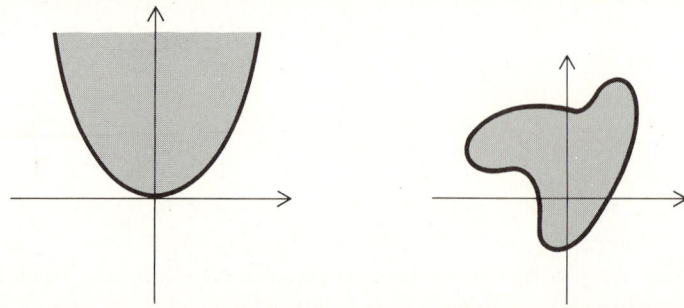

Figure 1.4

pictured in Figure 1.4 determines a relation on **R**. The restrictions stated above can be pictorially interpreted in the plane as follows.

(a) A relation on **R** is reflexive if the corresponding subset of $\mathbf{R}^2$ contains the diagonal line D of slope 1 through the origin. The relation in Figure 1.5 is thus reflexive, while the relations of Figure 1.4 are not.

(b) Symmetric relations on **R** correspond to subsets of $\mathbf{R}^2$ that are symmetric with respect to the diagonal line D in the usual sense of analytic geometry. The relation pictured in Figure 1.5 is symmetric, for example, but those in Figure 1.4 are not.

(c) Antisymmetric relations correspond to those subsets of $\mathbf{R}^2$ that do not contain any pairs of distinct points that are symmetric with

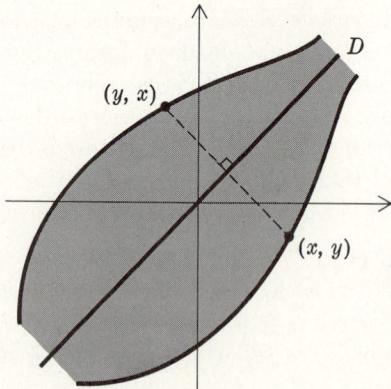

Figure 1.5

1.3. ORDER RELATIONS AND EQUIVALENCE RELATIONS 15

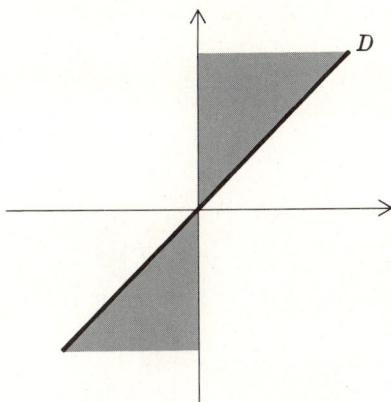

Figure 1.6

respect to the diagonal line D. The relation pictured in Figure 1.6 is antisymmetric.

(d) A relation R on **R** is transitive if it contains the fourth vertex of any rectangle parallel to the coordinate axes with one vertex on D and the other two vertices in R. (See Figure 1.7.) The relation determined in Figure 1.6 is thus seen to be transitive.

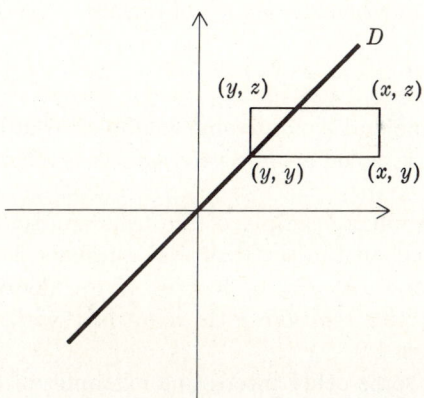

Figure 1.7

EXERCISES

1.17. Prove or disprove: A symmetric, transitive relation on a set A is always reflexive.

1.18. Describe all relations on **R** that are both symmetric and antisymmetric. Generalize your result to such a relation on an arbitrary set A.

1.19. Which of the above restrictions are satisfied by the relation on **R** pictured in Figure 1.8?

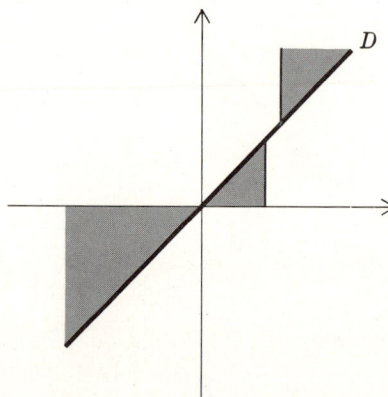

Figure 1.8

1.20. Let T denote the set of triangles in the plane and define the relation R to be the subset of $T \times T$ consisting of all (t_1, t_2) such that t_1 and t_2 share a common side. Which of the above restrictions are satisfied by R?

Order Relations

We now define and look at some examples of order relations. A relation R on a set A is called an *order relation* if R is reflexive, antisymmetric, and transitive.

The most familiar example of an order relation on a set A is the *usual order relation* on the set **R** of real numbers defined by the subset $R = \{(a,b) \in \mathbf{R} \times \mathbf{R}: a \leq b\}$ of $\mathbf{R} \times \mathbf{R}$. If we identify $\mathbf{R} \times \mathbf{R}$ with the Cartesian plane, the relation R is identified with the shaded region indicated in Figure 1.9.

We now list some other interesting examples of order relations.

Example 1.2. Suppose that X is a set. Define the relation R on the power set $P(X)$ of X by

$$R = \{(A,B) \in P(X) \times P(X): A \subset B\},$$

that is, an ordered pair (A,B) of subsets of X is in R if and only if A is a subset of B. R is clearly a reflexive relation on $P(X)$ and R is antisymmetric since two sets A and B are equal if and only if $A \subset B$ and $B \subset A$.

1.3. ORDER RELATIONS AND EQUIVALENCE RELATIONS 17

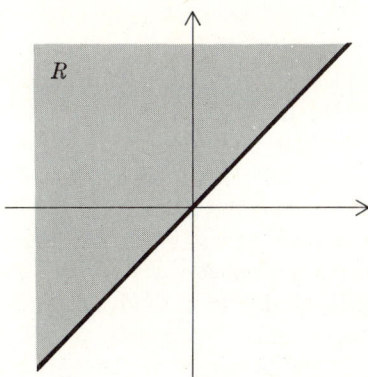

Figure 1.9

Also, it follows immediately from the definition of subset that if A, B, and C are subsets of X such that $A \subset B$ and $B \subset C$, then $A \subset C$. Therefore, R is transitive, and we conclude that R is an order relation on $P(X)$. R is referred to as the *inclusion relation* on $P(X)$.

Example 1.3. Suppose that $\mathbf{N}$ is the set of all natural numbers and that R is the relation on $\mathbf{N}$ defined by

$$R = \{(a,b) \in \mathbf{N} \times \mathbf{N} : a = kb \text{ for some } k \in \mathbf{N}\}.$$

Thus, if a, b are natural numbers, then $a \, R \, b$ if and only if a is a multiple of b. It is an easy matter to verify that R is an order relation on $\mathbf{N}$.

Example 1.4. Suppose that R is the relation defined on $\mathbf{R}^2$ by

$$R = \{((a,b),(c,d)) \in \mathbf{R}^2 \times \mathbf{R}^2 : a > c, \text{ or } a = c \text{ and } b \geq d\}.$$

Thus, two elements (a,b) and (c,d) of $\mathbf{R}^2$ satisfy $(a,b) \, R \, (c,d)$ if and only if either:

(1) The first component of (a,b) is greater than the first component of (c,d).

or

(2) The first components of (a,b) and (c,d) are equal and the second component of (a,b) is greater than or equal to the second component of (c,d).

For example, $(3,2) \, R \, (2,7)$ and $(3,7) \, R \, (3,5)$. R is an order relation on $\mathbf{R}^2$ which is sometimes referred to as the *dictionary* (or *lexicographic*) *order* on $\mathbf{R}^2$. A moment's thought will show why this terminology is appropriate.

18 SET THEORY

The important notion of strict inequality of real numbers is defined by the relation $R' = \{(a,b) \in \mathbf{R} \times \mathbf{R}: a < b\}$. R' is easily seen to be transitive, and R' is antisymmetric by default since $a < b$ and $b < a$ are never simultaneously true for any real numbers a and b. However, since R' is not reflexive, it is not an order relation on $\mathbf{R}$. Nevertheless, the relation R' obviously is intimately associated with the usual order relation R on $\mathbf{R}$ mentioned earlier, and the purpose of the discussion that follows is to clarify the connection between pairs of relations such as R and R'.

A relation R' on a set A is called a *strict order relation* if R' is transitive and if $(x,y) \in R'$ implies that $x \neq y$. (The latter restriction on R' may be thought of as an "antireflexive" property for R' since it implies that $(x,x) \notin R'$ for all $x \in A$.) It is easy to see that the relation R' just defined on the set $\mathbf{R}$ of real numbers is a strict order relation on $\mathbf{R}$. Another example of a strict order relation can be obtained by similarly modifying set inclusion. Suppose that X is a set and that $P(X)$ is the power set of X. If $A \in P(X)$ and $B \in P(X)$, then A is a *proper subset* of B if $A \subset B$ and if $A \neq B$. The relation R' defined on $P(X)$ by

$$R' = \{(A,B) \in P(X) \times P(X): A \subset B, A \neq B\}$$

is easily seen to be a strict order relation on $P(X)$.

The following result establishes the expected connection between order relations and strict order relations in general.

Proposition 1.3. If R is an order relation on a set A, then $R' = \{(x,y) \in R: x \neq y\}$ is a strict order relation on A. If R' is a strict order relation on A, then $R = \{(x,y) \in A \times A: (x,y) \in R' \text{ or } x = y\}$ is an order relation on A.

PROOF. Suppose that R is an order relation on the set A, and that $R' = \{(x,y) \in R: x \neq y\}$. If $(x,y) \in R'$ and $(y,z) \in R'$, then $(x,y) \in R$, $(y,z) \in R$, $x \neq y$, $y \neq z$. Since R is transitive, it follows that $(x,z) \in R$. If $x = z$, then $(z,x) \in R$; consequently, since $(x,y) \in R$, it would follow that $(z,y) \in R$ since R is transitive. But $(y,z) \in R$, so that $y = z$ since R is antisymmetric. This contradicts the fact that $y \neq z$; hence, it must be true that $x \neq z$. Therefore, $(x,z) \in R'$ and we see that R' is transitive. By definition of R', it is also true that $(x,y) \in R'$ implies $x \neq y$; consequently, R' is a strict order relation. The proof of the converse is left as an exercise.

EXERCISES

1.21. Define the relation R on $\mathbf{R}^2$ by

$$R = \{((a,b),(c,d)) \in \mathbf{R}^2 \times \mathbf{R}^2: a \geq c, b \geq d\}.$$

1.3. ORDER RELATIONS AND EQUIVALENCE RELATIONS

Prove that R is an order relation on $\mathbf{R}^2$. If $\mathbf{R}^2$ is identified with the Cartesian plane, sketch the following sets of points:
(a) $C = \{(a,b) \in \mathbf{R}^2 \colon (a,b) \; R \; (0,0)\}$.
(b) $O = \{(a,b) \in \mathbf{R}^2 \colon (a,b) \; R \; (1,1)$ and $(3,2) \; R \; (a,b)\}$.

1.22. Complete the proof of Proposition 1.3.
1.23. Is the relation $\{(x,y) \in \mathbf{R}^2 \colon 2y - x \geq 0\}$ an order relation on $\mathbf{R}$?
1.24. Suppose that R_1 and R_2 are order relations on a set A. Does it follow that $R_1 \cap R_2$ is also an order relation on A?
1.25. Verify that the dictionary order on $\mathbf{R}^2$ is an order relation. (See Example 1.4.)
1.26. Suppose that S is a sector in the Cartesian plane $\mathbf{R}^2$ with vertex at the origin and central angle θ where $0 \leq \theta < \pi$. (See Figure 1.10. It

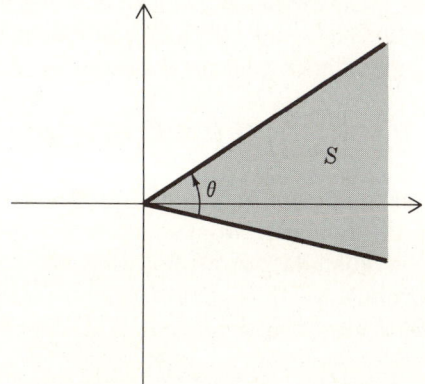

Figure 1.10

is assumed that the boundary rays for S as well as the origin are included in S.) Define a relation R on $\mathbf{R}^2$ by

$$R = \{((a,b),(c,d)) \in \mathbf{R}^2 \times \mathbf{R}^2 \colon (a - c, b - d) \in S\}.$$

Show that R is an order relation on $\mathbf{R}^2$.

Equivalence Relations

A relation R on a set A is called an *equivalence relation* on A if R is reflexive, symmetric, and transitive.

Probably the most familiar example of an equivalence relation is the equality relation on a set A, that is, the subset $\{(x,y) \in A \times A \colon x = y\}$ of $A \times A$. The reflexive, transitive, and symmetric properties of this relation are axiomatic features of our notion of equality. Not only is the equality relation on a set A an example of an equivalence relation, but also it is a fairly accurate general observation that all other interesting equivalence relations that occur in mathematics have been introduced for the purpose of establishing a notion of "equality" in the sets in question that is less stringent than the actual equality relation. For example, when we study the set of all plane triangles in high school geometry, we are ordinarily concerned with congruence or similarity of triangles, not with their equality (that is, identity). Congruence and similarity clearly define equivalence relations on the set T of all plane triangles. Also, in grade school when we study fractions (that is, indicated quotients of integers a and b with $b \neq 0$) we are not concerned with the equality of fractions as fractions, but rather with the equality of the rational numbers represented by two unequal fractions. For example, we wish to equate the fraction 3/4 with the fraction 6/8 because they both represent the same rational number; however, as fractions they are certainly not the same since $3 \neq 6$ and $4 \neq 8$. Nevertheless, we can proceed quite satisfactorily if we define two fractions a/b and c/d to be equivalent if $ad = bc$, that is, if we introduce an equivalence relation R on the set F of all fractions as

$$R = \left\{ \left(\frac{a}{b}, \frac{c}{d}\right) \in F \times F : ad = bc \right\}.$$

We shall now examine the nature of equivalence relations in some detail. Throughout the following discussion, if R is a fixed equivalence relation on a set A, we shall denote the fact that $(x,y) \in R$ by $x \sim y$ and read these latter symbols as "*x is equivalent to y.*" If $x \in A$, the *equivalence class* $[x]$ *of the element* x is the subset of A defined by

$$[x] = \{y \in A : y \sim x\}.$$

Each member y of $[x]$ is called a *representative of* $[x]$.

Example 1.5. If A is any set and if R is the equivalence relation defined by the equality relation on A, that is,

$$R = \{(x,y) \in A \times A : x = y\},$$

then for each $x \in A$, it is true that $[x] = \{x\}$ and that x is the only representative of $[x]$.

Example 1.6. If A is the set of all plane triangles and R is the equivalence relation on A defined by

$$R = \{(t_1, t_2) \in A \times A : t_1 \text{ is similar to } t_2\},$$

then for each plane triangle t, the equivalence class $[t]$ of t consists of all plane triangles t' that are similar to t, and any such plane triangle is a representative of $[t]$.

1.3. ORDER RELATIONS AND EQUIVALENCE RELATIONS 21

Example 1.7. The term "vector" is often defined in a rather vague way in many books on mathematics or physics. It is not uncommon to find the concept of vector (or free vector) described as "an entity with magnitude and direction." Other books define vectors to be directed line segments, but then stipulate that two distinct line segments are the same vector if they have the same magnitude and direction. Such vague and confusing descriptions can be made precise through the use of the notion of equivalence relation.

Suppose that D is the set of all directed line segments (either in a plane or in three dimensional space). If d_1, d_2 are members of D, define d_1 to be *equivalent* to d_2 if the directed line segments d_1 and d_2 have the same magnitude and direction. It is easy to verify that this defines an equivalence relation on D. We now define a *vector* to be any equivalence class with respect to this relation on D. The equivalence class containing the directed line segment d is traditionally denoted by $\vec{d}$ instead of $[d]$. Any directed line segment $d \in D$ is a representative of a vector, namely, the vector $\vec{d}$ consisting of all directed line segments with the same magnitude and direction as d. We define the *direction* and *magnitude* of a vector $\vec{d}$ to be the common direction and magnitude of all directed line segments that are members of $\vec{d}$.

Example 1.8. Suppose that **I** denotes the set of integers (that is, **I** is the set consisting of the number zero and all positive and negative whole numbers). For each positive integer m, define the relation R_m on **I** by

$$R_m = \{(a,b) \in \mathbf{I} \times \mathbf{I}: a - b \text{ is an integral multiple of } m\}.$$

The fact that $(a,b) \in R_m$ is denoted by

$$a \equiv b \pmod{m}$$

and we say that *a is congruent to b modulo m*. For example, $7 \equiv -1 \pmod{4}$ since $7 - (-1) = 8$ is an integral multiple of 4, but 7 is not congruent to 2 modulo 4 since $7 - 2 = 5$ is not an integral multiple of 4.

Since $a \equiv b \pmod{m}$ means that m divides the number $a - b$, the notion of congruence provides a useful framework in which divisibility properties of numbers can be studied. The language of congruences was first introduced by Gauss at the beginning of the nineteenth century and has proved to have important application in the theory of numbers.

We now show that congruence modulo m is an equivalence relation on **I**. R_m is reflexive since $a - a = m \cdot 0$ for any integer a. R_m is also symmetric, for if $a \equiv b \pmod{m}$, then $a - b = mk$ for some integer k, so that

$b - a = m(-k)$; that is, $b \equiv a \pmod{m}$. Finally, suppose that $a \equiv b \pmod{m}$ and $b \equiv c \pmod{m}$. This means that there exist integers k_1 and k_2 such that $a - b = mk_1$ and $b - c = mk_2$. Then $a - c = (a - b) + (b - c) = m(k_1 + k_2)$, which means that $a \equiv c \pmod{m}$. Hence, R_m is transitive, and we conclude that the congruence modulo m relation is an equivalence relation for each positive integer m.

It is traditional to refer to the equivalence classes determined by R_m as *residue classes modulo m*. Thus, two integers a and b belong to the same residue class modulo m if and only if m divides $a - b$. It follows that the residue class containing the integer a is the set

$$[a] = \{a + mk : k \in \mathbf{I}\}.$$

The set of residue classes modulo m is denoted by $\mathbf{I}_m$. For example, if $m = 2$, then there are just two residue classes, namely, [0] consisting of all even integers and [1] consisting of all odd integers. Thus, $\mathbf{I}_2 = \{[0],[1]\}$. Note that these two residue classes are disjoint subsets of $\mathbf{I}$ since no integer is both even and odd. If $m = 5$, then $\mathbf{I}_5$ has five members, namely,

$$[0] = \{5k : k \in \mathbf{I}\},$$
$$[1] = \{5k + 1 : k \in \mathbf{I}\},$$
$$[2] = \{5k + 2 : k \in \mathbf{I}\},$$
$$[3] = \{5k + 3 : k \in \mathbf{I}\},$$
$$[4] = \{5k + 4 : k \in \mathbf{I}\}.$$

Each particular integer belongs to one and only one of these residue classes.

In each of the preceding examples, the set $\mathcal{P}$ of equivalence classes determined by an equivalence relation R on a set A has the following properties:

(a) The union of all the sets in $\mathcal{P}$ is A.
(b) If $B \in \mathcal{P}$, then $B \neq \emptyset$.
(c) If $B \in \mathcal{P}$, $C \in \mathcal{P}$ and if $B \neq C$, then B and C are disjoint, that is, $B \cap C = \emptyset$.

Such a set $\mathcal{P}$ of subsets of a set A is called a *partition of A*. The following result shows that this is a feature of all equivalence relations.

Proposition 1.4. *If R is an equivalence relation on a set A and if $\mathcal{P} = \{[x] : x \in A\}$ is the set of all equivalence classes determined by R, then $\mathcal{P}$ is a partition of A.*

PROOF. For each $x \in A$, it is true that $x \in [x]$ since R is reflexive. Therefore, $A \subset \cup \{[x]: x \in A\}$. On the other hand, $[x] \subset A$ for each $x \in A$, so condition (a) for a partition is satisfied.

Also, since $x \in [x]$ for each $x \in A$, it is true that each set in $\mathcal{P}$ is not empty, which verifies condition (b).

Let $[x] \in \mathcal{P}$, $[y] \in \mathcal{P}$ be distinct elements of $\mathcal{P}$, that is, $[x] \neq [y]$. We wish to show that $[x]$ and $[y]$ are disjoint. If we suppose that $[x] \cap [y] \neq \varnothing$, then there is a $z \in A$ such that $z \in [x]$ and $z \in [y]$. By definition of equivalence class, this means that $z \sim x$ and $z \sim y$. Now, if w is an arbitrary member of $[x]$, then $w \sim x$. Also, $x \sim z$ since R is symmetric, and thus $w \sim z$ because R is transitive. But $z \sim y$, so that $w \sim y$ by the transitivity of R. This means $w \in [y]$, which implies that $[x] \subset [y]$ since w is an arbitrary member of $[x]$. A similar argument would show that $[y] \subset [x]$, so that $[x] = [y]$ contrary to our hypothesis. Thus, the supposition that $[x] \cap [y] \neq \varnothing$ leads to a contradiction, and we conclude that $[x]$ and $[y]$ are disjoint sets. This completes the proof of the fact that $\mathcal{P}$ is a partition of A.

The converse to Proposition 1.4 is also true; that is, any partition $\mathcal{P}$ of a set A determines an equivalence relation R on A in a natural way. (See Exercise 1.28.)

EXERCISES

1.27. Let $A = \{1,2,3,4,5\}$ and let $R = \{(1,1),(2,2),(3,3),(4,4),(5,5),(1,3), (2,5),(3,1),(5,2)\}$. Is R an equivalence relation on A? If so, what are the equivalence classes?

1.28. Prove that if $\mathcal{P}$ is any partition of a set A, then the relation R defined on A by

$$R = \{(x,y) \in A \times A: \text{ there is a } P \in \mathcal{P}$$
$$\text{such that } x \in P \text{ and } y \in P\}$$

is an equivalence relation on A and that $\mathcal{P}$ is the collection of equivalence classes determined by R.

1.29. How many equivalence relations are there on a set A consisting of 3 elements? [*Hint:* How many ways can A be partitioned?]

1.30. Show that if R_1 and R_2 are equivalence relations on a set A, then $R_1 \cap R_2$ is also an equivalence relation on A. Given an example to show that $R_1 \cup R_2$ need not be an equivalence relation.

1.31. Suppose that a relation R on a set A satisfies both of the following:
(a) For each $a \in A$, there exists $x \in A$ such that $(a,x) \in R$.
(b) If $(a,c) \in R$ and $(b,c) \in R$, then $(a,b) \in R$.
Must R be an equivalence relation on A?

1.32. Suppose S denotes the surface of the earth and consider the following relations on S:

$R_1 = \{(p,q) \in S \times S: p \text{ and } q \text{ have the same latitude}\}$,
$R_2 = \{(p,q) \in S \times S: p \text{ and } q \text{ have the same longitude}\}$.

Are R_1 and R_2 equivalence relations on S?

1.33. Define the relation R on the set **N** of natural numbers as follows: $(a,b) \in R$ if and only if the sum of the digits of a equals the sum of the digits of b. Show that R is an equivalence relation on **N** and describe the equivalence class containing the number 10.

1.34. Suppose that the relation R is defined on $\mathbf{R}^2$ as follows: $(x_1,y_1)\, R\, (x_2,y_2)$ if an only if $x_1 + y_1 = x_2 + y_2$. Show that R is an equivalence relation on $\mathbf{R}^2$. Sketch the subset in the Cartesian plane corresponding to the equivalence class with representative (1,3).

1.35. Show that the relation R on **I** defined by

$$R = \{(a,b) \in \mathbf{I} \times \mathbf{I}: a^2 - b^2 \text{ is an integral multiple of } 2\}$$

is an equivalence relation. Find the equivalence classes determined by this relation.

1.4. FUNCTIONS

A *function* (*or mapping*) f *from a set A to a set B* is a relation from A to B such that:

(a) For each $x \in A$, there is a $y \in B$ such that $(x,y) \in f$.
(b) If $(x,y) \in f$ and $(x,z) \in f$, then $y = z$.

The notation $f: A \to B$ will indicate that f is a function from A to B. The set A is called the *domain* of f and B is called the *range* of f.

It is customary to write $y = f(x)$ instead of $(x,y) \in f$ and to call y the *value* of the function f at x. Conditions (a) and (b) together assert that for each $x \in A$ there is a unique $y \in B$ such that $y = f(x)$, that is, such that y is the value of f at x.

An alternate and more traditional approach to the notion of function can be given in terms of the "rule" definition. In this version, a function from A to B is a rule that assigns to each $x \in A$ a unique $y \in B$ where $y = f(x)$ is written to denote the correspondence. This is essentially the same as the definition given above since the term "rule" is to be interpreted in such a way that each rule f determines the relation $\{(x,y) \in A \times B: y = f(x)\}$, and conversely. We shall discuss these two approaches to the definition of function from a pedagogical point of view in the Remarks and References section at the end of this chapter.

Frequently, the verbiage associated with functions is suppressed and reference is made, for example, to "the function $y = 1/x$." In such cases it is understood that the domain of the function consists of all elements of the set under discussion, say the set **R** of real numbers, for which the definition makes sense. In the case of $y = 1/x$ this would be the set of all $x \in \mathbf{R}$ such that $x \neq 0$. Usually this shorthand causes no confusion, but in certain instances, especially when there is some ambiguity about the domain, the full and detailed definition of the function must be given for clarity.

Some other examples of functions will now be considered.

Example 1.9. Suppose that **R** is the set of all real numbers. For each $x \in \mathbf{R}$, define $G(x)$ to be the *greatest integer in* x; that is, $G(x) = k$ where k is the integer satisfying $k \leq x < k + 1$. (For example, $G(\frac{1}{2}) = 0$, $G(\pi) = 3$, $G(-7.1) = -8$.) The rule that assigns to each real number x the greatest integer $G(x)$ in x defines a function $G: \mathbf{R} \to \mathbf{R}$. In terms of relations, we can write

$$G = \{(x,y) \in \mathbf{R} \times \mathbf{R}: y = G(x)\}.$$

Example 1.10. Suppose that T is the set of all plane triangles and that B is the set of all positive real numbers. Define

$$f = \{(t,a) \in T \times B: a = \text{area of } t\}.$$

Then f is a function from T to B, namely, the function determined by the rule that assigns to each triangle t its area a.

Example 1.11. Suppose that X is a set and that A is a fixed subset of X. The relation

$$g = \{(B,C) \in P(X) \times P(X): C = B \cap A\}$$

defines a function $g: P(X) \to P(A)$. That is to say, $g(B) = B \cap A$ for each subset B of X.

If f is a function from A to B and if S is a subset of A, then the *image of S under f* is the set $f(S) = \{f(x): x \in S\}$. If $T = f(S)$, we often say that "*f* maps S onto T." For example, if G is the "greatest integer" function defined in Example 1.9 and if $S = \{x \in \mathbf{R}: 0 < x < 4\}$, then $G(S) = \{0,1,2,3\}$. A convenient visualization of a function f from a set A to a set B is shown in Figure 1.11 where the shaded portions represent a subset S of A and its image.

If f is a function from A into B and if P is a subset of B, then the *preimage* of P under f is the set $f^{-1}(P) = \{x \in A: f(x) \in P\}$. For example, if $g: P(X) \to P(A)$ is the function defined in Example 1.11 above, then

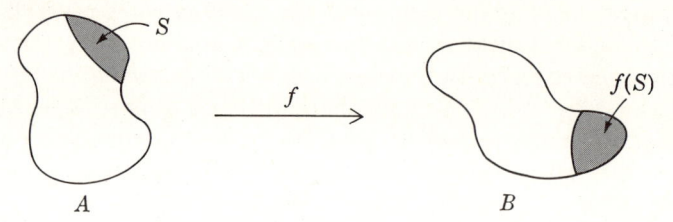

Figure 1.11

$g^{-1}(\emptyset)$ consists of all subsets B of X such that A and B are disjoint. The symbol f^{-1} is *not* used here to indicate a function. It is used only as a convenient notation to designate the set of points in the domain A that have their images in the specified subset of B. The notion of inverse function will be discussed shortly.

EXERCISES

1.36. Let f be a function from A to B. Show that if S and T are subsets of A, then $f(S \cup T) = f(S) \cup f(T)$. Show that $f(S \cap T) \subset f(S) \cap f(T)$ and give an example showing that inclusion may be proper.

1.37. Show that if $f: A \to B$ and P and Q are subsets of B, then $f^{-1}(P \cup Q) = f^{-1}(P) \cup f^{-1}(Q)$. Is the analogous equality for intersections valid?

1.38. Which of the relations in the examples of relations given in the previous section are functions?

1.39. Suppose a relation f on A is both a function from A to A and an equivalence relation on A. What can be said about f?

If $f: A \to B$ and $g: B \to C$, then the *composite function* $g \circ f: A \to C$ is defined by $g \circ f(x) = g(f(x))$ for all $x \in A$; that is, $g \circ f = \{(x,z) \in A \times C: \text{for some } y \in B, (x,y) \in f, (y,z) \in g\}$.

Example 1.12. Suppose that A is the set of all triangles and that B is the set of all circles in the coordinate plane C. Define functions $f: A \to B$ and $g: B \to C$ by

$$f = \{(x,y) \in A \times B: \text{the vertices of } x \text{ lie on } y\},$$
$$g = \{(y,z) \in B \times C: z \text{ is the center of } y\}.$$

In other words, f is the function that assigns to each plane triangle x its circumscribing circle y and g is the function that assigns to each circle in the plane the center of that circle. (We leave the verification of the fact that f and g are functions to the reader as an exercise.) The composite function $g \circ f \colon A \to C$ is obviously defined by

$$g \circ f = \{(x,z) \in A \times C \colon z \text{ is the center of the circumscribing circle for } x\}.$$

We now turn to a discussion of inverse functions. For this purpose, we introduce the following elementary function on a set. The *identity* function $i_A \colon A \to A$ on a set A is defined by $i_A(x) = x$ for all $x \in A$.

Proposition 1.5. The following conditions on a function $f \colon A \to B$ are equivalent:

(1) There exists a function $g \colon B \to A$ such that $g \circ f = i_A$, that is, $g(f(x)) = x$ for $x \in A$.
(2) If $f(x_1) = f(x_2)$ for x_1, x_2 in A, then $x_1 = x_2$.

A function $g \colon B \to A$ satisfying (1) is called a *left inverse* of f.

PROOF. We first prove that (1) implies (2). If $f(x_1) = f(x_2)$, then

$$\begin{aligned} x_1 &= i_A(x_1) = g(f(x_1)) = g(f(x_2)) \\ &= i_A(x_2) = x_2. \end{aligned}$$

Conversely, suppose condition (2) holds. Then for each $y \in f(A)$, $f^{-1}(\{y\})$ consists of a single point in A. Thus, for $y \in f(A)$, we define $g(y) = x$ where $f^{-1}(\{y\}) = \{x\}$. If $f(A) = B$, then $g \circ f = i_A$. If $f(A) \neq B$, then g is not defined on all of B. In this case, choose any $x_0 \in A$ and define $g(y) = x_0$ for all $y \in B$ such that $y \notin f(A)$. Then $g \colon B \to A$ and $g \circ f = i_A$.

A function $f \colon A \to B$ that satisfies condition (2) of Proposition 1.5 is called a *one-to-one* function. The proposition states that f has a left inverse if and only if f is a one-to-one function.

A function $f \colon A \to B$ is called an *onto* function if $f(A) = B$; that is, if for each $y \in B$ there exists at least one element $x \in A$ such that $f(x) = y$. This is the condition that f must satisfy to possess a right inverse as the next result states.

Proposition 1.6. The following conditions on a function $f \colon A \to B$ are equivalent:

(1) There exists $g \colon B \to A$ such that $f \circ g = i_B$.
(2) f is an onto function.

A function $g \colon B \to A$ satisfying (1) is called a *right inverse* of the function $f \colon A \to B$.

PROOF. To see that (1) implies (2), we simply observe that for each $y \in B$ we have $y = i_B(y) = (f \circ g)(y) = f(g(y))$; that is, $g(y)$ is an element x of A such that $f(x) = y$. Therefore, since y is an arbitrary element of B, f is an onto function.

Conversely, if $f: A \to B$ is an onto function and if $y \in B$, choose an element $x \in A$ such that $y = f(x)$ and define $g(y) = x$. This defines a function $g: B \to A$. By definition of g, $f(g(y)) = y$ for each $y \in B$; that is, $f \circ g = i_B$.

If $f: A \to B$ is a function from A to B and if there exists a function $g: B \to A$ from B to A such that $f \circ g = i_B$ and $g \circ f = i_A$, then g is called the *inverse function* of f and we write f^{-1} in place of g. For any function $f: A \to B$, at most one function $g: B \to A$ of the above sort can exist. For, if $g_1: B \to A$ and $g_2: B \to A$ satisfy $f \circ g_1 = f \circ g_2 = i_B$ and $g_1 \circ f = g_2 \circ f = i_A$, respectively, then for each $y \in B$ we have

$$g_1(y) = g_1(i_B(y)) = g_1(f(g_2(y))) = i_A(g_2(y)) = g_2(y)$$

and therefore $g_1 = g_2$. Thus, we are justified in speaking of *the* inverse function.

Proposition 1.7. A function $f: A \to B$ has an inverse function $f^{-1}: B \to A$ if and only if f is a one-to-one, onto function.

PROOF. If f is a one-to-one, onto function and if $g = \{(y,x) \in B \times A: (x,y) \in f\}$, then g is a function from B into A. Moreover, $f \circ g = i_B$ and $g \circ f = i_A$; hence, $g = f^{-1}$. If f has an inverse function $g: B \to A$, then g is both a left inverse and a right inverse for f, and Propositions 1.5 and 1.6 imply that f is a one-to-one, onto function.

A one-to-one, onto function $f: A \to B$ is usually called a *one-to-one correspondence* between A and B. The phrase "between A and B" can be used in place of "from A to B" because in the case of a one-to-one correspondence $f: A \to B$, the inverse function $f^{-1}: B \to A$ is also a one-to-one correspondence (see Exercise 1.41). Thus, A and B may be treated symmetrically. Two sets A and B are said to be in *one-to-one correspondence* if there is a one-to-one correspondence between them.

EXERCISES

1.40. Given a function $f: A \to B$, define a relation R on A by

$$R = \{(x_1, x_2) \in A \times A: f(x_1) = f(x_2)\}.$$

Prove that R is an equivalence relation on A. If $A = B = \mathbf{R}$, the set of all real numbers, and if f is the function whose value at the real number x is sin x, describe the equivalence classes determined by the relation R.

1.41. If $f: A \to B$ is a one-to-one correspondence, prove that $f^{-1}: B \to A$ is a one-to-one correspondence.

1.42. Show that if $f: A \to B$ and $g: B \to C$ are both one-to-one correspondences, then the composite $g \circ f: A \to C$ is also a one-to-one correspondence and $(g \circ f)^{-1} = f^{-1} \circ g^{-1}$.

1.43. If $A = \mathbf{R}$, $B = \{x \in \mathbf{R}: x \geq 0\}$, and $f: A \to B$ is the function defined by $f(x) = x^2$, find two different functions from B to A that are right inverses of f.

1.44. Show that the function $f: \mathbf{R} \to \mathbf{R}$ defined by $f(x) = x^3 + 1$ is a one-to-one, onto function and find f^{-1}.

1.45. Suppose that $f: \mathbf{R} \to \mathbf{R}$ and $g: \mathbf{R} \to \mathbf{R}$ are both one-to-one functions. Does it follow that $f + g$ and fg are also one-to-one functions? ($f + g$ and fg are defined for all x in $\mathbf{R}$ by $(f + g)(x) = f(x) + g(x)$ and $(fg)(x) = f(x)g(x)$, respectively.) If f and g are onto functions, does it follow that $f + g$ and fg are onto functions?

1.5. ALGEBRAIC SYSTEMS

Two particular classes of functions that play an important role in mathematics are the so-called unary and binary operations. Given a set A, a *unary operation on A* is simply a function $u: A \to A$, and a *binary operation on A* is a function $b: A \times A \to A$. Thus, a unary operation u on A associates with each element x of A some element $u(x)$ belonging to A, while a binary operation b on A associates with each ordered pair (x,y) of elements of A an element $b(x,y)$ of A. For example, the function $u: \mathbf{R} \to \mathbf{R}$ defined by $u(x) = |x|$ is a unary operation (called "absolute value") on $\mathbf{R}$, and the function $b: \mathbf{R} \times \mathbf{R} \to \mathbf{R}$ defined by $b(x,y) = x + y$ is a binary operation (called "addition") on $\mathbf{R}$. Also, if $P(X)$ is the power set of a set X, the function $u: P(X) \to P(X)$ defined by $u(A) = A'$ is a unary operation (called "complementation") on $P(X)$, and the function $b: P(X) \times P(X) \to P(X)$ defined by $b(A,B) = A \cap B$ is a binary operation (called "intersection") on $P(X)$.

Although we are inclined to think of such operations as "square root," that is, $s(x) = \sqrt{x}$, and "reciprocation," that is, $r(x) = 1/x$, as unary operations and of such operations as "subtraction," that is, $m(x,y) = x - y$, and "division," that is, $d(x,y) = x/y$, as binary operations, it should be realized that such operational interpretations do not make sense technically unless suitable domains and ranges are specified for the functions in question. For example, square root and reciprocation cannot legitimately be regarded as unary operations on the set $\mathbf{N}$ of natural numbers since neither $s(x) = \sqrt{x}$ nor $r(x) = 1/x$ define functions that map $\mathbf{N}$ into $\mathbf{N}$ (for example, $s(2) = \sqrt{2} \notin \mathbf{N}$ and $r(2) = 1/2 \notin \mathbf{N}$). Similarly, neither subtraction nor division are binary operations on $\mathbf{N}$ since neither $m(x,y) = x - y$ nor $d(x,y) = x/y$ define functions mapping $\mathbf{N} \times \mathbf{N}$ into $\mathbf{N}$ (for example, $m(2,3) = -1 \notin \mathbf{N}$ and $d(2,3) = 2/3 \notin \mathbf{N}$). On the other hand, $s(x) = \sqrt{x}$ and $r(x) = 1/x$ do define unary operations on appropriately selected sets such as the set $\mathbf{R}^+$ of positive real numbers. Similarly, $m(x,y) = x - y$ defines a binary operation on the set $\mathbf{I}$ of all integers since $m(\mathbf{I} \times \mathbf{I}) \subset \mathbf{I}$, and $d(x,y) = x/y$ defines a binary operation on either the set $\mathbf{Q}_0$ of nonzero rational numbers or the set $\mathbf{R}_0$ of nonzero real numbers.

In discussions of unary and binary operations where the domains of definition are ignored, there is an attempt to compensate for this rather casual attitude by introducing the notion of a set being "closed" with respect to an operation. One says that a set S is closed with respect to an operation if the result of performing the operation on an element (or elements) of S is always an element of S. This rather vague description is usually given meaning by referring to some examples that exhibit the desired behavior; for example, the set $\mathbf{I}$ is closed with respect to subtraction while $\mathbf{N}$ is not. This type of discussion puts the "cart before the horse," since the very definition of the operation requires some specification of the domain and range of the operation. These imprecisions can be simply avoided by stating the domain on which the operation in question is defined at the outset. The notion of a set being closed with respect to an operation then has the following clear and precise meaning.

If $u: A \to A$ is a given unary operation on a set A, then a subset S of A is said to be *closed with respect to u* if $u(S) \subset S$. Similarly, if $b: A \times A \to A$ is a given binary operation on A, then a subset S of A is said to be *closed with respect to b* if $b(S \times S) \subset S$.

For example, if $u: \mathbf{R} \to \mathbf{R}$ is the unary operation on $\mathbf{R}$ defined by $u(x) = |x|$ and if $b: \mathbf{R} \times \mathbf{R} \to \mathbf{R}$ is the binary operation on $\mathbf{R}$ defined by $b(x,y) = x + y$, then the subset $\mathbf{Q}$ of rational numbers in $\mathbf{R}$ is closed with respect to both u and b. However, the subset P of irrational real numbers is not closed with respect to addition since there exist pairs of irrational numbers whose sums are rational.

We now wish to define the rather general notion of an algebraic system. Very often, the sets that are of interest in mathematics have natural operations and relations associated with them. For example, the set $\mathbf{R}$ of real numbers is ordinarily studied as a system with the binary operations of addition and multiplication, the unary operations of absolute value or square root, and relations such as $\leq$ or $<$. Also, the power set $P(X)$ of a set X is naturally equipped with the binary operations of union and intersection, the unary operation of complementation, and the relation of inclusion.

At times, it is advantageous to study a set with certain operations and/or relations singled out for particular attention. Such studies ordinarily have as their objective the characterization or classification of the set in question in terms of these operations and/or relations. Specific results in this direction will be discussed in Chapters 2 and 6. With these situations in mind, we define an *algebraic system* to be a set A together with certain specified unary or binary operations and certain relations defined on A. For example, if we denote the binary operation of multiplication on the set $\mathbf{R}$ of real numbers by $\cdot$ and the usual order relation on $\mathbf{R}$ by $\leq$, then we can regard $\mathbf{R}$ as an algebraic system with the binary operation $\cdot$ and the relation $\leq$. In this case, we would denote the resulting algebraic system by $(\mathbf{R}, \cdot, \leq)$. Similarly, if we denote the binary operation of addition on $\mathbf{R}$ by $+$ and the unary operation of absolute value on $\mathbf{R}$ by $|\cdot|$, then $(\mathbf{R},+,|\cdot|)$ and $(\mathbf{R},+,\cdot,|\cdot|,\leq)$ are also algebraic systems. Even though the three algebraic systems mentioned above are defined with respect to the same set, namely the set $\mathbf{R}$ of real numbers, the operations and relations that we choose to single out for attention are different in each case, so these are regarded as different algebraic systems. Similarly, if X is a set, if $\subset$ is the relation of set inclusion on $P(X)$, if $\cup$ and $\cap$ are the binary operations of union and intersection respectively and if $'$ denotes the unary operation of complementation, then $(P(X),\subset)$ and $(P(X),\cup,\cap,')$ are distinct algebraic systems.

The following example describes an algebraic system that we shall encounter in our study of number systems and the theory of numbers.

Example 1.13. The set $\mathbf{I}_m$ of residue classes modulo m was introduced in Section 1.3 during the discussion of equivalence relations. It is useful to define binary operations corresponding to addition and multiplication on $\mathbf{I}_m$. In order to do this, we must first prove the following statement:

(*) If $a \equiv b \pmod{m}$ and $c \equiv d \pmod{m}$,
then $a + c \equiv b + d \pmod{m}$ and $ac \equiv bd \pmod{m}$.

For example, since $7 \equiv 2 \pmod 5$ and $6 \equiv -4 \pmod 5$, we may conclude that $13 \equiv -2 \pmod 5$ and $42 \equiv -8 \pmod 5$. To prove (*), we first note that from the definition of congruence modulo m, we have $a - b = k_1 m$ and $c - d = k_2 m$ for some integers k_1 and k_2. Then $(a + c) - (b + d) = (a - b) + (c - d) = (k_1 + k_2)m$; hence, $a + c \equiv b + d \pmod{m}$. Also, $ac - bd = (a - b)c + (c - d)b = (k_1 c + k_2 d)\, m$; hence, $ac \equiv bd \pmod{m}$.

We now define two binary operations $+$ and $\cdot$ on $\mathbf{I}_m$ as follows: Given residue classes $[a]$ and $[b]$ in $\mathbf{I}_m$, define $[a] + [b]$ to be the residue class containing $a + b$, and define $[a] \cdot [b]$ to be the residue class containing ab. That is,
$$[a] + [b] = [a + b], \qquad [a] \cdot [b] = [ab].$$

The statement (*) assures us that these definitions do not depend on the particular choice of representatives for $[a]$ and $[b]$ and thus $+$ and $\cdot$ are well-defined binary operations on $\mathbf{I}_m$. We shall refer to binary operations $+$ and $\cdot$ on $\mathbf{I}_m$ as *addition modulo m* and *multiplication modulo m*.

For example, let us compute the sum and product of $[3]$ and $[4]$ in $\mathbf{I}_5$. Since the number $3 + 4 = 7$ lies in the residue class $[2]$, we see that $[3] + [4] = [2]$ in $\mathbf{I}_5$. Similarly, since $3 \cdot 4 = 12 \equiv 2 \pmod 5$, we have $[3] \cdot [4] = [2]$ in $\mathbf{I}_5$. To say that addition modulo 5 and multiplication modulo 5 are well-defined means that if different members of $[3]$ and $[4]$ had been selected, the resulting sum and product would have been no different. For example, since 13 is a representative of $[3]$ and 19 is a representative of $[4]$ for congruence modulo 5, the number $13 + 19 = 32$ determines the same residue class as the number $3 + 4 = 7$, namely $[2]$.

The basic arithmetic properties of the algebraic system $(\mathbf{I}_m, +, \cdot)$ will be studied in our discussion of number systems in Chapter 2.

In certain instances, formally different algebraic systems have the same basic mathematical structure, and for all practical purposes there is no need to study them separately. For example, consider the algebraic system $(\mathbf{R}^+, \cdot, \sqrt{\ }, <)$ determined by the set $\mathbf{R}^+$ of positive real numbers, the binary operation $\cdot$ of multiplication on $\mathbf{R}^+$, the unary operation $\sqrt{\ }$ of square root on $\mathbf{R}^+$, and the usual strict order relation $<$ for $\mathbf{R}^+$. Also consider the algebraic system $(\mathbf{R}, +, t, <)$ determined by the set $\mathbf{R}$ of *all* real numbers, the binary operation $+$ of addition on $\mathbf{R}$, the unary operation t on $\mathbf{R}$ defined by $t(x) = x/2$ for each $x \in \mathbf{R}$, and the usual strict order relation $<$ for $\mathbf{R}$. Although these two systems are by no means identical since $\mathbf{R} \neq \mathbf{R}^+$ and since the corresponding operations are not the same, there is a natural algebraic similarity between these two systems. For if $L: \mathbf{R}^+ \to \mathbf{R}$ is the function defined by
$$L(x) = \log x$$
for all $x \in \mathbf{R}^+$, then L is a one-to-one function from $\mathbf{R}^+$ onto $\mathbf{R}$ such that
$$L(x \cdot y) = L(x) + L(y)$$
$$L(\sqrt{x}) = \frac{L(x)}{2} = (t \circ L)(x)$$
$$x < y \text{ if and only if } L(x) < L(y)$$

for all x and y in $\mathbf{R}^+$. Thus, the operations and the relations on the two sets correspond to one another under the mapping L. This sort of similarity between algebraic systems is of fundamental importance in mathematics since it reflects the structural identity of two algebraic systems that are formally distinct. When two algebraic systems exhibit this sort of structural identity they are termed "isomorphic," a word whose Greek root means "same form." We shall now provide a formal definition of this term for algebraic systems.

Suppose (A, u_i, b_j, r_k) and (A', u'_i, b'_j, r'_k) are two algebraic systems determined by certain unary operations u_i, u'_i, $i = 1, \cdots, \ell$, and/or certain binary operations b_j, b'_j, $j = 1, \cdots, m$, and/or certain relations r_k, r'_k, $k = 1, \cdots, n$, defined on A and A' respectively. Note that this requires that both systems have the same number of unary operations, the same number of binary operations, and the same number of relations. Then (A, u_i, b_j, r_k) is *isomorphic* with (A', u'_i, b'_j, r'_k) if there is a one-to-one function φ from A onto A' such that

$$\varphi(u_i(x)) = u'_i(\varphi(x)), \qquad i = 1, \cdots, \ell$$
$$\varphi(b_j(x,y)) = b'_j(\varphi(x), \varphi(y)), \quad j = 1, \cdots, m$$
$$(x,y) \in r_k \text{ if and only if } (\varphi(x), \varphi(y)) \in r'_k, k = 1, \cdots, n$$

for all x, y in A. In this case, φ is called an *isomorphism* of (A, u_i, b_j, r_k) onto (A', u'_i, b'_j, r'_k). In the above example, $A = \mathbf{R}^+$, $u_1 = \sqrt{}$, $b_1 = \cdot$, $r_1 = <$, and $A' = \mathbf{R}$, $u'_1 = t$, $b'_1 = +$, $r_2 = <$ (that is, there is only one unary operation, one binary operation, and one relation specified for each system), and the function $L: \mathbf{R}^+ \to \mathbf{R}$ is an isomorphism of $(\mathbf{R}^+, \cdot, \sqrt{}, <)$ onto $(\mathbf{R}, +, t, <)$.

Example 1.14. For each real number r, define a function $f_r: \mathbf{R} \to \mathbf{R}$ by

$$f_r(x) = x + r$$

for all $x \in \mathbf{R}$. If $A = \{f_r: r \in \mathbf{R}\}$, and if $\circ$ is the binary operation of composition of functions, and if R is the relation on A defined by

$$R = \{(g,h) \in A \times A : g(x) \leq h(x) \text{ for all } x \in \mathbf{R}\},$$

then the algebraic system $(A, \circ, R)$ is isomorphic with $(\mathbf{R}, +, \leq)$ where $+$ denotes the binary operation of addition on $\mathbf{R}$ and $\leq$ denotes the usual order relation on $\mathbf{R}$. The mapping $\varphi: A \to \mathbf{R}$, defined by $\varphi(f_r) = r$ for all $f_r \in A$, provides the desired isomorphism. To verify this, it must be shown that φ is a one-to-one mapping of A onto $\mathbf{R}$ and that

$$\varphi(f_r \circ f_s) = \varphi(f_r) + \varphi(f_s)$$
$$f_r R f_s \text{ if and only if } \varphi(f_r) \leq \varphi(f_s).$$

The details are left to the reader as an exercise.

EXERCISES

1.46. If $f: \mathbf{R} \times \mathbf{R} \to \mathbf{R}$ is the binary operation on $\mathbf{R}$ defined by $f(x,y) = xy/2$, determine which of the following subsets of $\mathbf{R}$ are closed with respect to f.
   (a) $\mathbf{N}$, the set of natural numbers.
   (b) E, the set of even integers.
   (c) $\mathbf{Q}$, the set of rational numbers.
   (d) P, the set of irrational numbers.

1.47. A binary operation b on a set A is called *commutative* if $b(x,y) = b(y,x)$ for all $(x,y) \in A \times A$. Which of the following binary operations are commutative?
   (a) $b: \mathbf{R} \times \mathbf{R} \to \mathbf{R}$ defined by $b(x,y) = x - y$.
   (b) $b: \mathbf{R} \times \mathbf{R} \to \mathbf{R}$ defined by $b(x,y) = x^2 + y^2$.
   (c) $b: \mathbf{R} \times \mathbf{R} \to \mathbf{R}$ defined by $b(a,b) = \max\{a,b\}$ where $\max\{a,b\}$ denotes the larger of the numbers a and b. (For example, max $\{5,3\} = 5$, max $\{-6,2\} = 2$.)
   (d) $b: \mathbf{R}_0 \times \mathbf{R}_0 \to \mathbf{R}_0$ defined by $b(x,y) = y/x$ (where $\mathbf{R}_0$ denotes the set of nonzero real numbers).
   (e) $b: \mathbf{R} \times \mathbf{R} \to \mathbf{R}$ defined by $b(x,y) = (x+y)/2$.

1.48. A binary operation b on a set A is called *associative* if $b(x,b(y,z)) = b(b(x,y),z)$ for all x, y, z in A. (For example, addition on $\mathbf{R}$ is associative since $x + (y + z) = (x + y) + z$ for all x, y, z in $\mathbf{R}$.) Which of the binary operations of Exercise 1.47 are associative?

1.49. Verify that the mapping φ defined in the Example 1.14 is an isomorphism of $(A, \circ, R)$ onto $(\mathbf{R}, +, \leq)$.

1.50. Let $\mathbf{I}$ denote the set of all integers, let $E = \{2k: k \in \mathbf{I}\}$ denote the set of even integers, and let $+, \cdot, \leq$ denote the usual operations on $\mathbf{I}$ and E of addition and multiplication and the usual order relation, respectively. Show that the mapping $\varphi: \mathbf{I} \to E$ defined by $\varphi(k) = 2k$ for all $k \in \mathbf{I}$ is an isomorphism of $(\mathbf{I}, +, \leq)$ onto $(E, +, \leq)$. Is φ also an isomorphism of $(\mathbf{I}, +, \cdot, \leq)$ onto $(E, +, \cdot, \leq)$?

1.51. Show that if the sets X and Y can be put in one-to-one correspondence, then the algebraic systems $(P(X), \cup, \cap, ', \subset)$ and $(P(Y), \cup, \cap, ', \subset)$ are isomorphic.

APPENDIX A: INFINITE SETS

The term "infinite" occurs frequently in mathematics and has different meanings in different contexts. For example, the phrases "becoming infinite" or "tending to infinity" are sometimes used in connection with a variable or function that attains large values without bound; this notion can be made precise in the theory of limits. There is also the notion of the "point at infinity" that arises in geometry (see Chapter 8). Another sense in which the term is used is in describing a set having more than a finite number of elements; it is this usage that we wish to discuss in this section. Among Georg Cantor's major contributions to the theory of sets were the realization that infinite sets occur in different "sizes" and the introductions of techniques for distinguishing different types of infinite sets.

Georg Cantor
David Smith Collection

To motivate the basic notion of comparing two sets, we first consider the situation for finite sets. Consider the procedure by which we customarily count the objects of a given set, say a sack of marbles. We remove the marbles from the sack one at a time while saying "one, two, three, $\cdots$." When the last marble is removed, it will correspond to some natural number n and we conclude that there were n marbles in the sack. Thus, we arrive at the number of marbles in the sack by establishing a one-to-one correspondence between the set of marbles and the set $\{1,2, \cdots ,n\}$ of natural numbers from 1 to n. In a similar manner, shepherds of old kept track of their flocks by transferring pebbles from one container to a second empty container as the sheep were let out to graze, one pebble for each sheep. The pebbles were replaced in the first container as the sheep returned and if the second container was empty at the end of the day, the shepherd knew that no sheep had strayed.

These considerations suggest the following mathematical formulation of the concepts of finite and infinite sets. A set S is defined to be *finite* if it is empty or if there is a one-to-one correspondence between S and a set of the type $\{1, 2, \cdots, n\}$ for some natural number n, in which case we say that S has n elements. A set is called *infinite* if it is not finite. One type of infinite set is delineated by extending the counting process to the set **N** of all natural numbers. More precisely, a set S is called *denumerable* (or *countably infinite*) if there exists a one-to-one correspondence between S and **N**. In this sense, denumerable sets can be counted, but there is no "last" element.

Example 1.15. The set E of even natural numbers is denumerable. This is proved by exhibiting a one-to-one function f from **N** onto E. The function defined by $f(n) = 2n$ for each $n \in$ **N** is clearly a one-to-one function. Also, $f(\mathbf{N}) = E$ since each even natural number is of the form $2n$ for some $n \in$ **N**. In a similar fashion, it can be proved that the set O of odd natural numbers is also denumerable. Thus, we have two disjoint subsets E and O of **N** such that $E \cup O = \mathbf{N}$, and yet each of these subsets can be put into one-to-one correspondence with **N**. This situation is presented more colorfully by relating the story of Cantor's hotel. This hotel has denumerably many rooms numbered $1, 2, 3, \cdots$. On a certain day, every room in the hotel is filled. Then, a denumerable number of guests arrive unannounced and in need of accommodations. The desk clerk solves the housing problem by moving each occupant into a room whose number is twice the number of the room he currently occupies. This leaves the odd numbered rooms empty and the new guests can move in without difficulty!

As the above example shows, the fact that a set is a proper subset of another set does not preclude the existence of a one-to-one correspondence between them. E and **N** can be placed into one-to-one correspondence even though E is a proper subset of **N**. It is readily seen that this situation is impossible for finite sets; that is, a finite set cannot be put into one-to-one correspondence with a proper subset of itself. However, as the next proposition shows, an infinite set can always be put in one-to-one correspondence with a proper subset of itself, and this property distinguishes infinite sets from finite sets. Before proving this assertion, we observe that any infinite set contains a denumerable subset. In fact, suppose S is an infinite set and choose any element a_1 of S. Then successively select $a_2 \in S$ from the complement of $\{a_1\}$, $a_3 \in S$ from the complement of $\{a_1, a_2\}$, $a_4 \in S$ from the complement of $\{a_1, a_2, a_3\}$, and so on. This process cannot terminate, for if the elements of S were exhausted at any stage it would follow that S would be a finite set, contrary to our assumption. Thus we obtain a subset $A = \{a_1, a_2, a_3, a_4, \cdots\}$ of S which is denumerable.

Proposition 1.8. A set is infinite if and only if it can be put into one-to-one correspondence with a proper subset of itself.

PROOF. Suppose that S is an infinite set and that $A = \{a_1, a_2, a_3, \cdots, a_n, \cdots\}$ is a denumerable subset of S. Define a one-to-one correspondence between A and the proper subset $B = \{a_2, a_3, a_4, \ldots\}$ of A by setting $f(a_n) = a_{n+1}$ for each $n = 1, 2, \ldots$. This correspondence is illustrated below.

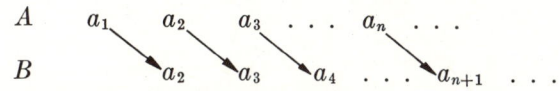

$$A \quad a_1 \quad a_2 \quad a_3 \quad \ldots \quad a_n \quad \ldots$$
$$B \quad\quad a_2 \quad a_3 \quad a_4 \quad \ldots \quad a_{n+1} \quad \ldots$$

Now if $x \in S$, but $x \notin A$, we simply set $f(x) = x$. Thus we shift the elements of A and leave the elements of S not in A untouched. This defines a one-to-one function f of S onto the proper subset $\{x \in S : x \neq a_1\}$ of S. The converse is clear.

Probably the most famous results of Cantor in the theory of infinite sets are the statements that the set of rational numbers is denumerable and the set of real numbers is not denumerable. Our next objective will be to establish these results. The proofs of these assertions should be known to every serious student of mathematics. The ideas contained in the proofs have been utilized in many areas of mathematics.

In the proof of the next proposition we will make use of the fact that a subset of a denumerable set is either finite or denumerable. This can be seen as follows. Let $S = \{s_1, s_2, \cdots, s_n, \cdots\}$ be a denumerable set where s_n is the element of S corresponding to the natural number n, and let A be a subset of S. If A is finite there is nothing to show. If A is infinite, then $A = \{s_{n_1}, s_{n_2}, \cdots, s_{n_k}, \cdots\}$ and $f(k) = s_{n_k}$ for $k \in \mathbf{N}$ defines a one-to-one correspondence between $\mathbf{N}$ and A.

Proposition 1.9. The set $\mathbf{Q}$ of rational numbers is denumerable.

PROOF. We will show that the set $\mathbf{Q}^+$ of positive rational numbers is denumerable. Since the union of two denumerable sets is again denumerable (see Exercise 1.52), the result will follow once this is established.

Instead of giving a specific formula for the one-to-one correspondence between $\mathbf{N}$ and $\mathbf{Q}^+$, we shall describe a counting technique from which such a correspondence can be obtained. We list all the positive fractions a/b as shown below where the first row contains all fractions with denominator 1, the second row contains all fractions with denominator 2, and so forth. The pictured zig-zag line then tells us how to establish a one-to-one correspondence between the set of positive fractions and **N**.

$$\begin{array}{cccccc}
\frac{1}{1} & \frac{2}{1} & \frac{3}{1} & \frac{4}{1} & \frac{5}{1} & \frac{6}{1} \cdots \\
\frac{1}{2} & \frac{2}{2} & \frac{3}{2} & \frac{4}{2} & \frac{5}{2} & \frac{6}{2} \cdots \\
\frac{1}{3} & \frac{2}{3} & \frac{3}{3} & \frac{4}{3} & \frac{5}{3} & \frac{6}{3} \cdots \\
\frac{1}{4} & \frac{2}{4} & \frac{3}{4} & \frac{4}{4} & \frac{5}{4} & \frac{6}{4} \cdots \\
\frac{1}{5} & \frac{2}{5} & \frac{3}{5} & \frac{4}{5} & \frac{5}{5} & \frac{6}{5} \cdots
\end{array}$$

We simply follow the line and count each time we pass through a fractoin. Every fraction gets counted in this way and since the set of fractions is not finite, every natural number will be used in the count. Thus, the set of fractions is denumerable. Now the elements of $\mathbf{Q}^+$ appear many times in the above list since each rational number has many equivalent representations as fractions. If we take only those fractions that are in lowest terms, then we can view $\mathbf{Q}^+$ as this subset of the set of all fractions. But since a subset of a denumerable set is either finite or denumerable, and since $\mathbf{Q}^+$ is obviously not finite, we conclude that $\mathbf{Q}^+$ is denumerable, and therefore that $\mathbf{Q}$ is denumerable.

Thus we have shown that in a sense there are as many natural numbers as there are rational numbers. We next prove that this cannot be said about the set of all real numbers. The proof makes use of Cantor's famous diagonal procedure.

Proposition 1.10. The set **R** of all real numbers is not denumerable.

PROOF. It evidently suffices to show that the set $\mathbf{R}_1 = \{x \in \mathbf{R} : 0 < x < 1\}$ is not denumerable. Suppose that $\mathbf{R}_1$ is denumerable. Then there exists a one-to-one function f from $\mathbf{N}$ onto $\mathbf{R}_1$. For each $n \in \mathbf{N}$, let us write down the decimal representation of $f(n)$.

$$f(1) = .x_{11}x_{12}x_{13}\cdots$$
$$f(2) = .x_{21}x_{22}x_{23}\cdots$$
$$f(3) = .x_{31}x_{32}x_{33}\cdots$$
$$\cdots$$
$$f(n) = .x_{n1}x_{n2}x_{n3}\cdots$$
$$\cdots$$

In case a number has two different decimal representations, which happens if a representation is eventually constantly 0 or constantly 9, we choose the one that is eventually constantly 0 in order to have a uniquely associated decimal representation for each x in $\mathbf{R}_1$. The assumption that f is an onto function means that every element of $\mathbf{R}_1$ is in the above list. The proof is completed by contradicting this assertion. This is done by constructing an x in $\mathbf{R}_1$ that is not in the above list. Define $x = .x_1x_2x_3\cdots x_n\cdots$ by

$$x_n = \begin{cases} 5, & \text{if } x_{nn} \neq 5 \\ 6, & \text{if } x_{nn} = 5. \end{cases}$$

This defines a real number x in terms of a decimal representation. x is an element of $\mathbf{R}_1$, but $x \neq f(n)$ for any choice of n since x and $f(n)$ disagree in the nth decimal place. Hence, the assumption that f is an onto function has been contradicted and the proof is complete.

Corollary. The set P of irrational numbers is not denumerable.

PROOF. Suppose P is denumerable. Since the set $\mathbf{Q}$ of rational numbers is denumerable, it would follow that $\mathbf{R} = P \cup \mathbf{Q}$ is also denumerable, for we can place P in one-to-one correspondence with the set of odd natural numbers and $\mathbf{Q}$ in one-to-one correspondence with the set of even natural numbers. But this contradicts Proposition 1.10 and thus the corollary follows.

We have already seen that an infinite set may be placed in one-to-one correspondence with certain subsets of itself. Therefore, it should not be surprising that any two intervals on the real line have the "same number" of points, that is, any two such intervals can be placed into one-to-one correspondence. One way to exhibit a one-to-one correspondence between intervals of different lengths is shown in Figure 1.12. The points x and y correspond if they lie on the same line drawn through the fixed reference point P. One could also define a specific one-to-one function from an

40 SET THEORY

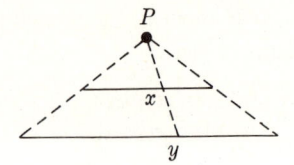

Figure 1.12

interval (a,b) onto an interval (c,d). For example, the function f defined by

$$f(x) = c + \frac{d-c}{b-a}(x-a), \quad a < x < b$$

does this as Figure 1.13 illustrates. Similar considerations can be used to

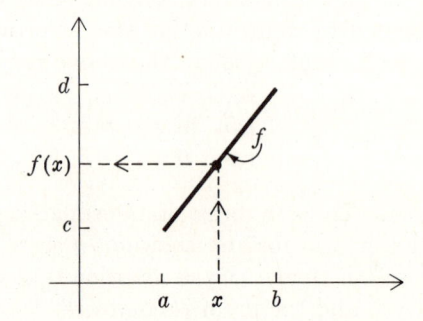

Figure 1.13

show that the entire number line can be placed in one-to-one correspondence with a given interval. One method of defining such a correspondence is shown in Figure 1.14. Here the interval $(-1,1)$ is "bent" in the middle

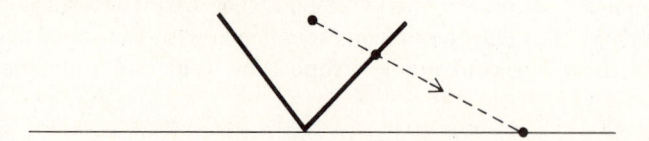

Figure 1.14

and the points projected down onto the line as illustrated. Again, a specific function can be defined, if desired. For example, $f(x) = x/(1 - x^2)$, $-1 < x < 1$, defines a one-to-one function from $(-1,1)$ onto **R**. (See Figure 1.15.)

In an attempt to obtain a set having "more" elements than the set **R** of real numbers, one might think of the Cartesian plane **R**². But a some-

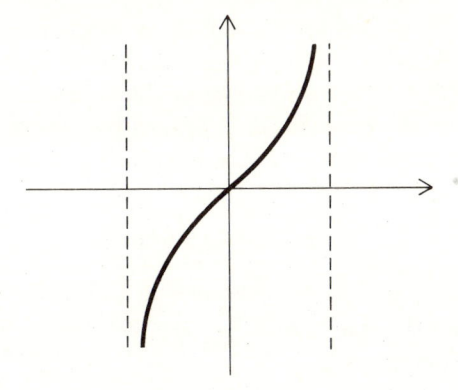

Figure 1.15

what surprising fact is that these two sets can actually be put in one-to-one correspondence. The basic idea is to use decimal representations. If the decimal representations of the coordinates of a point in the plane are given, we can "interlace" the digits to obtain a single real number. For example, the point $(\frac{1}{2}, \frac{1}{3})$ would correspond to the number $.530303 \cdots$. Conversely, given the decimal representation of a single real number, we can use the digits in the odd places and the digits in the even places to define a pair of real numbers. In this situation, there is no simple formula for a function yielding the one-to-one correspondence. We will not pursue the details.

It is natural to ask if there are any sets that are properly "larger" than **R**, that is, sets that cannot be put into one-to-one correspondence with **R** but that contain subsets that can be put into one-to-one correspondence with **R**. The following result, also due to Cantor, shows that the answer is affirmative. The set of all subsets of **R** is larger than **R** in this sense.

Proposition 1.11. If S is any set, then there does not exist a one-to-one correspondence between S and the set $P(S)$ of all subsets of S. There does exist a one-to-one correspondence between S and a proper subset of the set $P(S)$.

PROOF. The proof of the first assertion is by contradiction. Suppose there is a one-to-one function f from S onto $P(S)$. Consider the set

$$B = \{x \in S : x \notin f(x)\}.$$

Since B is a subset of S, the assumption implies that there exists an $x_0 \in S$ such that $f(x_0) = B$. Now x_0 must either be an element of B or of the complement B'. However, each of these possibilities leads to a contradiction. For if $x_0 \in B$, then from the definition of B, we see that $x_0 \notin f(x_0) = B$, a contradiction. On the other hand, if $x_0 \notin B$, then $x_0 \notin f(x_0)$ which means that $x_0 \in B$, another contradiction. This proves the first assertion. The second part of the proposition is proved by noting that the function $g: S \to P(S)$ defined by $g(x) = \{x\}$ is obviously a one-to-one function.

EXERCISES

1.52. Prove that if A and B are denumerable, then $A \cup B$ is denumerable.
1.53. Prove that if $A_1, A_2, \cdots, A_n, \cdots$ are all denumerable sets, then $\bigcup_{n \in \mathbf{N}} A_n$ is denumerable.
1.54. A real number is called *algebraic* if it is the root of a polynomial equation with rational coefficients. Show that the set of algebraic numbers is denumerable. [*Hint:* Prove that the set of roots of all polynomial equations of a fixed degree with rational coefficients is denumerable and apply Exercise 1.53.]
1.55. A real number is called *transcendental* if it is not algebraic. Prove that the set of transcendental numbers is not denumerable.
1.56. Show that if A and B are finite sets of n and m elements respectively, then $A \times B$ has mn elements.
1.57. Suppose that A and B are denumerable sets. Is $A \times B$ also denumerable?
1.58. Describe a specific one-to-one correspondence between the open interval (0,1) and the half-open interval [0,1) on the real number line.
1.59. Let $F(X)$ denote the set of all functions from a set X to $\mathbf{R}$. Show that if X is nonempty, then it is impossible to find a one-to-one correspondence between X and $F(X)$. [*Hint:* To each $A \in P(X)$, associate the function $f_A: X \to \mathbf{R}$ defined by $f_A(x) = 1$ for $x \in A$, $f_A(x) = 0$ for $x \notin A$. Apply Proposition 1.11.]

APPENDIX B: A GLIMPSE AT AXIOMATIC SET THEORY

During the last quarter of the nineteenth century, informal set theory evolved rapidly from the founding efforts of Georg Cantor (1845–1918) to the set-theoretic development of arithmetic by Gottlob Frege (1848–1925) at the turn of this century. All of this initial work was based on Cantor's original description of a set as "any collection into a whole of definite and separate objects of our intuition or our thought." In 1902, the foundations of informal set theory were shaken by an observation due to Bertrand Russell (1872–1970) regarding the meaning that had been ascribed to the term "set" up to that time. This observation, which has come to be known as Russell's paradox, provided the stimulus for the axiomatic development of set theory that continues to the present day.

Russell's paradox can be rather simply described as follows: First of all, Russell noted that all sets are of one of two basic types, sets that are

Bertrand Russell
The Granger Collection

members of themselves and sets that are not members of themselves. For example, "the set of all sets that can be defined in fewer than twenty English words" as well as "the set of all concepts" are sets that are members of themselves since the former set is defined above using fifteen English words while the latter set is certainly a concept itself. On the other hand, most sets that come to mind such as "the set of positive integers" or "the set of all plane triangles" are not members of themselves. We shall refer to sets that are not members of themselves as *ordinary sets* while sets that are members of themselves will be referred to as *extraordinary sets*.

Now consider the collection O of all ordinary sets. Observe that O itself cannot be an ordinary set since $O \notin O$ would imply that $O \in O$ by definition of O. On the other hand, O itself cannot be an extraordinary set since it would then follow that $O \in O$, which in turn would imply that O is an ordinary set by definition of O. Thus, the collection O can neither be an ordinary set nor an extraordinary set. Consequently, even though the collection O qualifies as a set according to Cantor's original definition, it cannot be regarded as a set.

There are several less formal ways of describing the basic idea behind Russell's paradox which may serve to clarify the reasoning used above. For example, consider the small town (male) barber who shaves *precisely* those men in town who do not shave themselves. Does the barber shave himself or not? If he does, then he should not, and if he does not, then he should! Thus, such a barber cannot exist.

Another interesting formulation proceeds as follows: A librarian notes that certain catalogues in his library list themselves as entries while others do not. He decides to compose a new catalogue for his library that will list all of the catalogues that do not list themselves. The question is: Should the librarian list his new catalogue as an entry in his new catalogue? If he does, then the catalogue becomes a catalogue that lists itself and therefore should not be listed in the new catalogue. On the other hand, if the new catalogue does not list itself, then it should be listed as an entry in the new catalogue. Thus, the librarian cannot compose a catalogue that will list all catalogues that do not list themselves!

Russell's paradox forced mathematicians to carefully examine the foundations of informal set theory so that a suitable, more restrictive concept of set could be found which would exclude the collection O mentioned above and yet provide a workable basis for the theory of sets. Before and after the discovery of Russell's paradox, other paradoxes in informal set theory were found. Though these paradoxes seemed on the surface to be less fundamental than Russell's paradox, they contributed significantly to the subsequent and continuing drive to place the theory of sets on a sound axiomatic foundation.

The first axiomatic development for set theory was provided by Ernst Zermelo (1871–1956) early in this century. Other axiomatic systems were later developed by Russell and Alfred North Whitehead (1861–1947) and still later by W. V. Quine (1908–   ). We shall now outline a modified version of the Zermelo axiom system for set theory.

The primitive undefined terms in this axiomatic theory of sets are the terms "set" and "set membership." All of the eight axioms stated below are concerned either with the specification of the basic relations between these undefined terms or with the existence of certain special sets. In stating these basic axioms we shall employ lower case letters x, y, z, and so on, for sets, the symbol $\in$ for set membership, and $\notin$ for the negation of this relation. The reader should note that this system of axioms does not distinguish between sets and elements; that is, it is not necessary to regard the concept of "element" as a separate undefined term of the theory.

(1) The Axiom of Extensionality. If x and y have the property that for all z it is true that $z \in x$ if and only if $z \in y$, then $x = y$.

Less formally, this axiom states that two sets are identical if they contain the same elements. The converse statement is also true; that is, if two sets are identical, then they contain the same elements. However, this converse statement is a consequence of the fact that equality means identity (that is, $x = y$ means that x and y name the same object) so it need not be included as an axiom.

In order to see that the Axiom of Extensionality is specifying an important property of the membership relation, it is useful to consider a situation in which the membership relation $\in$ is given a concrete meaning quite different from its intended meaning in informal set theory and for which the statement corresponding to the Axiom of Extensionality does not hold. For example, if the letters x and y denote people and if $x \in y$ is interpreted to mean that x is an ancestor of y, then the statement corresponding to the Axiom of Extensionality in this setting is "Two people x and y are identical if they have the same ancestors." This statement is obviously false since brothers and sisters have the same ancestors. However, the converse statement is true; that is, x and y have the same ancestors if $x = y$ because, as we have already noted, the converse depends only on the fact that equality means identity.

(2) Axiom of the Empty Set. There is a set denoted by $\emptyset$ and called the *empty set* such that the statement $z \in \emptyset$ is false for all z.

There is only one set that can serve as the set $\emptyset$ in the preceding axiom. For if $\emptyset'$ is also a set satisfying the property stated in axiom (2), then it would be true that $z \in \emptyset$ if and only if $z \in \emptyset'$. (In fact, since the statements $z \in \emptyset$ and $z \in \emptyset'$ are actually false for all z the condition "$z \in \emptyset'$ if and only if $z \in \emptyset'$" would be satisfied by default.) But then, by Axiom (1), it would follow that $\emptyset = \emptyset'$.

(3) Axiom of Union. Given a set x there exists a set y called the *union of the set* x such that $z \in y$ if and only if $z \in w$ for some $w \in x$.

The informal set-theoretic basis for the preceding axiom is easy to clarify. The set x mentioned in the axiom is to be regarded as a set of sets, that is, $x = \{S_\alpha : \alpha \in I\}$ where S_α is a set for each α in some indexing set I. With this interpretation for x, the union y of x mentioned in the axiom is just the set $y = \bigcup_{\alpha \in I} S_\alpha$.

EXERCISE

1.60. Use the Axiom (1) to prove that there is only one set y with the properties stated in Axiom (3), that is, the union of the set x is unique.

(4) Axiom of Unordered Pairs. If x and y are sets, there is a set z such that $w \in z$ if and only if $w = x$ or $w = y$.

The interpretation of the preceding axiom within the context of informal set theory is obtained by regarding the set z mentioned in the axiom to be the set $\{x,y\}$ consisting of the two elements x and y. For this reason, the set z is usually referred to as the *unordered pair with components x, y*. As in the case of the empty set and the union, Axiom (1) implies that for given x and y, there is only one set z with the properties described in Axiom (4), that is, the unordered pair with components x, y is unique.

In connection with the above definition of unordered pairs, it should be noted that for each set x, the *singleton* $\{x\}$ (that is, the set consisting of x alone) can be defined as the unordered pair $\{x,x\}$ with components x and x. Note that $z \in \{x\}$ if and only if $z = x$.

It is not necessary to include another axiom stipulating the existence of ordered pairs. For, as we have already noted in Section 1.3, we can construct ordered pairs on the basis of unordered pairs. In fact, we define the *ordered pair (x,y) with first component x and second component y* to be the unordered pair whose components are $\{x\}$ and the unordered pair $\{x,y\}$, that is, $(x,y) = \{\{x\},\{x,y\}\}$.

It is also not necessary to include special axioms asserting the existence of ordered triples (x,y,z), ordered quadruples (x,y,z,w), and so on, because, for example, we could define (x,y,z) to be the ordered pair $((x,y),z)$ whose first component is the ordered pair (x,y) and whose second component is z. Unordered triples $\{x,y,z\}$, unordered quadruples $\{x,y,z,w\}$, and so on, can also be defined on the basis of unordered pairs with the aid of Axiom (3). For example, we can define $\{x,y,z\}$ to be the union of the set $\{\{x,y\},z\}$ while $\{x,y,z,w\}$ can be defined as the union of the set $\{\{x,y,z\},w\}$.

EXERCISE

1.61. Provide an appropriate definition for the phrase "the set z is the Cartesian product of the sets x and y" and prove that the Cartesian product of two sets always exists using only Axioms (1)–(4).

(5) Axiom of the Power Set. If x is a set, there is a set y called the power set of x such that $z \in y$ if and only if $w \in z$ implies $w \in x$.

In our description of informal set theory, we denoted the power set y of x by $P(x)$. Also, when the condition "$w \in z$ implies $w \in x$" was satisfied, we said that z was a *subset* of x and we wrote $z \subset x$. Thus, the set y described in Axiom (5) is the "set of all subsets of the set x" which we referred to as the power set of x in Section 1.2.

(6) Axiom of Infinity. There exists a set y such that $\emptyset \in y$ and such that if $x \in y$, then $\{x\} \in y$.

The primary concern of this axiom is the existence of the infinite sets of informal set theory. To establish the existence of such sets, we first note that it is possible to prove on the basis of Axioms (1) through (5) alone that the sets $\emptyset$, $\{\emptyset\}$, $\{\{\emptyset\}\}$, $\cdots$ are all distinct. In fact, $\emptyset \neq \{\emptyset\}$ since $\emptyset \in \{\emptyset\}$ while $\emptyset \notin \emptyset$ by definition of $\emptyset$. By the same reasoning $\emptyset \neq \{\{\emptyset\}\}$. Moreover, it is easy to conclude from the definition of singleton and Axiom (1) that $\{x\} = \{y\}$ if and only if $x = y$. Consequently $\{\emptyset\} \neq \{\{\emptyset\}\}$ since $\emptyset \neq \{\emptyset\}$, so that $\emptyset$, $\{\emptyset\}$, $\{\{\emptyset\}\}$ are three distinct sets. Proceeding in this way, we can verify that the sets $\emptyset$, $\{\emptyset\}$, $\{\{\emptyset\}\}$, $\cdots$ are all distinct. However, one cannot prove on the basis of Axioms (1) through (5) alone that there is a set y such that $\emptyset \in y$, $\{\emptyset\} \in y$, $\{\{\emptyset\}\} \in y$, and so on; the existence of such a set is implied by Axiom (6).

(7) Axiom of Specification. If x is a set and if $S(\cdot)$ is an open statement in which the variable has the domain x, then there is a set y such that $z \in y$ if and only if $S(z)$ is true.

Informally speaking, this axiom states that any open statement whose domain is a set determines a subset of the given set, namely, the subset consisting of those elements in the given set for which the statement is true. Moreover, this subset is uniquely determined by the given set x and the open statement $S(\cdot)$ by virtue of Axiom (1).

By making use of Axiom (7), we can establish the following result: *Given any set x, there is always a set y such that $y \notin x$.* To verify this assertion, we proceed as follows. If x is any set and if z is a variable with domain x, the open statement $S(\cdot)$ which is the negation of $z \in z$, that is,

$$S(z): z \notin z,$$

defines a subset y of x according to *Axiom* (7). We now observe that $y \notin x$. For if $y \in x$, then either $S(y)$ is true or $S(y)$ is false. If $S(y)$ is true, then $y \in y$ which, by definition of y, implies that $S(y)$ is false—a contradiction. On the other hand, if $S(y)$ is false, then $y \in y$ by definition of $S(\cdot)$, so $S(y)$ is true by definition of y—again a contradiction. Thus we see that $y \notin x$. (Note the similarity between the reasoning applied here and that used in the discussion of Russell's Paradox.)

The assertion proved in the preceding paragraph shows us that there is no such thing as "the set of all sets." This fact is intimately related to our earlier discussion of Russell's Paradox. In that discussion, the sentence "z is an ordinary set" was treated as an open statement in which the variable z had the collection of all sets as its domain. However, since this collection is not a set, as we have just seen, this sentence is really not an open statement at all. In view of this, it is not surprising that the collection of all ordinary sets is not itself a set. Thus, there is no conflict between Axiom (7) and the fact that the sentence "z is an ordinary set" does not define a set.

EXERCISE

1.62. Provide an appropriate definition of the phrase "y is the *intersection* of the set x" and prove that for each set x there is a unique set y which is the intersection of x by using Axioms (1) through (7).

(8) Axiom of Choice. If x is a set, if y is the union of x and if $z \in x$ implies $z \neq \emptyset$, then there is a function $f: x \to y$ such that for each $z \in x$, it is true that $f(z) \in z$.

In the preceding axiom, we regard x to be a "set of nonempty sets," say $x = \{A_\alpha : \alpha \in I\}$ where for each α in the indexing set I, A_α is a nonempty set. The function $f: x \to y$ mentioned in the axiom assigns to each A_α an element $f(A_\alpha) \in A_\alpha$. In other words, the function f "chooses" an element from each set $A_\alpha \in x$. For this reason f is sometimes called a "choice function" for the set x.

If x is a finite set, the function $f: x \to y$ required by Axiom (8) can be shown to exist by making use of Axiom (3) and Axiom (4) since the existence of such a function is guaranteed by the existence of ordered n-tuples [see the discussion after Axiom (4)]. Also, for certain infinite sets x, it may be possible to prove the existence of such a function directly without appeal to Axiom (8). To illustrate this latter point, let x be the set of all nonempty subsets of the set of positive integers. Each $z \in x$ contains a unique integer n_z which is less than all other integers in z; that is, each $z \in x$ has a unique least element n_z. If y is the union of x, then the function $f: x \to y$ defined by $f(z) = n_z$ for $z \in x$ is a function of the sort required by Axiom (8). However, if we take x to be the set of all nonempty subsets of the set of real numbers, then it was shown in 1963 by P. J. Cohen (1934–) that no function $f: x \to y$ such that $f(z) \in z$ for all $z \in x$ can be proved to exist without utilizing the Axiom of Choice. In particular, his result shows that Axiom (8) is not a consequence of the other axioms of set theory.

Russell used the following less mathematical but more colorful example to illustrate the difficulty of finding a choice function for some infinite sets. Consider an infinite collection x of pairs of shoes. The function which assigns to each $z \in x$ the left shoe in z is clearly a choice function for x. However, if we replace x by an infinite collection of pairs of socks, no obvious choice function presents itself since the two socks in any given pair are identical. However, by Axiom (8), a choice function does exist in this case also.

This completes our list of axioms essentially embodying the set theory proposed by E. Zermelo in 1908. This axiom system is quite satisfactory since most of mathematics, certainly all of mathematics on an elementary level, can be formulated within the framework of Zermelo's set theory.

Various refinements and extensions of the axiomatic theory of sets followed Zermelo's pioneering efforts. In particular, in 1922, A. Fraenkel (1891–1965) replaced the Axiom of Specification by a stronger axiom usually referred to as the Axiom of Replacement. Although this axiom is technically preferable, it is difficult to describe informally. This axiom system was further augmented in 1925 by John von Neumann (1903–1957) with the introduction of an axiom known as the Axiom of Regularity. The main purpose of this axiom is to rule out certain types of irregularities such as a set being an element of itself. The resulting axiom system for set theory is now generally referred to as the Zermelo-Fraenkel or the Zermelo-Fraenkel–von Neumann system.

Of all the axioms descussed above, the Axiom of Choice is the most difficult to accept as a "self-evident" truth. It has proved to be of great value in mathematics, but its use can lead to rather surprising and unintuitive conclusions. However, in 1938, K. Gödel (1906–   ) proved that if no contradictions can be derived in the Zermelo-Fraenkel system without making use of the Axiom of Choice, then no contradictions can be derived if this axiom is added. In other words, the inclusion of the Axiom of Choice does not increase the possibility that the resulting theory will lead to contradictions. This raises the question of whether or not there may be contradictions within Zermelo-Fraenkel set theory. The answer to this question is not known. However, Gödel did prove that *if*, in fact, there are no contradictions in this set theory, then it is impossible to *prove* within the context of the theory that there are no such contradictions!

REMARKS AND REFERENCES

As we pointed out in the introductory remarks for this chapter, set theory provides a language that may be used to formalize and communicate mathematical ideas. Although set theory is certainly a subject that can be studied for its own sake, the purpose for its inclusion in school mathematics is to enable textbook, teacher, and student to express themselves more clearly and precisely and to organize mathematical content and procedures.

Virtually all of the mathematical subject matter that is taught from kindergarten through college can be formalized in terms of set theory. However, mathematical concepts exist independently of their set theoretic formulation, and the process of doing and learning mathematics certainly involves much more than simply mastering and applying the techniques of set theory. The decision regarding the extent to which set theoretic considerations should be stressed in the classroom treatment of a given topic should be based on the mathematical maturity of the students involved and the relevance of the set theoretic formalism to a proper understanding of that topic.

Most students now begin their secondary school mathematics program with some intuitive background concerning sets. This background is ordinarily quite modest, usually amounting to little more than an acquaintance with set notation and some experience with computing unions, intersections, and complements. Nevertheless, after some review and reinforcement, this background can provide an adequate set-theoretic basis for the first-year algebra course.

As we have seen, the definitions and basic properties of the set operations of union, intersection, and complementation depend on the meaning of the logical connectives "and," "or," "not." This might suggest that the classroom discussion of these concepts ought to be preceded by some work in logic. However, from a pedagogical point of view, this neither seems necessary nor advisable. Students at the secondary school level generally accept the meaning and properties of these connectives as being rather "obvious" and any attempt to formalize this material may actually confuse rather than clarify matters. Venn diagrams can be very useful and instructive in the presentation of the basic properties of the set operations, particularly if they are used to discover as well as illustrate set equations.

It is quite worthwhile to ask students to compare the properties of the set operations with the properties of the more familiar arithmetic operations. Such a comparison will point out a number of similarities and differences, and thereby enable the student to appreciate the significance of these properties. For example, if the arithmetic operation of addition is compared with the set operation of union, it can be seen that both share the commutative and associative properties and that the arithmetic property $a + 0 = a$ is analogous to the set-theoretic property $A \cup \emptyset = A$.

On the other hand, while $A \cup A = A$ for *any* set A, the corresponding arithmetic equation $a + a = a$ is true only if a is zero.

The concepts of variable, constant, and open statement are important to the development of algebra. These ideas are most easily grasped by the beginner if they are introduced by means of well-chosen examples drawn from the students' mathematical and nonmathematical backgrounds. The teacher should take care to specify explicitly the domains of the variables and the truth sets for the open statements in such examples. When the open statement used is an equation or inequality involving a variable whose domain is a subset of the real number line **R**, the domain of the variable as well as the truth set (that is, solution set) for the equation or inequality should be graphed on the real number line. After some appreciation of the meaning of these terms has been gained through the consideration of examples, the formal definitions should be studied so that the student can interpret new situations involving these terms for himself.

The content of secondary school mathematics provides an opportunity for the discussion of a number of order and equivalence relations. The order relations defined by inequality for real numbers and inclusion for sets and the equivalence relations defined by equivalence of fractions, congruence and similarity of triangles are perhaps the most significant of these. In the classroom, a careful study of the properties of such concrete examples of relations can lead to the proper conception of a relation as a set of ordered pairs and to an appreciation of the meaning of such restrictions as "reflexive," "transitive," "symmetric," and "antisymmetric." Graphing relations on the set **R** of real numbers (that is, plotting such relations as subsets of the Cartesian plane $\mathbf{R}^2$) should be stressed.

The function concept is fundamental in algebra, geometry and, for that matter, in almost all areas of mathematics. The less formal "rule" definition of function seems to be the best starting point for a classroom discussion of this topic even though the term "rule" must be left rather vague at the beginning. After the scope of this definition has been clarified with a variety of examples, considerable emphasis should be placed on the graphical representation of functions $f: \mathbf{R} \to \mathbf{R}$ (where **R** is the set of real numbers). Students have ordinarily encountered graphing problems for special types of functions by the time they begin their secondary school training. Further work on graphing leads quite naturally to the consideration of functions as special sets of ordered pairs since it is natural to identify a function $f: \mathbf{R} \to \mathbf{R}$ with its graph which is a subset of $\mathbf{R}^2$. In this way, one can lay the groundwork for the more abstract definition of function in terms of ordered pairs.

Composite functions and inverse functions do not begin to play a significant role in the secondary school program until the elementary trigonometric functions and the exponential and logarithmic functions are considered. For this reason, these notions as well as the related ideas of one-to-one and onto functions are not usually considered in the first course in algebra. However, one-to-one correspondences between finite sets are considered early (even before secondary school) because of their relevance to the counting process.

Very little consideration is given to one-to-one correspondences between infinite sets in most secondary school texts. Although this material is both interesting and accessible to good students at this level, it is ordinarily omitted because it is rather independent of the remainder of the program and because it has little immediate application at this stage of the student's mathematical development. Nevertheless, the teacher will find this subject matter to be excellent enrichment material for independent study and mathematics club lectures by students inclined to pure mathematics.

We shall now mention some references for further reading. Hundreds of examples and exercises illustrating the basic concepts of set theory can be found in the supplementary problem book by Lipschutz [3]. The short book by Halmos [1] presents a very readable and well-motivated account of set theory without the trappings of symbolic logic. Hamilton and Landin [2] discuss the basic notions of set theory and then demonstrate how the theory of sets can be used to construct models for the fundamental number systems of mathematics. The book by Stoll [5] is a more extensive work that relates set theory, logic, number systems, Boolean algebras, and axiomatic systems in general. Several nontechnical articles on set theory and the foundations of mathematics can be found in [4].

1. Halmos, P. R., *Naive Set Theory*, Princeton, N.J.: D. Van Nostrand Company, Inc., 1960.
2. Hamilton, N. T. and J. Landin, *Set Theory, the Structure of Arithmetic*, Boston: Allyn and Bacon, Inc., 1961.
3. Lipschutz, S., *Set Theory and Related Topics*, Schaum's Outline Series, New York: McGraw-Hill Book Company, Inc., 1964.
4. *Mathematics in the Modern World*, Readings from *Scientific American* with introductions by Morris Kline, San Francisco: W. H. Freeman and Company, Inc., 1968.
5. Stoll, R. R., *Set Theory and Logic*, San Francisco: W. H. Freeman and Company, Inc., 1963.

Chapter 2

NUMBER SYSTEMS AND THEIR PROPERTIES

The development of number systems must certainly be regarded as one of mankind's greatest accomplishments. It has provided man with a practical means of dealing with economics as larger social units evolved and it has enabled him to make a scientific study of his environment so that he could make nature his servant. To be sure, this development was a long and arduous process; as a matter of fact, the story of man's effort to invent, use, and understand numbers seems to be a part of all of his recorded history.

Although the usual manner in which we are introduced to the number systems is quite natural and appropriate, it makes a serious study of number systems and their properties somewhat difficult when we are older. For we learn to compute with numbers long before we are sophisticated enough to wonder what numbers really are, and we ordinarily come to know (almost unconsciously) the many properties of the arithmetic operations with numbers through computational experience rather than deductive reasoning. However, if a student is ever to go beyond the stage of performing routine computations patterned after those performed by his teacher or displayed in his textbook, it is absolutely necessary that he develop some appreciation of the number systems as deductive systems. The structures found in the various number systems are typical of the important mathematical structures that arise when one studies nature by means of mathematical models, and an understanding of important mathematical concepts such as convergence, operator, matrix, and so on, must ultimately depend on an understanding of and an intuitive feeling for the properties and deductive character of the familiar number systems.

2.1. THE NATURAL NUMBER SYSTEM

The set $\mathbf{N} = \{1,2,3,4, \cdots, n, \cdots\}$ of all positive whole numbers will be referred to as the *natural number system* and the elements of $\mathbf{N}$ will be called *natural numbers*. We shall take the meaning of the elements of $\mathbf{N}$ for granted and concentrate on the task of displaying and studying a list of axioms about $\mathbf{N}$ that turn out to distinguish this set from all other sets that we might consider. These axioms are called the *Peano axioms*, named after Guiseppe Peano (1858–1932) whose efforts to place arithmetic on a sound deductive foundation gave impetus to the development of modern axiomatic mathematics.

The Peano axioms can be abstractly formulated as follows: The natural number system is a set $\mathbf{N}$ with a unary operation s that has the following properties:

(P-1) There is a unique element $e \in \mathbf{N}$ such that $e \neq s(x)$ for all $x \in \mathbf{N}$.

(P-2) s is a one-to-one function.

(P-3) If $M \subset \mathbf{N}$, if $e \in M$ and if $s(x) \in M$ whenever $x \in M$, then $M = \mathbf{N}$.

More explicitly, the function s is defined on $\mathbf{N}$ by

$$s(n) = n + 1$$

for $n \in \mathbf{N}$, and the element e referred to in (P-1) is the natural number 1. With these more explicit descriptions of s and e, we can paraphrase (P-1) through (P-3) as follows:

(P-1)' There is a unique natural number called 1 that is not equal to $n + 1$ for any natural number n.

(P-2)' If n and m are natural numbers such that $n + 1 = m + 1$, then $n = m$.

(P-3)' If M is a set of natural numbers such that
   (a) $1 \in M$,
   (b) the assumption that $n \in M$ implies that $n + 1 \in M$,
then M coincides with the set of all natural numbers.

If it was our purpose here to present a deductive development of the natural number system, we could start with the Peano axioms (P-1) through (P-3) and proceed to *define* operations of addition and multiplication and an order relation on $\mathbf{N}$ and then establish all the familiar arithmetic properties of $\mathbf{N}$ as theorems. The resulting theoretical development would be quite similar in spirit to that found in a traditional secondary school plane geometry course in that all of the theorems would be developed in a logical deductive fashion from a basic set of axioms and undefined terms. However, unlike plane geometry, we would not be likely to encounter many theorems regarding the natural numbers that would appear to be surprising or new in their content because of our great informal familiarity with the arithmetic properties of natural numbers. The object of such a development would be the logical organization of arithmetic rather than the development of unfamiliar arithmetical facts.

It is possible on the basis of set theory alone to construct a set S and define a unary operation s on S having properties (P-1) through (P-3). Although this construction leads to a specific model for the natural number system, it can be proved that any two models for the natural number system must be isomorphic as algebraic systems. Source material dealing with the axiomatic development of number systems and their construction via set theory can be found in the Remarks and References section at the end of this chapter.

It is not our purpose to present an extensive development of the natural number system in this book. Instead, we shall dwell on the Axiom (P-3)' because it is this axiom which embodies the important Principle of Mathematical Induction.

Textbooks often formulate the Principle of Mathematical Induction in terms of statements about natural numbers. One such formulation is as follows.

Principle of Mathematical Induction. Suppose that for each natural number n, $P(n)$ is a statement. If

(a') $P(1)$ is true,

(b') the assumption that $P(n)$ is true implies that $P(n + 1)$ is true,

then the statement $P(n)$ is true for each natural number n.

This form of the Principle of Mathematical Induction (which we shall refer to as (PMI) for the sake of convenience) is logically equivalent to Axiom (P-3)' in the following sense:

(1) (PMI) *may be derived as a theorem if we accept* (P-3)' *as a "given" or axiomatic property of* **N**.

(2) (P-3)' *may be derived as a theorem if we accept* (PMI) *as a "given" logical principle*.

The derivation mentioned in (1) can be carried out as follows: We are given that for each natural number n, $P(n)$ is a statement such that:

(a') $P(1)$ is true,

(b') the assumption that $P(n)$ is true implies that $P(n + 1)$ is true.

We are also *given* statement (P-3)' about **N**. We must *prove* that $P(n)$ is true for all n. Another way to put this conclusion is as follows: If $M = \{n \in \mathbf{N}: P(n) \text{ is true}\}$, then $M = \mathbf{N}$. To verify this conclusion, we first note that $1 \in M$ because of the *given* statement (a'). Moreover, the assumption that $n \in M$ for a natural number n means that $P(n)$ is assumed true for that n. Consequently, by the *given* statement (b'), it would follow that $P(n + 1)$ is also true. By definition of M, this implies that $n + 1 \in M$. Thus, the assumption that $n \in M$ implies that $n + 1 \in M$. Therefore, we have shown that M is a set of natural numbers satisfying the hypothesis (a), (b), of (P-3)'. Consequently, since we accept (P-3)' as a given property of **N**, we are forced to the conclusion that $\mathbf{N} = M$; that is, $P(n)$ is true for all natural numbers n, which is the desired result.

The derivation mentioned in (2) proceeds as follows: This time, we are *given* that M is a set of natural numbers satisfying (a) and (b) of (P-3)' and we are *given* (PMI) as an accepted logical principle. We must *prove* that $M = \mathbf{N}$; that is, we must prove that the statement $P(n)$, defined by

$$P(n)\colon n \text{ is an element of } M,$$

is true for all natural numbers n. But the given statement (a) about M implies that $P(1)$ is true, and the given statement (b) about M shows that the assumption that $P(n)$ is true implies that $P(n + 1)$ is true. Consequently, by applying the given logical principle (PMI), we are forced to conclude that $P(n)$ is true for each natural number n, that is, $\mathbf{N} = M$, which is the desired conclusion. Thus, we have demonstrated the logical equivalence of (P-3)' and the Principle of Mathematical Induction.

Some work with mathematical induction is quite generally considered to be a part of the content of the algebra courses in the secondary schools. It is safe to say that the majority of the students find mathematical induction to be one of the most difficult topics in the algebra program. For this reason, it seems worthwhile to explore some of the common difficulties and misconceptions that arise with this topic.

One frequent source of confusion regarding mathematical induction involves the hypothesis (b') of (PMI). In order to understand the real meaning of (b'), one should carefully note that in (b') we seek to *deduce* the truth of $P(n + 1)$ as a consequence of the *assumption* that $P(n)$ is true. It makes no difference as far as the verification of (b') is concerned whether or not $P(n)$ is actually true.

For example, if $P(n)$ is the statement "$n + 7 = n$," then $P(n)$ is evidently false for all natural numbers n. However, the *assumption* that $P(n)$ is true for a given natural number n does imply that $P(n + 1)$ is also true! (For if $n = n + 7$, then $n + 1 = (n + 7) + 1 = (n + 1) + 7$.) Therefore, (b') of (PMI) is satisfied. Fortunately, $P(1)$ is not true so that (PMI) does not enable us to conclude that $n = n + 7$ for all natural numbers n.

2.1. THE NATURAL NUMBER SYSTEM

The following example should serve to further clarify the meaning of condition (b′).

Example 2.1. A standard example illustrating the use of mathematical induction deals with the verification of the formula

$$1 + 2 + \cdots + n = \frac{n(n+1)}{2}$$

for every natural number n. The appropriate statement $P(n)$ in this case is

$$P(n): 1 + 2 + \cdots + n = \frac{n(n+1)}{2}.$$

The verification of (a′) in (PMI) presents no difficulty since

$$1 = \frac{1(1+1)}{2}$$

is clearly true. Condition (b′) of (PMI) can be verified by a computation such as

$$1 + 2 + \cdots + n + (n+1) = \frac{n(n+1)}{2} + (n+1) = \frac{(n+1)(n+2)}{2}$$

obtained by adding $n + 1$ to both sides of the equality assumed in $P(n)$. Therefore, (PMI) implies that $P(n)$ is true for every natural number n. Note that in the above computation used to verify (b′) in (PMI), we merely assumed $P(n)$ to be true in order to show that it would follow from this assumption that $P(n + 1)$ would also be true. In other words, we did not concern ourselves with the actual truth of $P(n)$ in verifying (b′), but rather only with the fact that $P(n + 1)$ is a consequence of the assumed truth of $P(n)$.

Another worthwhile observation concerning the use of mathematical induction is illustrated in the following example.

Example 2.2. Derive the formula for the sum of the cubes of the first n natural numbers.

Our first task is to arrive at a suitable conjecture for such a formula. We do this by examining the sum of the cubes of the first n natural numbers for small values of n.

$$\begin{align}
n = 1: &\quad 1^3 = 1, \\
n = 2: &\quad 1^3 + 2^3 = 9, \\
n = 3: &\quad 1^3 + 2^3 + 3^3 = 36, \\
n = 4: &\quad 1^3 + 2^3 + 3^3 + 4^3 = 100, \\
n = 5: &\quad 1^3 + 2^3 + 3^3 + 4^3 + 5^3 = 225.
\end{align}$$

We note that these sums are all perfect squares, and a bit more reflection shows that:

$$n = 1: \quad 1 = 1^2,$$
$$n = 2: \quad 9 = (1 + 2)^2,$$
$$n = 3: \quad 36 = (1 + 2 + 3)^2,$$
$$n = 4: \quad 100 = (1 + 2 + 3 + 4)^2,$$
$$n = 5: \quad 225 = (1 + 2 + 3 + 4 + 5)^2.$$

On the basis of these observations and the formula derived in the preceding example, we *conjecture* the following formula:

$$P(n): 1^3 + \cdots + n^3 = \frac{n^2(n+1)^2}{4}$$

for all natural numbers n.

We shall now *prove* this formula by making use of the Principle of Mathematical Induction. Since we have already noted that $P(1)$ is true, it is only necessary to prove that $P(n + 1)$ is true if we assume that $P(n)$ is true. However, this conclusion follows from the computation:

$$1^3 + \cdots + (n+1)^3 = (1 + \cdots + n^3) + (n+1)^3$$
$$= \frac{n^2(n+1)^2}{4} + (n+1)^3$$
$$= \frac{(n+1)^2[(n+1)+1]^2}{4}.$$

Consequently, (PMI) shows that the conjectured formula is valid for all natural numbers n.

Notice that mathematical induction played no role in the discovery of the above formula, but was used only to confirm that the formula was correct. If our conjecture had been incorrect, however, then an attempt to verify it by mathematical induction would have brought the error to light. For example, suppose we had conjectured that

$$1^3 + 2^3 + \cdots + n^3 = \frac{19n^2 - 41n + 24}{2},$$

a formula that is true for $n = 1, 2, 3$. If we try to prove this formula by mathematical induction, we obtain

$$1^3 + 2^3 + \cdots + n^3 + (n+1)^3 = \frac{19n^2 - 41n + 24}{2} + (n+1)^3$$
$$= \frac{2n^3 + 25n^2 - 35n + 26}{2}$$

which is in obvious disagreement with our conjecture. Thus, mathematical induction can be used to discard incorrect conjectures as well as to verify correct ones.

Mathematical induction is quite useful for verifying formulas such as those presented in the preceding examples. However, one should not draw the conclusion that it is necessary or even preferable to verify all such formulas in this way. An extreme example that serves to illustrate this point is provided by the formula

$$(n + 1)^2 = n^2 + 2n + 1 \qquad (n \in \mathbf{N}).$$

To be sure, the Principle of Mathematical Induction provides a proof of this formula for all natural numbers n. However, a direct verification is trivial, and it is clear that the assumption made in (b') of (PMI) is not really needed to verify the conclusion of that statement.

We also note that the formula

$$1 + 2 + \cdots + n = \frac{n(n + 1)}{2} \qquad (n \in \mathbf{N})$$

established in the first example considered above may be easily verified without appeal to the Principle of Mathematical Induction. In fact, if we define s_n to be the sum of the first n natural numbers, then

$$\begin{aligned} s_n &= 1 + \cdots + n \\ + s_n &= n + \cdots + 1 \\ \hline 2s_n &= (n + 1) + \cdots + (n + 1) = n(n + 1). \end{aligned}$$

Hence,

$$1 + \cdots + n = s_n = \frac{n(n + 1)}{2}$$

for each natural number n.

There is an amusing anecdote concerning the great mathematician K. F. Gauss (1777–1855) that relates to the above formula for the sum of the first n natural numbers. According to the story, one of his grade school teachers assigned the task of adding all the integers from 1 to 100 to Gauss and his classmates as a form of punishment. To his teacher's wonder, Gauss produced the correct answer almost immediately after the assignment was given! It is likely that he discovered the formula for the sum of the first 100 natural numbers by reasoning similar to that used above.

Although mathematical induction does provide a useful tool for the verification of formulas such as those considered above, its utility is by no means restricted to such applications. We shall now consider some further examples to illustrate the versatility of mathematical induction. Additional examples can be found in the exercises.

The following so-called Golden Needle Problem is an old but still rather fascinating and instructive problem which illustrates rather effectively the use of mathematical induction.

Example 2.3. According to a very old tale, there was a temple in the Orient in which, at the time of creation, 64 golden disks, no two the same size, were placed on one of three golden needles in such a way that no disk was placed above a smaller disk on the needle (see Figure 2.1). Since that

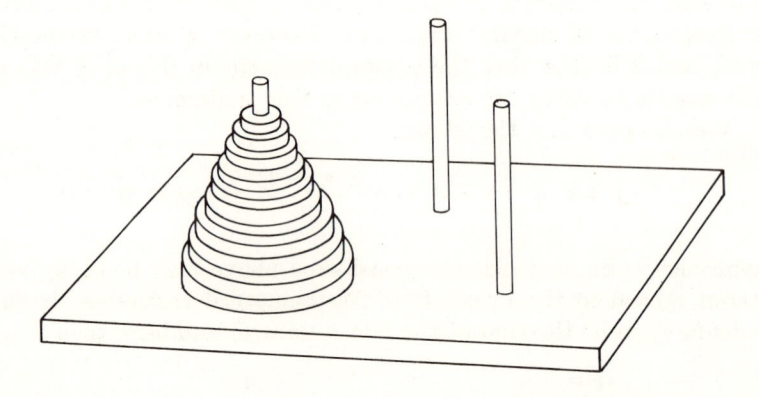

Figure 2.1

time, the priests of the temple have been engaged in moving the disks, one at a time and one per second, to any one of the needles subject only to the condition that no disk is ever placed above a smaller disk on a needle. The goal of their labors was to move all the disks to one of the other needles. When this goal was achieved, the faithful would be rewarded and the unfaithful would be doomed. Of course, the unfaithful were quite interested in knowing the minimum length of time in which the task could be completed. One of their leaders, Prince Math Indu, was able to compute this minimum time. Can you?

The key to the solution of this problem is generalization. Instead of trying to solve the problem for 64 disks, try to determine the minimum time required for moving any number n of disks from one needle to another. At first glance it might seem that such a generalization merely replaces the given problem with a more difficult one. However, after a systematic inspection of the minimum number of moves required for $n = 1, 2, 3, 4$, it should become clear that a reasonable *conjecture* for the minimum time required for n disks is $2^n - 1$. (See Figure 2.2 for the case

2.1. THE NATURAL NUMBER SYSTEM

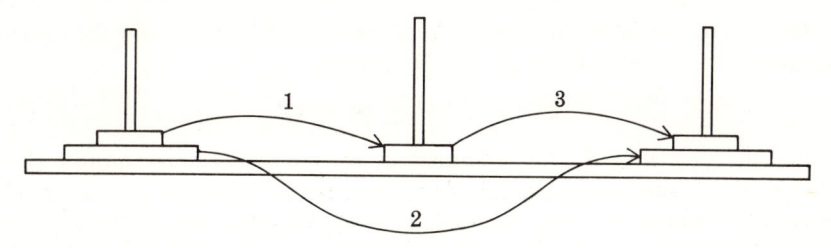

Figure 2.2

$n = 2$.) Again, by inspecting a systematic solution for small values of n, it should become clear that the solution for the problem for a given number $n + 1$ disks can be derived easily if the solution for n disks is known. Once this is done, a *proof* of the conjecture for all natural numbers n is simple. The reader should carry out the details of the solution. No doubt the unfaithful took great comfort from Prince Indu's answer of $2^{64} - 1$ seconds! If the priests worked unerringly (and without tea breaks), it would take them approximately 5,845,420,460,906 years to complete their task!

Example 2.4. Another role that mathematical induction plays in mathematics is the definition of objects that depend on natural numbers. For example, if x is a real number and n is a natural number, then the number x^n is frequently defined to be the number obtained by multiplying x by itself n times. Strictly speaking, however, this is not a satisfactory definition for $n \geq 3$ since multiplication is a binary operation and is therefore defined only on pairs of real numbers. A careful definition of x^n can be given in terms of mathematical induction as follows: We first set $x^1 = x$ and then specify that $x^{n+1} = x \cdot x$ for each natural number n. The complete and detailed verification of the fact that this uniquely defines the exponentiation function is not easy and we shall not pursue the logical difficulties here.

Example 2.5. For each $n \in \mathbf{N}$, consider the statement

$$P(n): 2^n < n!$$

An attempt to prove that $P(n)$ is true for all natural numbers n would be futile since it is readily seen that $P(1)$, $P(2)$, and $P(3)$ are false. However, $P(4)$ is true since $16 < 24$ and $P(5)$ is also true since $32 < 120$. It seems reasonable that the inequality should hold when $n \geq 4$; mathematical induction can be used to verify this conjecture.

In order to establish condition (b′) of (PMI), we must prove that $2^{n+1} < (n + 1)!$ on the basis of the assumption $2^n < n!$. Multiplying both sides of the latter inequality by 2 yields $2^{n+1} < 2 \cdot n!$ and the observation that $2 \leq n + 1$ for all $n \in \mathbf{N}$ gives us

$$2^{n+1} < 2 \cdot n! \leq (n+1)!$$

Thus, for any $n \in \mathbf{N}$, the assumption that $P(n)$ is true implies that $P(n+1)$ is true (even though $P(n)$ is actually false for $n = 1, 2, 3$). Since $P(4)$ is true, we may conclude from (PMI) that $P(n)$ is true for all natural numbers n satisfying $n \geq 4$. In this case, the Principle of Mathematical Induction applies with starting point 4 instead of 1. Notice that we could convert this situation into the formal context of (PMI) by considering the statements $P(n+3)$ in place of $P(n)$ for $n \in \mathbf{N}$.

Example 2.6. It is often instructive to be confronted with an obviously false "theorem" that has a "proof" in which the error is not completely apparent. The following example is one of this sort in which mathematical induction plays a role.

"Theorem." If n is any natural number and if the maximum of two natural numbers p and q is n, then $p = q$.

"Corollary." If p and q are any two natural numbers, then $p = q$.

"PROOF." For each natural number n, let $P(n)$ be the statement of the "Theorem." $P(1)$ is certainly true because if p and q are natural numbers and if the maximum of p and q is 1, then $p = q = 1$. If we assume that $P(n)$ is true and that the maximum of the natural numbers p and q is $n + 1$, then the maximum of $p - 1$ and $q - 1$ is n. Consequently, $p - 1 = q - 1$ since $P(n)$ is assumed true. Therefore, $p = q$ and the "proof" is complete.

Did you detect the error in the argument? You should study the proof until you discover the mistake.

Mathematical induction can be made to seem more plausible and legitimate if it is related to the following more intuitively acceptable property of the natural number system **N**.

Proposition 2.1. (Well-Ordering Property of the Natural Number System **N**) Every nonempty subset S of $\mathbf{N}$ contains a least element s.

PROOF. This result may be proved by making use of (P-3)'. Suppose, contrary to the assertion, that S is a nonempty subset of $\mathbf{N}$ that does not contain a least element. Define $M = \{n \in \mathbf{N} : 1 \notin S, 2 \notin S, \cdots, n \notin S\}$. Clearly, $1 \in M$; for if $1 \in S$, then 1 must be the least element of S since 1 is the least of all the natural numbers. If we assume that $n \in M$, then $1 \notin S, \cdots, n \notin S$ by definition of M. But then $n + 1 \in M$ also; otherwise $n + 1$ would be the least element of S. Therefore, $n \in M$ implies that $n + 1 \in M$. Thus, (P-3)$'$ can be applied to conclude that $n \in M$ for all $n \in \mathbf{N}$. But this contradicts the fact that S is nonempty; consequently, we conclude that S must contain a least element and the proof of the theorem is complete.

On the other hand, (P-3)$'$ may be derived as a theorem if the Well-Ordering Property for $\mathbf{N}$ is considered to be a given axiom about the natural number system $\mathbf{N}$. In fact, let M be a subset of $\mathbf{N}$ such that $1 \in M$ and such that the assumption that $n \in M$ implies that $n + 1 \in M$ for each natural number n. Suppose, contrary to the conclusion of (P-3)$'$, that $M \neq \mathbf{N}$. Define $S = \{n \in \mathbf{N}: n \notin M\}$, then S is a nonempty subset of $\mathbf{N}$ by the preceding supposition. By the Well-Ordering Property for $\mathbf{N}$, S contains a least element n_0. Since $1 \in M$, $n_0 \neq 1$. Therefore, $n_0 - 1$ *is a natural number* and $n_0 - 1 \in M$. The hypotheses on M imply that $n_0 \in M$ contrary to the definition of n_0. This contradiction proves that $M = \mathbf{N}$, that is, (P-3)$'$ is a valid theorem.

In view of our earlier results, we can summarize our conclusions as follows: *In conjunction with properties (P-1)$'$ and (P-2)$'$, the Well-Ordering Property, the Principle of Mathematical Induction* (PMI), *and the property (P-3)$'$ are all equivalent.*

In order to emphasize the equivalence of the Well-Ordering Property and the Principle of Mathematical Induction, we shall now present two proofs of the following theorem, one based on each of these two principles.

Proposition 2.2. The sum of the cubes of three consecutive natural numbers $n, n + 1, n + 2$ is always an integer multiple of 9.

FIRST PROOF. If $P(n)$ is the statement of the theorem, then $P(1)$ is certainly true since $1^3 + 2^3 + 3^3 = 36$ is an integer multiple of 9. If we assume that $P(n)$ is true for some n, then $n^3 + (n + 1)^3 + (n + 2)^3$ is an integer multiple of 9. Therefore, since

$$(n + 1)^3 + (n + 2)^3 + (n + 3)^3$$
$$= [n^3 + (n + 1)^3 + (n + 2)^3] + 9n^2 + 27n + 27,$$

it follows that $(n + 1)^3 + (n + 2)^3 + (n + 3)^3$ is also an integer multiple of 9. Thus, the assumption that $P(n)$ is true for a given n implies that $P(n + 1)$ is true. Therefore, by the Principle of Mathematical Induction, $P(n)$ is true for all n.

SECOND PROOF. Define S to be the set of all natural numbers n such that $n^3 + (n + 1)^3 + (n + 2)^3$ *is not* an integer multiple of 9. To prove the proposition, it would be sufficient to show that S is the empty set. Suppose to the contrary that S is not empty. Then, by the Well-Ordering Property, S contains a least natural number n_0. $n_0 \neq 1$ since $1^3 + 2^3 + 3^3 = 36$ is an integer multiple of 9. Therefore, $n_0 - 1$ is a natural number and $n_0 - 1 \notin S$ since n_0 is the least natural number in S. By definition of S, it follows that $(n_0 - 1)^3 + n_0^3 + (n_0 + 1)^3$ is an integer multiple of 9, and therefore

$$n_0^3 + (n_0 + 1)^3 + (n_0 + 2)^3 = [(n_0 - 1)^3 + n_0^3 + (n_0 + 1)^3] + 9n_0^2 + 9n_0 + 9$$

is also an integer multiple of 9. But this implies that $n_0 \notin S$ contrary to the definition of n_0. Hence, the assumption that S is not empty leads to a contradiction. Therefore, S must be empty, which completes the proof of the theorem.

EXERCISES

2.1. Prove that

$$\frac{1}{1 \cdot 2} + \frac{1}{2 \cdot 3} + \cdots + \frac{1}{n(n + 1)} = \frac{n}{n + 1}$$

for all natural numbers n.

2.2. Conjecture a simple formula for the sum $1 + 3 + \cdots + (2n - 1)$ of the first n odd integers, and check your conjecture by using the Principle of Mathematical Induction.

2.3. Prove or disprove:

$$1^2 + 3^2 + \cdots + (2n - 1)^2 = \frac{4n^3 - n}{3}$$

for all natural numbers n.

2.4. Prove or disprove:

$$1^2 + 2^2 + \cdots + n^2 = \frac{2n^3 + 9n^2 - 17n + 12}{6}$$

for all natural numbers n.

2.5. Find a natural number $k_0 > 1$ such that $n^3 - n$ is an integer multiple of k_0 for each natural number n.

2.6. For which natural numbers is it true that $n^2 < 2^n$?

2.7. Prove that if $x \geq -1$, then $(1 + x)^n \geq 1 + nx$ for each natural number n.

2.8. Find the error in the following "proof" that all natural numbers are even. Let $P(n)$ be the statement "all natural numbers less than n are even." $P(1)$ is true since there are no natural numbers smaller than 1. If $P(n)$ is assumed true and $n + 1$ is given, write $n + 1 = j + k$ where j and k are natural numbers such that $j < n$ and $k < n$. Then both j and k are even by assumption, and hence their sum $n + 1$ is even. Thus, the truth of $P(n + 1)$ follows from the assumed truth of $P(n)$. Therefore, $P(n)$ is true for all n by (PMI).

2.9. Consider the statement $P(n)$: For each natural number n, $n^2 - n + 41$ is a prime number (that is, $n^2 - n + 41$ is divisible only by itself and 1). Verify $P(n)$ for several small values of n, but show that $P(n)$ is false for some choices of n. [*Note:* In Chapter 3, we shall prove that if $f(x)$ is a polynomial of degree at least one with integer coefficients, then $f(n)$ cannot be prime for all natural numbers n.]

2.10. Prove that $a^n - b^n$ has $(a - b)$ as a factor for each natural number n. [*Note:* The factorization of $a^n - 1$ into irreducible factors with real number coefficients is interesting from the point of view of mathematical induction. For it was observed that in such factorizations, the absolute value of the coefficients did not exceed 1; for example,
$$a^2 - 1 = (a - 1)(a + 1),$$
$$a^3 - 1 = (a - 1)(a^2 + a + 1),$$
$$a^4 - 1 = (a - 1)(a + 1)(a^2 + 1),$$
$$a^5 - 1 = (a - 1)(a^4 + a^3 + a^2 + a + 1).$$

This property held for many other choices of n and it was conjectured that the property might actually hold for all natural numbers n. However, it was finally shown that $a^{105} - 1$ has an irreducible factor that has two coefficients of -2, so this property does not hold in general.]

2.11. Prove that any integer $n > 7$ can be written as $n = 3a + 5b$ where a and b are nonnegative whole numbers. [*Hint:* Use (PMI) and verify condition (b') by considering two cases according to whether $n = 3a + 5b$ with $b = 0$ or $b > 0$.]

2.12. Prove that for each natural number n, n straight lines in the plane always break up the plane into regions such that the entire plane may be colored with two colors in such a way that no two regions which are in contact along a straight line segment are colored with the same color.

2.13. Using only the Peano axioms (P-1) to (P-3), prove that:
   (a) For all $x \in \mathbf{N}$, $x \neq s(x)$.
   (b) If $x \neq 1$, then $x = s(y)$ for some $y \in \mathbf{N}$.

2.2. THE SYSTEM OF INTEGERS AND ORDERED DOMAINS

The purpose of this section is to identify a list of arithmetic properties for the system of integers that embodies all of the essential arithmetic information concerning this system. In other words, if the properties in this list are taken as *axioms* for the algebraic system of integers, then it is possible to deduce all other arithmetic properties of that system as *theorems*. In the course of the discussion of the axiomatic basis for this system, we shall investigate other mathematical systems that share some but not all of the basic properties of the system of integers.

The system **I** of all integers (positive and negative whole numbers and zero) can be axiomatically described as follows: The *system of integers* is an algebraic system $(\mathbf{I}, +, \cdot, \leq)$ consisting of a set **I**, two binary operations $+$ (called addition) and $\cdot$ (called multiplication) on **I**, and an order relation $\leq$ on **I** such that:

(A-1) For all a, b, c in **I**, $a + (b + c) = (a + b) + c$. (Associative Property for Addition)

(A-2) For all a and b in **I**, $a + b = b + a$. (Commutative Property for Addition)

(A-3) There is an element 0 in **I** (called the *zero* element of **I**) such that $a + 0 = a$ for all a in **I**. (Existence of a Zero Element)

(A-4) For each element a in **I**, there is an element of **I** denoted by $-a$ (called the *negative of* a) such that $a + (-a) = 0$. (Existence of Negative Elements)

(M-1) For all a, b, c in **I**, $a \cdot (b \cdot c) = (a \cdot b) \cdot c$. (Associative Property for Multiplication)

(M-2) For all a and b in **I**, $a \cdot b = b \cdot a$. (Commutative Property for Multiplication)

(M-3) There is an element 1 in **I** (called the *unit* element) such that $1 \neq 0$ and $a \cdot 1 = a$ for all a in **I**. (Existence of a Unit Element)

(D) For all a, b, c in **I**, $a \cdot (b + c) = a \cdot b + a \cdot c$. (Distributive Property for Multiplication over Addition)

(C) If a, b, c are in **I**, if $a \neq 0$ and if $a \cdot b = a \cdot c$, then $b = c$. (Cancellation Property)

(O-1) For all a and b in **I**, either $a \leq b$ or $b \leq a$. (Total Order Property)

(O-2) If a, b, c are in **I** and if $a \leq b$, then $a + c \leq b + c$. (Compatibility Property for Order and Addition)

(O-3) If a, b, c are in **I**, if $0 \leq c$ and if $a \leq b$, then $a \cdot c \leq b \cdot c$. (Compatibility Property for Order and Multiplication)

(O-4) If S is a nonempty subset of **I** and if $0 \leq s$, $s \neq 0$ for all $s \in S$ then S contains a least element. (Well-Ordering Property)

For notational convenience, we shall often write ab in place of $a \cdot b$ and $b \geq a$ in place of $a \leq b$.

The preceding properties of the system of integers should be familiar, and undoubtedly the reader could list other properties without difficulty.

2.2. THE SYSTEM OF INTEGERS AND ORDERED DOMAINS

The significance of the above list lies in the fact that these properties characterize the system of integers in the following sense: Any algebraic system $(A, +, \cdot, \leq)$ consisting of a set A, two binary operations $+$ and $\cdot$ on A and an order relation $\leq$ on A possessing all of the properties listed above (with the symbol **I** replaced by A) is isomorphic to the system of integers. A proof of this assertion will be outlined at the end of this section.

As a simple illustration of how arithmetic computations can be based on the above list of properties, let us solve the equation $2x + 3 = 7$ for x. We first write 7 as $4 + 3$ and add the negative -3 of 3 to both sides of the equation to obtain

$$(2x + 3) + (-3) = (4 + 3) + (-3).$$

By applying the Associative Property (A-1), we can write this as

$$2x + (3 + (-3)) = 4 + (3 + (-3)).$$

By property (A-4) for negatives, this becomes

$$2x + 0 = 4 + 0.$$

Using the property (A-3) for the zero element, we see this is the same as

$$2x = 4.$$

Finally, an application of the Cancellation Property (C) gives us the solution

$$x = 2.$$

It is certainly neither necessary nor appropriate to belabor this type of calculation by appealing to the axiomatic properties each time an equation is to be solved, but it is instructive to observe how each step can be justified on the basis of a relatively short list of fundamental properties.

Because of his familiarity with the arithmetic of integers, the reader may find it difficult to appreciate the significance of the properties listed above. For this reason we shall now discuss some other algebraic systems and compare their arithmetic features to those of the system of integers.

Example 2.7. The reader should note that all of the properties listed above for the system of integers with the exception of the Well-Ordering Property (O-4) correspond to familiar properties of the algebraic systems $(\mathbf{Q}, +, \cdot, \leq)$ of rational numbers and $(\mathbf{R}, +, \cdot, \leq)$ of real numbers. That the Well-Ordering Property fails to hold for the usual order relation on $\mathbf{Q}$ or $\mathbf{R}$ can be seen as follows. If S denotes the set of all positive numbers in $\mathbf{Q}$ (or $\mathbf{R}$), and if $x \in S$, then $x/2$ is also in S and $x/2 < x$. Therefore, S cannot contain a least element. In other words, there is no smallest positive rational (or real) number.

68 NUMBER SYSTEMS AND THEIR PROPERTIES

Example 2.8. Consider a set B consisting of two elements which we shall denote by 0 and 1. *Define* binary operations $+$ and $\cdot$ on B by means of the following tables:

+	0	1
0	0	1
1	1	0

$\cdot$	0	1
0	0	0
1	0	1

By a routine case consideration one can verify that the algebraic system $(B,+,\cdot)$ has properties corresponding to (A-1) through (A-4), (M-1) through (M-3), (D), and (C). At this point, it is natural to raise the question of whether or not it is possible to introduce an order relation $\leq$ on B in such a way that (O-1) through (O-4) are satisfied (with the set $\mathbf{I}$ replaced by B). It is easy to show that no such relation on B exists. In fact, if $\leq$ were such an order relation on B, then by (O-1) either $0 \leq 1$ or $1 \leq 0$. If $0 \leq 1$, then $1 = 0 + 1 \leq 1 + 1 = 0$ by (O-2), which implies $0 = 1$ since order relations are antisymmetric. However, $0 \neq 1$ by (M-3), and hence $0 \leq 1$ is not possible. Similarly, $1 \leq 0$ is excluded. Thus, no such order relation on B can exist. [*Note:* Equality is an order relation on B with properties (O-2) and (O-3), but not (O-1) nor (O-4).]

Example 2.9. In Section 1.3, we introduced an equivalence relation $\equiv$ on the system $\mathbf{I}$ of integers defined by congruence modulo a given natural number m. We introduced the notation $[a]$ for the residue class (that is, equivalence class for $\equiv$) containing the integer a, and we observed that $[a] = [b]$ if and only if $a \equiv b \pmod{m}$. In Section 1.5, we also introduced binary operations $+$ and $\cdot$ on the set $\mathbf{I}_m$ of residue classes modulo m by defining

$$[a] + [b] = [a + b], \qquad [a] \cdot [b] = [ab]$$

for all $[a]$ and $[b]$ in $\mathbf{I}_m$.

It is clear from these definitions and the definition of $\mathbf{I}_m$ that the mapping φ from $\mathbf{I}$ onto $\mathbf{I}_m$ defined by $\varphi(a) = [a]$ for each $a \in \mathbf{I}$ (that is, φ maps each integer into its residue class modulo m) has the following properties:

$$\varphi(a + b) = \varphi(a) + \varphi(b),$$
$$\varphi(ab) = \varphi(a) \cdot \varphi(b),$$
$$\varphi(0) = [0],$$
$$\varphi(1) = [1],$$

for all a and b in $\mathbf{I}$. Using these properties of φ together with the axiomatic properties for the system of integers, one can verify that the algebraic system $(\mathbf{I}_m,+,\cdot)$ satisfies (A-1) through (A-4), (M-1), (M-2), and (D) (with $\mathbf{I}$ replaced by $\mathbf{I}_m$). (For example, to verify (A-2) we simply observe

2.2. THE SYSTEM OF INTEGERS AND ORDERED DOMAINS

that for all $[a]$ and $[b]$ in $\mathbf{I}_m$, we have $[a] + [b] = \varphi(a) + \varphi(b) = \varphi(a + b) = \varphi(b + a) = \varphi(b) + \varphi(a) = [b] + [a]$.) Moreover, if $m > 1$, then $[1] \neq [0]$ and for all $[a] \in \mathbf{I}_m$, we have $[a] \cdot [1] = \varphi(a) \cdot \varphi(1) = \varphi(a \cdot 1) = [a]$. Thus, $[1]$ is a unit element and (M-3) is satisfied by $(\mathbf{I}_m, +, \cdot)$. However, for certain choices of the number m, the property corresponding to (C) is not satisfied by the system $(\mathbf{I}_m, +, \cdot)$. For example, if $m = 4$, then $[2] \cdot [2] = [2] \cdot [0]$, yet $[2] \neq [0]$ in $\mathbf{I}_4$. We shall show later in Chapter 3 that $(\mathbf{I}_m, +, \cdot)$ has the cancellation property if and only if m is a prime number.

Again we ask whether or not it is possible to introduce an order relation $\leq$ on $\mathbf{I}_m$ in such a way that (O-1) through (O-4) are satisfied. Actually, Example 2.8 shows that the answer to this question is negative for at least one choice of m, namely, $m = 2$. In fact, by comparing the definitions of the binary operations $+$ and $\cdot$ in $(B, +, \cdot)$ of Example 2.8 and $(\mathbf{I}_2, +, \cdot)$, we see that the algebraic systems $(B, +, \cdot)$ and $(\mathbf{I}_2, +, \cdot)$ are isomorphic. Therefore, the argument used in Example 2.8 can be applied to prove that no such order relation can exist on $(\mathbf{I}_2, +, \cdot)$. We shall see later (see the remarks following Proposition 2.6) that except for the trivial case $m = 1$ it is impossible to define an order relation on $\mathbf{I}_m$ that will have the properties corresponding to (O-1) through (O-4).

Example 2.10. Suppose that X is a nonempty set and that $P(X)$ is the set of all subsets of X. For A and B in $P(X)$, define $A + B$ and $A \cdot B$ as follows:
$$A + B = (A \cap B') \cup (A' \cap B),$$
$$A \cdot B = A \cap B.$$

Thus, $A \cdot B$ is simply the intersection of the sets A and B, while $A + B$ is the set depicted in the Venn diagram in Figure 2.3. $A + B$ is sometimes

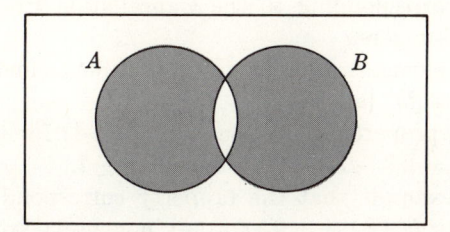

Figure 2.3

referred to as the *symmetric difference* or the *Boolean sum* of the sets A and B.

If the binary operations $+$ and $\cdot$ are defined on $P(X)$ as above and if $\leq$ is taken to be set inclusion (that is, $A \leq B$ if and only if $A \subset B$), then the algebraic system $(P(X), +, \cdot, \leq)$ evidently satisfies properties corresponding to (A-2), (M-1), (M-2), and (O-3). Also, the empty set $\emptyset$ provides the zero element for $P(X)$ [see (A-3)] since

$$A + \emptyset = (A \cap \emptyset') \cup (A' \cap \emptyset) = (A \cap X) \cup \emptyset = A.$$

Therefore, given $A \in P(X)$, the negative of A (see (A-4)) is provided by A itself since

$$A + A = (A \cap A') \cup (A' \cap A) = \emptyset \cup \emptyset = \emptyset.$$

The set X plays the role of the unit element in (M-3) since

$$A \cdot X = A \cap X = A$$

for all $A \in P(X)$.

It can also be shown that the system $(P(X), +, \cdot, \leq)$ has properties corresponding to (A-1) and (D) for the integers; however, the proofs of these two properties are somewhat intricate and will be omitted at this point since the corresponding results will be established in Chapter 6 in the more general context of Boolean algebras. Simple examples show that the properties corresponding to (C), (O-1), (O-2), and (O-4) do not hold for the system $(P(X), +, \cdot, \leq)$ whenever X contains more than a single element.

EXERCISES

2.14. Construct addition and multiplication tables analogous to those found in Example 2.8 for $(\mathbf{I}_5, +, \cdot)$ and $(\mathbf{I}_6, +, \cdot)$.

2.15. For each $[a]$ in $\mathbf{I}_5$ such that $[a] \neq [0]$, use the multiplication table constructed in Exercise 2.14 to verify that there is a $[b] \in \mathbf{I}_5$ such that $[b] \cdot [a] = [1]$. Use this fact to verify that $(\mathbf{I}_5, +, \cdot)$ satisfies the property corresponding to the Cancellation Property (C) for the system of integers.

2.16. For which elements $[a]$ in $\mathbf{I}_6$ is it impossible to find an element $[b] \in \mathbf{I}_6$ such that $[b] \cdot [a] = [1]$?

2.17. Verify the property corresponding to (A-4) for $(\mathbf{I}_m, +, \cdot)$ and show that $-[a] = [m - a] = [-a]$ for all $a \in \mathbf{I}$.

2.18. Show by example that the property corresponding to (O-2) does not hold for $(P(X), +, \cdot, \leq)$ for any nonempty set X, and that (C), (O-1), and (O-4) also fail to have analogues for $(P(X), +, \cdot, \leq)$ if X contains two or more elements.

2.19. Prove that if X is a set that contains two or more elements, then it is not possible to define an order relation $\leq$ on $P(X)$ so that $\leq$ satisfies the property corresponding to the Total Order Property (O-1). [*Hint:* Show that if A is a nonempty, proper subset of X, then neither $A \leq A'$ nor $A' \leq A$ can hold.]

2.2. THE SYSTEM OF INTEGERS AND ORDERED DOMAINS

2.20. Suppose that P denotes the set of all real polynomials with integer coefficients, that is, P is the set of all functions $p \colon \mathbf{R} \to \mathbf{R}$ of the form
$$p(x) = a_n x^n + \cdots + a_1 x + a_0 \qquad (x \in \mathbf{R})$$
where n is a natural number and $a_0, \cdots, a_n$ are integers. Define binary operations $+$ and $\cdot$ on P as follows: For any p and q in P, define the polynomials $p + q$ and $p \cdot q$ by
$$(p + q)(x) = p(x) + q(x)$$
$$(p \cdot q)(x) = p(x)q(x)$$
for all x in $\mathbf{R}$.
(a) Verify that the algebraic system $(P, +, \cdot)$ satisfies the properties corresponding to (A-1) through (A-4), (M-1) through (M-3), and (D).
(b) If, for p and q in P, we define $p \leq q$ to mean that $p(x) \leq q(x)$ for all $x \in \mathbf{R}$, prove that $\leq$ is an order relation on P that satisfies the analogues to (O-2) and (O-3) for $(\mathbf{I}, +, \cdot, \leq)$. Find examples to show that the relation $\leq$ does not have properties corresponding to (O-1) and (O-4).
(c) Define the relation $\lhd$ on P as follows: If p and q are polynomials in P, then $p \lhd q$ if the polynomial $q - p$ is either identically zero or has a positive leading coefficient. (For example, if $p(x) = x + 7$, $q(x) = 2x - 5$, and $r(x) = x^2 - 3x$, then $p \lhd q$ and $q \lhd r$.) Prove that $\lhd$ is an order relation on P with properties corresponding to (O-1), (O-2), (O-3), but not (O-4).

The examples given above represent a small sample of the great variety of algebraic systems which share some of the arithmetic features of the system of integers. Because of the importance of some of these other systems, mathematicians have made extensive studies of the structure of various classes of systems of this sort. Such studies comprise the subject matter of the broad field of mathematics known as abstract algebra. The ramifications of this field of study in mathematics and in science generally have become so great that almost all mathematics students and many students of science are given a broad introduction to this subject late in their undergraduate programs or near the beginning of their graduate study. We do not intend to delve deeply into abstract algebra in this text. However, it will be useful for us to identify certain classes of algebraic systems in order to clarify and unify our treatment of the properties of number systems.

It is possible to arbitrarily select several of the properties listed above for the system of integers and proceed to study algebraic systems in which the selected properties are regarded as axioms. However, this

72   NUMBER SYSTEMS AND THEIR PROPERTIES

would not be likely to result in a useful or significant theory. If such studies are to be fruitful, the selected axioms must have some cohesion and there must be significant concrete examples of the resulting algebraic systems to serve as guides for the development of the theory. We shall now define several important algebraic systems that are particularly relevant to our present discussion.

Definition. If $(A,+,\cdot)$ is an algebraic system consisting of a set A and two binary operations $+$ and $\cdot$ on A satisfying properties corresponding to (A-1) through (A-4), (M-1) through (M-3), and (D), then $(A,+,\cdot)$ is called a *commutative ring with unit*. If, in addition, the property corresponding to the Cancellation Property (C) is satisfied, then $(A,+,\cdot)$ is called an *integral domain*.

If $(A,+,\cdot,\leq)$ is an algebraic system consisting of a set A, two binary operations $+$ and $\cdot$ on A, and an order relation $\leq$ on A such that $(A,+,\cdot)$ is an integral domain and such that the relation $\leq$ satisfies properties corresponding to (O-1) through (O-3), then $(A,+,\cdot,\leq)$ is called an *ordered domain*.

Notation such as (A-2), (D), and terminology such as "the Cancellation Property (C)" will be used to refer to the defining axioms for a commutative ring with unit, an integral domain or an ordered domain even though these notations and descriptive terms were originally introduced for the system of integers.

The following diagram illustrates how the structures defined above are related:

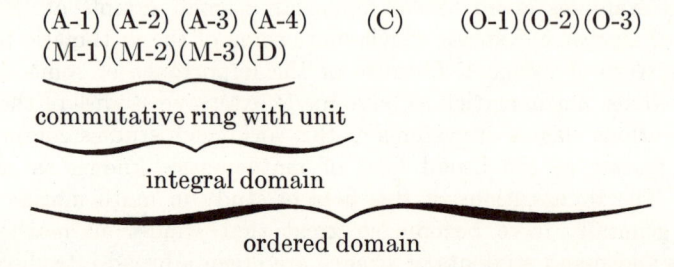

$$\underbrace{\underbrace{\underbrace{\text{(A-1) (A-2) (A-3) (A-4)} \quad\quad \text{(C)}}_{\text{commutative ring with unit}} \quad \text{(O-1)(O-2)(O-3)}}_{\text{integral domain}}}_{\text{ordered domain}}$$

(M-1)(M-2)(M-3)(D)

Note that the definitions of a commutative ring with unit and integral domain involve only the two binary operations $+$ and $\cdot$ while the definition of ordered domain concerns an order relation as well.

According to these definitions, we see that the systems $(\mathbf{R},+,\cdot,\leq)$ of all real numbers and $(\mathbf{Q},+,\cdot,\leq)$ of all rational numbers are ordered domains; however, neither of these systems has the Well-Ordering Property (O-4) (see Example 2.7 above). The two-element system $(B,+,\cdot)$ considered in Example 2.8 is an integral domain. However, we have shown

2.2. THE SYSTEM OF INTEGERS AND ORDERED DOMAINS

that it is not possible to define an order relation $\leq$ on B in such a way that $(B,+,\cdot,\leq)$ is an ordered domain. The system $(\mathbf{I}_m,+,\cdot)$ of residue classes modulo a natural number $m > 1$ is a commutative ring with unit. In view of our remarks in Example 2.9, $(\mathbf{I}_m,+,\cdot)$ is an integral domain if and only if m is a prime number. The algebraic system $(P(X),+,\cdot)$ discussed in Example 2.10 is a commutative ring with unit, but not an integral domain. The order relation $\leq$ defined by set inclusion on $P(X)$ satisfies (O-3) but not (O-1) or (O-2). (See Exercise 2.18.) Thus, $(P(X),+,\cdot,\leq)$ is not an ordered domain.

We now illustrate how the axioms for the types of algebraic systems defined above can be used to deduce further properties of these systems. The following pair of propositions provides a sample of the familiar properties of addition and multiplication for the systems of integers, rational numbers, and real numbers that can be deduced *solely* from the fact that these systems are commutative rings with unit.

Proposition 2.3. Suppose $(A,+,\cdot)$ is a commutative ring with unit. Then:

(1) If $c \in A$ is such that $a + c = a$ for some $a \in A$, then $c = 0$. (Uniqueness of Zero Element)

(2) If $a \in A$ and if $b \in A$ is such that $a + b = 0$, then $b = -a$. (Uniqueness of Negatives)

(3) If $d \in A$ is such that $a \cdot d = a$ for all $a \in A$, then $d = 1$. (Uniqueness of Unit Element)

(4) For any $a \in A$, $a \cdot 0 = 0$.

PROOF. (1) By (A-4), we have $a + (-a) = 0$. Therefore, adding $-a$ to both sides of $a = a + c$ and using the Commutative Property for Addition (A-2) and the Associative Property for Addition (A-1), we have

$$0 = a + (-a) = (a + c) + (-a) = (c + a) + (-a)$$
$$= c + (a + (-a)) = c + 0.$$

Since $c + 0 = c$ by (A-3), we conclude that $c = 0$.

(2) Since $a + b = 0 = a + (-a)$, an argument similar to that used in (1) shows that $b = -a$.

(3) If $a \cdot d = a$ for all $a \in A$, then taking $a = 1$ gives us $1 \cdot d = 1$. Since $1 \cdot d = d \cdot 1 = d$ by (M-2) and (M-3), it follows that $d = 1$.

(4) Since $a + 0 = a$ for any $a \in A$, we can apply the Distributive Property (D) to obtain

$$a \cdot a + a \cdot 0 = a \cdot (a + 0) = a \cdot a.$$

Hence, $a \cdot 0 = 0$ by (1).

Proposition 2.4. If $(A,+,\cdot)$ is a commutative ring with unit, then for all a and b in A we have:

(1) $-(a+b) = (-a) + (-b)$,
(2) $-(-a) = a$,
(3) $(-a) \cdot b = -(ab) = a \cdot (-b)$,
(4) $(-a) \cdot (-b) = a \cdot b$.

PROOF. (1) We wish to show that $(-a) + (-b)$ is the negative of $a + b$. Since the negative of an element of A is unique by (2) of Proposition 2.3, we need only show that the sum of $(-a) + (-b)$ and $a + b$ is 0. This is done as follows:

$$(a+b) + [(-a) + (-b)] = [a + (-a)] + [b + (-b)]$$
$$= 0 + 0 = 0.$$

(The reader should check to see which properties are used in this calculation.) Hence, $(-a) + (-b) = -(a+b)$.

(2) Since $(-a) + a = a + (-a) = 0$, it follows from (2) of Proposition 2.3 that a is the negative of $-a$. That is, $a = -(-a)$.

(3) and (4) are left as exercises.

If $(A,+,\cdot)$ is a commutative ring with unit, then an element $a \in A$ is called a *zero divisor* if $a \neq 0$ and $a \cdot b = 0$ for some $b \neq 0$. For example, if X is a set with two or more elements, then any nonempty proper subset of X is a zero divisor in the commutative ring with unit $(P(X),+,\cdot)$ discussed in Example 2.10 above. For if a denotes any such subset of X and b is the complement of a, then $a \neq 0$, $b \neq 0$ and $a \cdot b = 0$.

Zero divisors cannot exist in integral domains because of the Cancellation Property (C) which states that if $a \cdot b = a \cdot c$ and $a \neq 0$, then $b = c$. Therefore, if $a \neq 0$ and $a \cdot b = 0$, then $a \cdot b = a \cdot 0$ and hence $b = 0$. As the next proposition shows, the absence of zero divisors is equivalent to the Cancellation Property for commutative rings with unit.

Proposition 2.5. A commutative ring with unit $(A,+,\cdot)$ is an integral domain if and only if $a \cdot b = 0$ for a and b in A implies that either $a = 0$ or $b = 0$.

PROOF. In the remarks preceding this proposition, we observed that integral domains possess the stated property. For the converse, we must show that if $(A,+,\cdot)$ is a commutative ring with unit having the stated property, then the Cancellation Property (C) holds. Suppose $a \cdot b = a \cdot c$ for a, b, c in A where $a \neq 0$. Then $a \cdot (b - c) = 0$, where $b - c = b + (-c)$. Since this implies that either $a = 0$ or $b - c = 0$ by hypothesis, and since $a \neq 0$, we conclude that $b - c = 0$, that is, $b = c$.

EXERCISES

2.21. Prove (3) and (4) of Proposition 2.4.

2.22. If $(A,+,\cdot)$ is a commutative ring with unit, show that for a, b, c, d in A,
   (a) $(a - b) + (c - d) = (a + c) - (b + d)$
   (b) $(a - b)(c - d) = (ac + bd) - (ad + bc)$.

2.23. Find the zero divisors, if any, in $(\mathbf{I}_6,+,\cdot)$.

2.24. Prove or disprove: If a is a zero divisor in a commutative ring with unit $(A,+,\cdot)$, then $a \cdot b = 0$ for all $b \in A$.

2.25. Prove that the axioms (A-3) and (A-4) for a commutative ring with unit $(A,+,\cdot)$ are together equivalent to the following property:
   For each a and b in A there exists a $c \in A$ such that $a + c = b$.

2.26. Prove that if a is an element of an integral domain such that $a^2 = a$ (where $a^2 = a \cdot a$), then either $a = 0$ or $a = 1$. Show that this conclusion is not valid for arbitrary commutative rings with unit. [*Hint:* Consider $(P(X),+,\cdot)$ where X has two or more elements.]

2.27. Let F denote the set of all functions from $\mathbf{R}$ to $\mathbf{R}$. For f and g in F, define $f + g$ and $f \cdot g$ in F by
$$(f + g)(x) = f(x) + g(x)$$
$$(f \cdot g)(x) = f(x)g(x)$$
for all x in $\mathbf{R}$.
   (a) Show that $(F,+,\cdot)$ is a commutative ring with unit. What functions play the role of the zero and unit elements?
   (b) Show that $(F,+,\cdot)$ is not an integral domain by exhibiting a zero divisor in F.
   (c) Find an $f \in F$ other than the zero or unit element such that $f^2 = f$.

While the above results by no means exhaust the arithmetic consequences of the axioms for a commutative ring with unit or an integral domain, they are adequate to indicate the flavor of the arguments involved. We now turn to the derivation of some elementary consequences of the axioms for an ordered domain. We shall use the notation $a < b$ to mean $a \leq b$ and $a \neq b$.

Proposition 2.6. Suppose that $(A,+,\cdot,\leq)$ is an ordered domain:

(1) If $a \in A$ and if $a \leq 0$, then $0 \leq -a$.
(2) For all $a \in A$, it is true that $0 \leq a^2$ (where $a^2 = a \cdot a$). If $a \neq 0$, then $0 < a^2$.

76 NUMBER SYSTEMS AND THEIR PROPERTIES

(3) If 1 is the unit element of A, then $0 < 1$.
(4) A must be an infinite set.

PROOF. (1) If $a \leq 0$, then from the Compatibility of Order and Addition (O-2), we conclude that $0 = a + (-a) \leq 0 + (-a) = -a$, that is, $0 \leq -a$.

(2) For all $a \in A$ either $0 \leq a$ or $a \leq 0$ by the Total Order Property (O-1). If $0 \leq a$, then $0 = a \cdot 0 \leq a \cdot a = a^2$ by (4) in Proposition 2.3 and the Compatibility Property for Order and Multiplication (O-3). If $a \leq 0$, then $0 \leq -a$ by (2), so that $0 \leq (-a)^2 = a^2$ by (4) in Proposition 2.4. If $a \neq 0$, then $a^2 \neq 0$ and hence $0 < a^2$.

(3) Since $1 = 1 \cdot 1 = (1)^2$, (3) follows immediately from (2) and the fact that $0 \neq 1$.

(4) In view of (3), we know $0 < 1$. Therefore, the Compatibility Property for Order and Addition (O-2) implies that $1 = 1 + 0 \leq 1 + 1$ and $1 + 1 \neq 1$. Proceeding inductively, we conclude that the elements $0, 1, 1 + 1, 1 + 1 + 1, \cdots$ of A are all distinct. Therefore, A cannot be a finite set.

As a consequence of (4) in Proposition 2.6 we note that if m is a prime number, there does not exist an order relation $\leq$ on the integral domain $\mathbf{I}_m$ such that $(\mathbf{I}_m, +, \cdot, \leq)$ is an ordered domain.

It is interesting that the basic arithmetic properties of the absolute value operation for the systems of integers, rational numbers, and real numbers can be deduced from the fact that these systems are ordered domains. To show this, we proceed as follows: Suppose that $(A, +, \cdot, \leq)$ is an ordered domain. For each $a \in A$, define the *absolute value* $|a|$ of a by

$$|a| = \begin{cases} a, & \text{if } a \geq 0 \\ -a, & \text{if } a \leq 0. \end{cases}$$

Proposition 2.7. If $(A, +, \cdot, \leq)$ is an ordered domain, the following assertions are valid:

(1) $|a| \geq 0$ for all $a \in A$, and $|a| = 0$ if and only if $a = 0$.
(2) $|a \cdot b| = |a| \cdot |b|$ for all a and b in A.
(3) If $c \geq 0$ and if $a \in A$, then $|a| \leq c$ if and only if $-c \leq a \leq c$.
(4) If a and b are in A, then $|a + b| \leq |a| + |b|$. (The Triangle Inequality)

PROOF. (1) If $a \geq 0$, then $|a| = a \geq 0$. If $a \leq 0$, then $|a| = -a \geq 0$ by (1) of Proposition 2.6. Therefore, $|a| \geq 0$ for all $a \in A$. It is clear from the definition of absolute value that $|a| = 0$ if and only if $a = 0$.

(2) The assertion is proved by examining each of the following four

cases: $a \geq 0$ and $b \geq 0$, $a \geq 0$ and $b \leq 0$, $a \leq 0$ and $b \geq 0$, $a \leq 0$ and $b \leq 0$. For example, if $a \leq 0$ and $b \geq 0$, then $-a \geq 0$ and by (3) of Proposition 2.4 and (O-3) we have $-(a \cdot b) = (-a) \cdot b \geq 0$. Therefore, $|a \cdot b| = -(a \cdot b) = (-a) \cdot b = |a| \cdot |b|$. The other cases are handled in a similar manner.

(3) From the definition of absolute value, it follows that $|a| \leq c$ if and only if both of the inequalities $a \leq c$ and $-a \leq c$ hold. It is easily verified that $-a \leq c$ is equivalent to $-c \leq a$. Thus, $|a| \leq c$ if and only if $-c \leq a$ and $a \leq c$, that is, $-c \leq a \leq c$.

(4) If a and b are in A, then $|a| \geq a$, $|a| \geq -a$, $|b| \geq b$, $|b| \geq -b$. Therefore, by (O-2), we have $-(|a| + |b|) = (-|a|) + (-|b|) \leq a + b \leq |a| + |b|$. By virtue of (3), this implies that $|a + b| \leq |a| + |b|$.

In the remarks following the list of axiomatic properties for the system $(\mathbf{I}, +, \cdot, \leq)$ of integers, we pointed out that these properties served to characterize this system in the sense that any other algebraic system $(A, +, \cdot, \leq)$ with these properties must be isomorphic to the system of integers. We shall now outline a verification of this remark and also show that the system of integers is in a certain sense the "smallest" ordered domain.

First of all, let us suppose that $(A, +, \cdot, \leq)$ is an ordered domain. For the sake of clarity, we shall denote the zero and unit elements of A by $\bar{0}$ and $\bar{1}$ to distinguish these elements from the integers 0 and 1. We define a mapping $\varphi \colon \mathbf{I} \to A$ as follows:

(a) $\varphi(0) = \bar{0}$, $\varphi(1) = \bar{1}$.

(b) If n is a natural number and if $\varphi(n)$ has been defined, then define $\varphi(n + 1) = \varphi(n) + \bar{1}$. (In view of (a), this would imply that $\varphi(2) = \bar{1} + \bar{1}$, $\varphi(3) = \bar{1} + \bar{1} + \bar{1}$, and so on.)

(c) For each negative integer $-n$, define $\varphi(-n) = -\varphi(n)$.

By making use of the Principle of Mathematical Induction, it can be shown that the mapping φ has the following properties:

(1) φ is a one-to-one mapping.
(2) $\varphi(a + b) = \varphi(a) + \varphi(b)$ for all a and b in $\mathbf{I}$.
(3) $\varphi(a \cdot b) = \varphi(a) \cdot \varphi(b)$ for all a and b in $\mathbf{I}$.
(4) $a \leq b$ in $\mathbf{I}$ if and only if $\varphi(a) \leq \varphi(b)$ in A.

It is not necessarily true that $\varphi(\mathbf{I}) = A$. For example, if $(A, +, \cdot, \leq)$ is the system of rational numbers, then $r \notin \varphi(\mathbf{I})$ for any rational number r that is not an integer.

The mapping φ defined above permits us to regard the system $(\mathbf{I}, +, \cdot, \leq)$ of integers to be a "part" of any ordered domain $(A, +, \cdot, \leq)$. More precisely, if for each integer a we define the *integral element* $\bar{a}$ in A

by $\bar{a} = \varphi(a)$, then φ is a one-to-one correspondence between the set **I** of integers and the subset $\bar{\mathbf{I}}$ of integral elements in A. Moreover, properties (2), (3), (4) of φ show that φ is an isomorphism between the algebraic systems $(\mathbf{I},+,\cdot,\leq)$ and $(\bar{\mathbf{I}},+,\cdot,\leq)$. Thus, if $(A,+,\cdot,\leq)$ is an ordered domain, the system of integers may be regarded as a "part" of A if we identify each integer a with the corresponding integral element $\bar{a} = \varphi(a)$ in A.

Now let us assume in addition that the ordered domain $(A,+,\cdot,\leq)$ has the Well-Ordering Property (O-4). We shall now show that φ is an onto mapping, that is, that the set of integral elements of A coincides with A. In view of (c) in the definition of φ and (O-1), it suffices to show that the set

$$S = \{a \in A : \bar{0} < a, a \neq \varphi(n) \text{ for any natural number } n\}$$

is empty. If we suppose to the contrary that $S \neq \emptyset$, then the Well-Ordering Property (O-4) implies that S contains a least element a_0. It follows from (a) in the definition of φ that $a_0 \neq \bar{0}$ and $a_0 \neq \bar{1}$. Therefore, $\bar{0} < a_0 + (-\bar{1}) < a_0$, so that $a_0 + (-\bar{1}) \notin S$ since a_0 is the least element of S. But then there is a natural number n_0 such that $\varphi(n_0) = a_0 + (-\bar{1})$. Since $\varphi(n_0 + 1) = a_0$ by virtue of (b), we have arrived at a contradiction to the definition of a_0. We conclude that $\varphi(\mathbf{I}) = A$.

Combining the result of the preceding paragraph with our earlier conclusions concerning φ, we obtain the following result: *Any ordered domain $(A,+,\cdot,\leq)$ with the Well-Ordering Property (O-4) is isomorphic to the system $(\mathbf{I},+,\cdot,\leq)$ of integers.*

We conclude this section with a proof of the division algorithm, one of the most basic consequences of the Well-Ordering Property (O-4) for the system $(\mathbf{I},+,\cdot,\leq)$ of integers.

Proposition 2.8. (The Division Algorithm) If b is a positive integer and a is an integer, then there exist unique integers q (the quotient) and r (the remainder) satisfying $a = qb + r$ and $0 \leq r \leq b - 1$.

PROOF. We first establish the existence of the desired q and r. Define $S = \{s \in \mathbf{I} : s \geq 0, a = tb + s \text{ for some } t \in \mathbf{I}\}$. If $0 \in S$, then $a = tb$ for some $t \in \mathbf{I}$ and we can take $q = t$ and $r = 0$. If $0 \notin S$, then $s > 0$ for each $s \in S$. In order to apply (O-4) we must know that S is not empty. If $a \geq 0$, then $a + b \geq 0$ and $a = (-1)b + (a + b)$, so that $a + b \in S$. If $a < 0$, then $a - ab = (-a)(b - 1) \geq 0$. Thus $a = ab + (a - ab)$, so that $a - ab \in S$. Therefore, S is not empty in either case, and (O-4) implies that S has a least element. If this least element is denoted by r, then $r \geq 0$ and $a = qb + r$ for some $q \in \mathbf{I}$. Moreover since $a = (q +$

2.2. THE SYSTEM OF INTEGERS AND ORDERED DOMAINS

$1)b + (r - b)$, it must be true that $r - b < 0$, that is, $r \leq b - 1$, since r is the least element of S.

To prove the uniqueness of q and r, suppose that $a = q'b + r'$ for some $q' \in \mathbf{I}$ and $0 \leq r' \leq b - 1$. Then $r' \in S$, and if $r \neq r'$, then $r' = r + m$ for some natural number $m \leq b - 1$. It would then follow that $qb + r = a = q'b + r' = q'b + r + m$, so that $m = (q - q')b$. Therefore, m would be a multiple of b, contrary to the fact that $m \leq b - 1$. Consequently, $r = r'$ and since $a = q'b + r' = qb + r$, it also follows that $q' = q$. Thus, q and r are the unique integers such that $a = qb + r$ and $0 \leq r \leq b - 1$.

A geometric interpretation of the division algorithm can be given as follows. Consider the real number line with the multiples of b marked off as shown in Figure 2.4. A given integer a must fall into one of the intervals

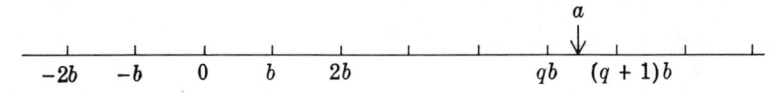

Figure 2.4

$[qb, (q + 1)b) = \{x \in \mathbf{R} : qb \leq x < (q + 1)b\}$ for some $q \in \mathbf{I}$. Thus, the quotient in $a = qb + r$ can be regarded as that integer q for which $qb \leq a < (q + 1)b$. The remainder $r = a - qb$ is simply the distance between the left-hand endpoint qb of the interval and the given integer a. Since the interval has length b, we have $0 \leq r < b$.

EXERCISES

2.28. Suppose that $(A, +, \cdot, \leq)$ is an ordered domain. Prove that:
 (a) If $a \leq b$ and if $c \leq 0$, then $a \cdot c \geq b \cdot c$.
 (b) If $a \leq 0$ and if $b \geq 0$, then $a \cdot b \leq 0$.
 (c) $a + |a| \geq 0$ for all $a \in A$.
 (d) $||a| - |b|| \leq |a - b|$ for all a, b in A.
 (e) $2ab \leq a^2 + b^2$ for all a and b in A.
 (f) If $0 < a < b$, then $a^2 < b^2$.

2.29. (a) If x and y are real numbers, show that $xy > 0$ if and only if $x > 0, y > 0$ or $x < 0, y < 0$.
 (b) Find all real numbers x satisfying $x^2 - 2x - 3 > 0$. (Note that $x^2 - 2x - 3 = (x - 3)(x + 1)$.)

2.30. Suppose that $(A, +, \cdot, \leq)$ is an ordered domain. Define the *positive cone* C in $(A, +, \cdot, \leq)$ by

$$C = \{x \in A : x \geq 0\}.$$

80 NUMBER SYSTEMS AND THEIR PROPERTIES

If $C + C = \{x + y : x \in C, y \in C\}$, if $C \cdot C = \{x \cdot y : x \in C, y \in C\}$ and if $-C = \{-x : x \in C\}$, then show that

(C-1) $C + C \subset C$,
(C-2) $C \cdot C \subset C$,
(C-3) $C \cap (-C) = \{0\}$,
(C-4) $C \cup (-C) = A$.

2.31. Prove that if $(A, +, \cdot)$ is an integral domain and if C is a subset of A with properties (C-1) through (C-4) in Exercise 2.30, then

$$x \leq y \text{ if } y - x \in C$$

defines an order relation $\leq$ on A such that $(A, +, \cdot, \leq)$ is an ordered domain with positive cone C.

2.32. The division algorithm was proved under the assumption that $b > 0$. Extend it to the case $b \leq 0$. That is, show that if a and b are arbitrary integers, then there exist unique integers q and r such that $a = qb + r$ and $0 \leq r \leq |b| - 1$.

2.3. THE SYSTEMS OF RATIONAL AND REAL NUMBERS AND ORDERED FIELDS

We now proceed to discuss some arithmetic features of the rational and real number systems that serve to distinguish these systems from other ordered domains.

With regard to the operation of multiplication, the rational and real number systems differ from the system of integers in the following fundamental respect: For each nonzero rational (or real) number a there is a rational (or real) number c (usually denoted by a^{-1} or $1/a$) such that $a \cdot c = 1$. It is this fact that constitutes the basis of the division process for each of these systems.

Definition. If $(A, +, \cdot)$ is a commutative ring with unit element that has the following additional property:

(M-4) For each $a \in A$ such that $a \neq 0$, there is an element a^{-1} (called the *inverse* of a) such that $a \cdot a^{-1} = 1$,

then $(A, +, \cdot)$ is called a *field*.

Note that a field $(A, +, \cdot)$ is always an integral domain. In fact, if a, b, c are elements of A, if $a \neq 0$, and if $a \cdot b = a \cdot c$, then since a^{-1} exists by (M-4), we have

$$b = (a^{-1} \cdot a) \cdot b = a^{-1} \cdot (a \cdot b) = a^{-1} \cdot (a \cdot c) = (a^{-1} \cdot a) \cdot c = c$$

by virtue of (M-1) and (M-3).

2.3. THE SYSTEMS OF RATIONAL AND REAL NUMBERS

As we have already pointed out in the remarks preceding the above definitions, the systems $(\mathbf{Q},+,\cdot)$ of rational numbers and $(\mathbf{R},+,\cdot)$ of real numbers are fields. The two-element integral domain discussed in Example 2.8 is a field in which $1^{-1} = 1$. As an application of Proposition 2.10 below, we shall see that if m is a prime number, the integral domain $(\mathbf{I}_m,+,\cdot)$ of residue classes modulo m is a field. If m is not a prime number, then $(\mathbf{I}_m,+,\cdot)$ is not a field since it is not even an integral domain. For the same reason, the commutative ring with unit $(P(X),+,\cdot)$ discussed in Example 2.10 is not a field unless X consists of a single point.

We shall now consider some other examples of fields.

Example 2.11. Consider the set $\mathbf{Q}(\sqrt{2})$ of all real numbers of the form $r + s\sqrt{2}$ where r and s are rational numbers; that is, $\mathbf{Q}(\sqrt{2}) = \{r + s\sqrt{2} : r \in \mathbf{Q}, s \in \mathbf{Q}\}$. Define binary operations $+$ and $\cdot$ on $\mathbf{Q}(\sqrt{2})$ as follows: If $r_1 + s_1\sqrt{2} \in \mathbf{Q}(\sqrt{2})$ and $r_2 + s_2\sqrt{2} \in \mathbf{Q}(\sqrt{2})$, then set:

$$(r_1 + s_1\sqrt{2}) + (r_2 + s_2\sqrt{2}) = (r_1 + r_2) + (s_1 + s_2)\sqrt{2}$$
$$(r_1 + s_1\sqrt{2}) \cdot (r_2 + s_2\sqrt{2}) = (r_1 r_2 + 2 s_1 s_2) + (r_1 s_2 + r_2 s_1)\sqrt{2}.$$

The fact that the algebraic system $(\mathbf{Q}(\sqrt{2}),+,\cdot)$ satisfies (A-1), (A-2), (M-1), (M-2), and (D) follows immediately from these definitions and the corresponding properties of the rational number system $(\mathbf{Q},+,\cdot)$. The element $0 + 0\sqrt{2} \in \mathbf{Q}(\sqrt{2})$ plays the role of the zero element needed in (A-3). The element $1 + 0 \cdot \sqrt{2}$ is the unit element required by (M-3). Given an element $r + s\sqrt{2}$ in $\mathbf{Q}(\sqrt{2})$, the element $(-r) + (-s)\sqrt{2}$ satisfies (A-4). Therefore, $(\mathbf{Q}(\sqrt{2}),+,\cdot)$ is a commutative ring with unit element. Finally, if $r + s\sqrt{2}$ is a nonzero element of $\mathbf{Q}(\sqrt{2})$, then the definition of multiplication and the description of the unit element given above show that $[r/(r^2 - 2s^2)] + [-s/(r^2 - 2s^2)]\sqrt{2}$ provides the inverse of $r + s\sqrt{2}$ required by (M-4). (Note that $r^2 - 2s^2 \neq 0$ since $\sqrt{2} \notin \mathbf{Q}$.) Therefore, $\mathbf{Q}(\sqrt{2})$ is a field.

Example 2.12. Consider the set $Q(t)$ of all rational functions of a real variable t with rational coefficients. That is, $Q(t)$ is the set of all functions r defined by $r(t) = p(t)/q(t)$ for those $t \in \mathbf{R}$ such that $q(t) \neq 0$, where p and q are polynomials in t with rational coefficients and where q is not the zero polynomial. If r_1 and r_2 are elements of $Q(t)$, then $r_1 = r_2$ if $r_1(t) = r_2(t)$ for all t for which both $r_1(t)$ and $r_2(t)$ are defined. If r_1 and r_2 are in $Q(t)$ and if $r_1 = p_1/q_1$, $r_2 = p_2/q_2$ where p_1, p_2, q_1, q_2 are polynomials in t with rational coefficients and q_1 and q_2 are not the zero polynomial, we define the *sum* $r_1 + r_2$ and the *product* $r_1 \cdot r_2$ of r_1 and r_2 by

$$r_1 + r_2 = \frac{p_1 q_2 + p_2 q_1}{q_1 q_2} \qquad r_1 \cdot r_2 = \frac{p_1 p_2}{q_1 q_2}.$$

We leave it to the reader to verify that these definitions are independent of the particular choices of polynomials p_1, p_2, q_1, q_2 such that $r_1 = p_1/q_1$, $r_2 = p_2/q_2$ and that under these definitions $(Q(t), +, \cdot)$ is a field.

We shall now establish some simple arithmetic properties of fields.

Proposition 2.9. Suppose that $(A, +, \cdot)$ is a field.

(1) If $a \in A$, $c \in A$, and $a \cdot c = 1$, then $c = a^{-1}$. (Uniqueness of Inverses)
(2) If a, b are nonzero elements of A, then $(a \cdot b)^{-1} = b^{-1} \cdot a^{-1}$.
(3) If a is a nonzero element of A, then $(a^{-1})^{-1} = a$.

PROOF. (1) In view of (4) in Proposition 2.3, we see that $a \neq 0$; consequently, (M-4) implies that $a \cdot c = 1 = a \cdot a^{-1}$. Therefore, by the Cancellation Property (C) we conclude that $c = a^{-1}$.

(2) To prove (2), it suffices to show that $(a \cdot b) \cdot (b^{-1} \cdot a^{-1}) = 1$ by virtue of (1) since $a \cdot b \neq 0$ by Proposition 2.5. This equality is easily established as follows:

$$(a \cdot b) \cdot (b^{-1} \cdot a^{-1}) = a \cdot (b \cdot b^{-1}) \cdot a^{-1} = a \cdot 1 \cdot a^{-1} = a \cdot a^{-1} = 1.$$

(3) Since $a^{-1} \cdot a = a \cdot a^{-1} = 1$, it follows from (1) that $(a^{-1})^{-1} = a$.

We have already observed that any commutative ring with unit element $(A, +, \cdot)$ that satisfies (M-4) also satisfies (C); that is, every field is an integral domain. The following result shows that the converse is true if A is finite.

Proposition 2.10. If $(A, +, \cdot)$ is an integral domain and if A is a finite set, then $(A, +, \cdot)$ is a field.

PROOF. If $a \in A$, $a \neq 0$, then the set $aA = \{ab : b \in A\}$ is a subset of A that has the same number of elements as A by property (C). Therefore, $aA = A$; in particular, there is an element $c \in A$ such that $a \cdot c = 1$, that is, (M-4) is satisfied. We conclude that $(A, +, \cdot)$ is a field.

In particular, it follows from the preceding proposition that the commutative ring with unit $(\mathbf{I}_m, +, \cdot)$ of residue class modulo m is a field if and only if m is a prime number.

We have already mentioned in Example 2.7 that the system $(\mathbf{Q}, +, \cdot, \leq)$ of rational numbers and the system $(\mathbf{R}, +, \cdot, \leq)$ of real numbers are ordered domains and we have just remarked that the systems $(\mathbf{Q}, +, \cdot)$ and $(\mathbf{R}, +, \cdot)$ are fields. This leads us to the following definition.

Definition. An algebraic system $(A,+,\cdot,\leq)$ consisting of a set A, two binary relations $+$ and $\cdot$ on A and an order relation $\leq$ on A such that $(A,+,\cdot,\leq)$ is an ordered domain and such that $(A,+,\cdot)$ is a field is called an *ordered field*.

In particular, the system $(\mathbf{Q},+,\cdot,\leq)$ of rational numbers and the system $(\mathbf{R},+,\cdot,\leq)$ of real numbers are examples of ordered fields. Also, if $\leq$ is the usual order relation for real numbers, then $(\mathbf{Q}(\sqrt{2}),+,\cdot,\leq)$ is an ordered field (see Example 2.11). An example of an ordered field whose elements are not real numbers is provided in Exercise 2.37.

The following result establishes some basic properties of ordered fields. As in Section 2.2, we shall use $a < b$ to mean $a \leq b$ and $a \neq b$. Also, $a \leq b$ and $b \geq a$ will be used interchangeably.

Proposition 2.11. Suppose that $(A,+,\cdot,\leq)$ is an ordered field:

(1) If $a > 0$, then $a^{-1} > 0$.
(2) If $a < b$, then $a < 2^{-1}(a+b) < b$ (where 2 is the element $1 + 1$ of A).

PROOF. (1) By the Total Order Property (O-1), we know that either $a^{-1} \geq 0$ or $a^{-1} \leq 0$. If $a^{-1} \leq 0$, then $a \cdot a^{-1} = 1 \leq 0$ which contradicts the fact that $0 < 1$ by (3) of Proposition 2.6. Consequently, $a^{-1} \geq 0$. Clearly, $a^{-1} = 0$ is impossible, hence $a^{-1} > 0$.

(2) If $a < b$, then $2a = (1+1)a = a + a < a + b$. Also, since $0 < 1$, we have $0 < 1 + 1 = 2$, and hence $0 < 2^{-1}$ by (1). Therefore, $a = 2^{-1}(2a) \leq 2^{-1}(a+b)$. Equality is not possible since it would imply that $a = b$. Thus, $a < 2^{-1}(a+b)$. The inequality $2^{-1}(a+b) < b$ is proved in a similar manner.

EXERCISES

2.33. Compute the inverse of each nonzero element in the fields $(\mathbf{I}_3,+,\cdot)$ and $(\mathbf{I}_5,+,\cdot)$.
2.34. If $(A,+,\cdot)$ is a field, prove that:
   (a) For each nonzero element $a \in A$, it is true that $(-a)^{-1} = -(a^{-1})$.
   (b) If a, b, c, d are elements of A and if b, d are nonzero, then $(ab^{-1}) + (cd^{-1}) = (ad + cb) \cdot (bd)^{-1}$.
2.35. If $(A,+,\cdot,\leq)$ is an ordered field, prove that:
   (a) If $0 < a < b$, then $0 < b^{-1} < a^{-1}$.
   (b) $|a^{-1}| = |a|^{-1}$ for all nonzero a in A.
2.36. Show that an ordered field cannot have the Well-Ordering Property (O-4).

2.37. Define a relation $\leq$ on the set $Q(t)$ of rational functions (see Example 2.12) as follows: If r_1, r_2 are rational functions in $Q(t)$ and if p_1, p_2, q_1, q_2 are polynomials such that $r_1 = p_1/q_1$, $r_2 = p_2/q_2$, then $r_1 \leq r_2$ if and only if the leading coefficient of the polynomial $(p_2q_1 - p_1q_2)q_1q_2$ is nonnegative.
  (a) If $r_1 = -t^4/t$, $r_2 = (t^3 + 1)/(-2t^2)$, prove that $r_1 \leq r_2$.
  (b) Prove that this definition of the relation $\leq$ on $Q(t)$ does not depend on the particular choice of polynomials p_1, p_2, q_1, q_2 for which $r_1 = p_1/q_1$, $r_2 = p_2/q_2$.
  (c) Prove that $(Q(t), +, \cdot, \leq)$ is an ordered field.

In Section 2.2, we defined an isomorphism φ between the system $(\mathbf{I}, +, \cdot, \leq)$ of integers and the system $(\bar{\mathbf{I}}, +, \cdot, \leq)$ of integral elements of a given ordered domain $(A, +, \cdot, \leq)$. If $(A, +, \cdot, \leq)$ is actually an ordered field, it is possible to extend the definition of φ to obtain a mapping from the set $\mathbf{Q}$ of rational numbers into the set A. In fact, if $r \in \mathbf{Q}$ and if $r = a/b$ where a, b are in $\mathbf{I}$ and $b \neq 0$, then define $\varphi(r) = \varphi(a) \cdot \varphi(b)^{-1}$. Note that this definition depends only on the rational number r and not on the particular choice of integers a, b such that $r = a/b$. For if c, d are also integers such that $r = c/d$, then $a \cdot d = b \cdot c$. Consequently, since φ is an isomorphism between the system of integers and the system of integral elements of A, it follows that $\varphi(a) \cdot \varphi(d) = \varphi(b) \cdot \varphi(c)$, and hence, $\varphi(a) \cdot \varphi(b)^{-1} = \varphi(c) \cdot \varphi(d)^{-1}$. Thus, we have unambiguously defined the mapping $\varphi \colon \mathbf{Q} \to A$.

If $(A, +, \cdot, \leq)$ is an ordered field, then an element $\bar{r} \in A$ is called a *rational element* of A if $\bar{r} = \varphi(r)$ for some rational number r. We shall denote the set of all rational elements of A by $\bar{\mathbf{Q}}$. Note that every integral element of A is a rational element of A, that is, $\bar{\mathbf{I}} \subset \bar{\mathbf{Q}}$.

The mapping $\varphi \colon \mathbf{Q} \to A$ is easily verified to be an isomorphism between $(\mathbf{Q}, +, \cdot, \leq)$ and $(\bar{\mathbf{Q}}, +, \cdot, \leq)$. For example, if $r_1 = a_1/b_1$, $r_2 = a_2/b_2$ where a_1, b_1, a_2, b_2 are integers and b_1, b_2 are nonzero, then

$$\varphi(r_1 + r_2) = \varphi\left(\frac{a_1b_2 + a_2b_1}{b_1b_2}\right) = \varphi(a_1b_2 + a_2b_1)\varphi(b_1b_2)^{-1}$$
$$= [\varphi(a_1)\varphi(b_2) + \varphi(a_2)\varphi(b_1)]\varphi(b_1)^{-1}\varphi(b_2)^{-1}$$
$$= \varphi(a_1)\varphi(b_1)^{-1} + \varphi(a_2)\varphi(b_2)^{-1}$$
$$= \varphi(r_1) + \varphi(r_2)$$

by virtue of the fact that φ is an isomorphism between the system $(\mathbf{I}, +, \cdot, \leq)$ of integers and the system $(\bar{\mathbf{I}}, +, \cdot, \leq)$ of integral elements of A. The other properties of an isomorphism are verified in a similar fashion.

We can now summarize the above conclusions as follows: *If $(A, +, \cdot, \leq)$ is an ordered field, then the ordered field $(\mathbf{Q}, +, \cdot, \leq)$ of rational numbers is* *isomorphic to the ordered field* $(\bar{\mathbf{Q}}, +, \cdot, \leq)$ *of rational elements of* A. In this sense, the system of rational numbers is the smallest ordered field since it may be regarded as a part of any ordered field $(A, +, \cdot, \leq)$. Because of this isomorphism between the system of rational numbers and the system of rational elements of $(A, +, \cdot, \leq)$, we shall often denote a rational element of A by the corresponding rational number for the sake of convenience. For example, we can use $\frac{3}{4}$ to denote the element $\varphi(3) \cdot \varphi(4)^{-1}$ of A.

The following order property of the systems of rational and real numbers is quite important: Given any rational or real number a, there exists a natural number n such that $a \leq n$. While this property is quite acceptable intuitively, it cannot be proved as a consequence of the axioms for an ordered field. For this reason we introduce the following restriction for ordered fields.

Definition. An ordered field $(A, +, \cdot, \leq)$ is called an *Archimedean ordered field* if it has the following additional property:

(O-5) For each $a \in A$, there is an integral element $n \in A$ such that $a \leq n$ and $0 \leq n$. (Archimedean Property)

An example of an ordered field which is not an Archimedean ordered field is given in Exercise 2.46 below. The requirement that an ordered field be Archimedean is quite restrictive; in fact, it is shown in the Appendix to this chapter that any such field may be regarded as a part of the ordered field $(\mathbf{R}, +, \cdot, \leq)$ of real numbers. This fact is anticipated by the following result.

Proposition 2.12. If $(A, +, \cdot, \leq)$ is an Archimedean ordered field and if $x \in A$, there is a unique integral element n_0 such that $n_0 \leq x < n_0 + 1$.

PROOF. We shall only consider the case in which $x \geq 0$. By virtue of the Archimedean Property (O-5), the set $S = \{a \in \bar{\mathbf{I}}: a \geq 0, a > x\}$ is nonempty. Since the ordered domain $(\bar{\mathbf{I}}, +, \cdot, \leq)$ of integral elements of A is isomorphic to the system of integers, it has the Well-Ordering Property (O-4). Consequently, S contains a least element a_0. Since $0 \leq x < a_0$, it follows that $0 \leq a_0 - 1$. Moreover, $a_0 - 1 \leq x$ since $a_0 - 1 \notin S$. Therefore, if we take $n_0 = a_0 - 1$, then $n_0 \leq x < n_0 + 1$. The case $x < 0$ is left to the reader as an exercise.

Our next objective will be to discuss the important Completeness Property for ordered fields. The following concepts will be needed in order to formulate this property.

Definition. Suppose that $(A, +, \cdot, \leq)$ is an ordered field and that B is a subset of A. An element $a \in A$ is an *upper bound* for B if $b \leq a$ for all $b \in B$. If there is an upper bound for B in A, B is said to be *bounded above*.

An element $u \in A$ is the *least upper bound* (or *supremum*) of B, written $u = \mathrm{lub}(B)$ (or $u = \sup(B)$), if:

(1) u is an upper bound for B,
(2) $u \leq a$ for each upper bound a for B.

The terms *lower bound*, *bounded below*, and *greatest lower bound* (or *infimum*) are defined in an analogous manner. If a subset B of A has a greatest lower bound $v \in A$, we sometimes write $v = \mathrm{glb}(B)$ (or $v = \inf(B)$).

If B is a subset of an ordered field $(A, +, \cdot, \leq)$, then B can have at most one least upper bound and at most one greatest lower bound in A (see Exercise 2.40). However, not all subsets of an ordered field have a least upper bound or a greatest lower bound.

Example 2.13. The subset $\mathbf{N}$ of natural numbers is not bounded above in the ordered field $(\mathbf{R}, +, \cdot, \leq)$ of real numbers. In particular, the least upper bound of $\mathbf{N}$ in $\mathbf{R}$ does not exist. However, it is clear that $1 = \mathrm{glb}(\mathbf{N})$.

Example 2.14. The subset $B = \{x \in \mathbf{R}: 0 < x \leq 1\}$ of the ordered field of real numbers is bounded above and bounded below. In fact, $1, \frac{3}{2}, \pi$ are several upper bounds for B, while $0, -2$, and $-\sqrt{2}$ are lower bounds for B. It is easy to see that $1 = \mathrm{lub}(B)$ while $0 = \mathrm{glb}(B)$. Note that B contains its least upper bound, but not its greatest lower bound.

If we replace the ordered field of real numbers by the ordered field of rational numbers throughout the above example, the same conclusions hold except that π is no longer an upper bound for B and $-\sqrt{2}$ is no longer a lower bound for B because π and $\sqrt{2}$ are not rational numbers.

Example 2.15. The subset $B = \{m/n: m \in \mathbf{N}, n \in \mathbf{N}, m < n\}$ of the system of rational numbers $(\mathbf{Q}, +, \cdot, \leq)$ has greatest lower bound 0 and least upper bound 1. Note that B contains neither its least upper bound nor its greatest lower bound in this case. If, in the definition of the subset B, we allowed m to be an arbitrary integer less than n, the resulting set would not be bounded below and hence would not have a greatest lower bound; however, the least upper bound would remain the same.

Example 2.16. The subset B of the ordered field $(\mathbf{Q}, +, \cdot, \leq)$ of rational numbers defined by

$$B = \{r \in \mathbf{Q}: r \geq 0, r^2 < 2\}$$

obviously has greatest lower bound 0. B is clearly bounded above; in fact, $\frac{3}{2}$, 2, 4 are all upper bounds for B. However, B does not have a least upper bound *in* $\mathbf{Q}$. To see this, we first note that if r is a least upper bound for B in $\mathbf{Q}$ then either $r^2 < 2$ or $r^2 > 2$ since $\sqrt{2}$ is not a rational number. If $r^2 < 2$, then choose a natural number n such that $1/n < \sqrt{2} - r$. Since $(r + 1/n)^2 < (r + \sqrt{2} - r)^2 = 2$, we conclude that $r + 1/n \in B$. This contradicts the fact that r is the least upper bound of B since $r < r + 1/n$. If $r^2 > 2$, choose a natural number m such that $1/m < r - \sqrt{2}$. Then $(r - 1/m)^2 > 2 > b^2$ for all $b \in B$ so that $r - 1/m$ is an upper bound for B. This also contradicts the fact that r is the least upper bound for B since $r > r - 1/m$. Therefore, we conclude that B does not have a least upper bound in $\mathbf{Q}$.

We now formulate the important Completeness Property for ordered fields.

Definition. An ordered field $(A, +, \cdot, \leq)$ is *complete* if it satisfies the following condition:

(O-6) Every nonempty subset of A that is bounded above has a least upper bound in A. (The Completeness Property)

If $(A, +, \cdot, \leq)$ is a complete ordered field, then every subset of A that is bounded below has a greatest lower bound (see Exercise 2.41). Example 2.16 above shows that the ordered field $(\mathbf{Q}, +, \cdot, \leq)$ of rational numbers is not complete.

The ordered field $(\mathbf{R}, +, \cdot, \leq)$ is a complete ordered field. Intuitively, this can be seen by making the usual identification of the set of real numbers with the set of points on the number line. When this is done, the statement that B is a nonempty subset of $\mathbf{R}$ which is bounded above in $\mathbf{R}$ means geometrically that there is a point P on the number line with the property that it lies to the right of all points b in B (see Figure 2.5). The

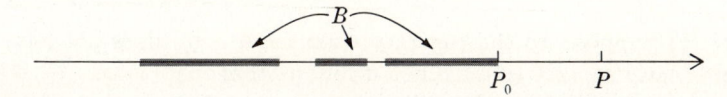

Figure 2.5

least upper bound P_0 of B is then the point farthest to the left with the preceding property. Even though it seems quite plausible that such a point P_0 exists, we have certainly not proved that it does and so this property of the number line, which is just the Completeness Property (O-6), will be regarded as an axiom concerning the real number system.

The Completeness Property (O-6) for ordered fields is very stringent.

As a matter of fact, it can be shown that *every complete ordered field must be isomorphic to the ordered field of real numbers.* We shall not prove this result in this book; however, a proof could be based on the concept of decimal representation. (See the Appendix to this chapter.) From the point of view of our work in this chapter, the importance of this isomorphism theorem lies in the fact that it shows that the Completeness Property (O-6) serves to distinguish the real number system from other ordered fields just as the Well-Ordering Property (O-4) distinguished the system of integers from other ordered domains.

We shall conclude this section by deducing two important consequences of the Completeness Property (O-6) for ordered fields.

Proposition 2.13. If $(A, +, \cdot, \leq)$ is a complete ordered field, then:

(1) $(A, +, \cdot, \leq)$ has the Archimedean Property (O-5).

(2) If, for each natural number n, $B_n = \{x \in A : a_n \leq x \leq b_n\}$ where a_n and b_n are elements of A such that $a_n \leq a_{n+1} < b_{n+1} \leq b_n$, then $\bigcap_{n \in \mathbb{N}} B_n$ is not empty. (The Nested Interval Property)

The condition on the sequence of intervals $\{B_n\}$ in (2) means that $B_{n+1} \subset B_n$ for all n. This nesting property is illustrated in Figure 2.6.

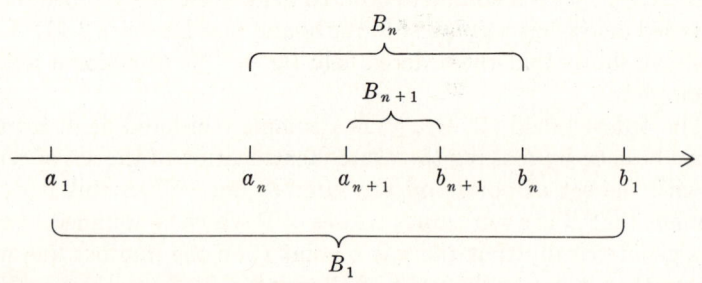

Figure 2.6

PROOF. (1) Suppose, to the contrary, that $(A, +, \cdot, \leq)$ does not have the Archimedean Property (O-5). Then it follows that there is an $x \in A$ such that $n \leq x$ for every positive integral element $n \in A$. But then the set $\bar{I}$ of integral elements is bounded above by x. Consequently, since $(A, +, \cdot, \leq)$ is complete, $\bar{I}$ has a least upper bound u. Since $u - 1 < u$, it follows from the definition of least upper bound that $u - 1$ is not an upper bound for $\bar{I}$. Hence, there is an integral element n_0 such that $u - 1 < n_0$. But then $u < n_0 + 1$ and $n_0 + 1 \in \bar{I}$, which contradicts the fact that u is an upper bound for $\bar{I}$. Therefore, we conclude that $(A, +, \cdot, \leq)$ must have the Archimedean Property (O-5).

(2) Since $a_n \leq a_{n+1} < b_{n+1} \leq b_n$ for all n, it follows that the set

$L = \{a_n : n = 1,2,3, \cdots\}$ is bounded above. If u is the least upper bound of L, then certainly $a_n \leq u$ for all n. Moreover, since each b_n is an upper bound for L, it follows that $u \leq b_n$ for all n. Therefore, we conclude that $u \in B_n$ for all n, that is, $u \in \bigcap_{n \in \mathbf{N}} B_n$.

Note that if the intervals B_n in the Nested Interval Property are altered so as to exclude either a_n or b_n, then the conclusion that the intersection $\bigcap_{n \in \mathbf{N}} B_n$ is not empty is no longer valid. For example, if $B_n = \{x \in \mathbf{R} : 0 < x \leq 1/n\}$, then $\bigcap_{n \in \mathbf{N}} B_n = \emptyset$.

EXERCISES

2.38. Find $\text{lub}(B)$ and $\text{glb}(B)$ when
  (a) $B = \{1, \frac{1}{2}, \frac{1}{3}, \cdots\}$,
  (b) $B = \{(-1)^n/n : n \in \mathbf{N}\}$,
  (c) $B = \{1/n - 1/m : n \in \mathbf{N}, m \in \mathbf{N}\}$.
2.39. Show that any nonempty finite subset of an ordered field has a least upper bound and a greatest lower bound.
2.40. Show that if u_1 and u_2 are both least upper bounds of a subset B of an ordered field, then $u_1 = u_2$.
2.41. Show that each subset of a complete ordered field that is bounded below has a greatest lower bound.
2.42. Show that if $(A, +, \cdot, \leq)$ is an ordered field with the property that each nonempty subset that is bounded below has a greatest lower bound, then it is complete.
2.43. Suppose B is a subset of $\mathbf{R}$ that is bounded above and that U denotes the set of all upper bounds for B. Show that $\text{lub}(B) = \text{glb}(U)$.
2.44. Suppose B is a nonempty subset of $\mathbf{R}$ that is bounded above and below. Show that $\text{glb}(B) \leq \text{lub}(B)$. What can be said about B if $\text{glb}(B) = \text{lub}(B)$?
2.45. Suppose A and B are subsets of $\mathbf{R}$ that are bounded above and let $C = \{a + b : a \in A, b \in B\}$. Show that if $x = \text{lub}(A)$ and $y = \text{lub}(B)$, then $x + y = \text{lub}(C)$.
2.46. Prove that the ordered field $(Q(t), +, \cdot, \leq)$ introduced in Exercise 2.37 does not have the Archimedean Property.

2.4. THE COMPLEX NUMBER SYSTEM

The first systematic studies of the arithmetic properties of complex numbers were carried out during the sixteenth century in connection with research into the solution of algebraic equations. The so-called "quadratic formula"

$$x = \frac{-b}{2a} \pm \frac{\sqrt{b^2 - 4ac}}{2a}$$

for determining the solutions of the quadratic equation $ax^2 + bx + c = 0$ was well known at that time. It had also been observed that the solution of some quadratic equations (for example, $x^2 + 1 = 0$) by means of the quadratic formula led to the consideration of square roots of negative real numbers. At first, such solutions were regarded as meaningless and were summarily discarded. Later it was observed that if expressions of the form $a + \sqrt{b}$, where a and b are *any* real numbers, were formally manipulated as binomials, valid arithmetic results could be obtained. Mathematicians of that period found it useful to consider such expressions in other connections, including the solution of cubic and quartic equations, and such expressions soon acquired the name complex numbers.

By the end of the sixteenth century, the arithmetic of complex numbers was in common use among mathematicians, although a certain aura of mystery continued to surround their nature until the nineteenth century. During the nineteenth century, the geometry of complex numbers was developed and exploited. The research of this period not only served to clarify the meaning of complex numbers, but also established the importance of complex numbers for the study of mathematical analysis, geometry, and abstract algebra.

In this section, we shall limit our discussion of the complex number system to a description of its definition and basic arithmetic properties. In Chapter 8, we shall consider the geometry of the complex number system in some detail.

Traditionally, a complex number was defined as a number of the form $z = x + iy$ where x and y are real numbers and i is the so-called imaginary unit $\sqrt{-1}$. This definition is quite natural for two basic reasons. First of all, if quadratic equations of the form

$$ax^2 + bx + c = 0, \quad a, b, c \text{ real}$$

are solved through formal application of the quadratic formula, the two roots r_1 and r_2 may be written in the form

$$r_1 = s_1 + it_1, \quad r_2 = s_2 + i(-t_1),$$

where s_1, s_2, and t_1 are real numbers. Secondly, the definition of the basic operations that turn out to be useful for the complex number system are suggested by treating the complex numbers algebraically as binomials. For example, if $z_1 = x_1 + iy_1$ and $z_2 = x_2 + iy_2$ are complex numbers, formal algebraic manipulation of these binomial expressions and use of the fact that $i^2 = -1$ would suggest the following appropriate definitions for the sum $z_1 + z_2$ and product $z_1 z_2$:

$$z_1 + z_2 = (x_1 + x_2) + i(y_1 + y_2),$$
$$z_1 z_2 = (x_1 x_2 - y_1 y_2) + i(x_1 y_2 + x_2 y_1).$$

Nevertheless, the traditional definition of a complex number has the following logical shortcoming. The expression $z = x + iy$ involves an indicated product iy and an indicated sum $x + (iy)$ of terms that are not all real numbers. Thus, strictly speaking, this definition seems to include an undefined product and sum. In reality, however, these indicated operations need not be regarded as operations for the purpose of defining complex numbers. They are merely to indicate that a complex number is to be considered as a noncommutative binomial of real numbers; that is, $x + iy$ is to be regarded as distinct from $y + ix$. Thus, a complex number z is determined by a pair of real numbers x and y in which the order in which x and y are written is important. These considerations suggest the following definition of complex number.

Definition. A *complex number* z is an ordered pair (x,y) of real numbers.

The first component x is the *real part* of z (denoted Re(z)) and the second component y of z is the *imaginary part* of z (denoted Im(z)). If Re(z) = 0, then z is sometimes called an *imaginary number*. The imaginary number $z = (0,1)$ will often be denoted by i. The set of all complex numbers will be denoted by **C**.

We shall often use the traditional notation $z = x + iy$ for the complex number $z = (x,y)$ in order to take advantage of the suggestive features of this "binomial" notation. In view of the discussion preceding the above definition, the indicated sum and product in the traditional notation should not be a source of confusion to the reader.

We now proceed to define some basic algebraic operations on the set **C** of complex numbers. If $z_1 = (x_1, y_1) \in \mathbf{C}$, $z_2 = (x_2, y_2) \in \mathbf{C}$, define:

$$z_1 + z_2 = (x_1 + x_2, y_1 + y_2),$$
$$z_1 \cdot z_2 = (x_1 x_2 - y_1 y_2, x_1 y_2 + x_2 y_1),$$
$$|z_1| = \sqrt{x_1^2 + y_1^2},$$
$$\bar{z}_1 = (x_1, -y_1).$$

The first two definitions given above define binary operations $+$ and $\cdot$ on **C** which we shall refer to as *addition* and *multiplication*, respectively. $|z|$ is called the *absolute value* or *modulus* of z, while $\bar{z}$ is referred to as the *complex conjugate* of z.

Proposition 2.14. The algebraic system $(\mathbf{C}, +, \cdot)$ is a field. There does not exist an order relation $\leq$ on **C** for which $(\mathbf{C}, +, \cdot, \leq)$ is an ordered field.

PROOF. Properties (A-1), (A-2), (M-1), (M-2), and (D) follow from the definitions of addition and multiplication for **C** and the corresponding properties of the real number system. The complex number (0,0) provides the zero element required by (A-3) while (1,0) is easily seen to provide the unit element in (M-3). Given $z = (x,y) \in C$, we have $(x,y) + (-x,-y) = (0,0)$ and

$$(x,y) \cdot \left(\frac{x}{x^2+y^2}, \frac{-y}{x^2+y^2}\right) = (1,0)$$

if $(x,y) \neq (0,0)$. Therefore, $-z = (-x,-y)$ is the negative of z so that (A-4) holds, while the inverse z^{-1} needed for (M-4) is

$$z^{-1} = \left(\frac{x}{x^2+y^2}, \frac{-y}{x^2+y^2}\right).$$

Therefore, (**C**,+,·) is a field.

Suppose that $\leq$ is an order relation on **C** for which (**C**,+,·,$\leq$) is an ordered field. Then, as we have already noted in Proposition 2.6, the unit element (1,0) of **C** must satisfy $(1,0) \geq (0,0)$ and for any $z = (x,y) \in$ **C**, we must have $z^2 = (x,y) \cdot (x,y) \geq (0,0)$. In particular, $i^2 = (0,1)(0,1) = (-1,0) \geq (0,0)$. Thus, we have $(1,0) \geq (0,0)$ and $(-1,0) = -(1,0) \geq (0,0)$ which is impossible since $(1,0) \neq (0,0)$. Therefore, there does not exist an order relation $\leq$ on **C** for which (**C**,+,·,$\leq$) is an ordered field.

Strictly speaking, a real number is not a complex number since a complex number is, by definition, an ordered pair of real numbers. However, the real number system is isomorphic to a certain subsystem of (**C**,+,·) which we shall now describe. The set $\tilde{\mathbf{R}} = \{z \in \mathbf{C}: z = (x,0), x \in \mathbf{R}\}$ is a subset of **C** that is closed with respect to the binary operations $+$ and $\cdot$ on **C** since for all $z_1 = (x_1,0) \in \tilde{\mathbf{R}}$, $z_2 = (x_2,0) \in \tilde{\mathbf{R}}$, we have $z_1 + z_2 = (x_1 + x_2, 0) \in \tilde{\mathbf{R}}$ and $z_1 \cdot z_2 = (x_1 x_2, 0) \in \tilde{\mathbf{R}}$. The zero element (0,0) and the unit element (1,0) for (**C**,+,·) belong to $\tilde{\mathbf{R}}$; moreover, if $z = (a,0) \in \tilde{\mathbf{R}}$ and $a \neq 0$, then $z^{-1} = (a^{-1},0) \in \tilde{\mathbf{R}}$. From these observations, it is a simple matter to conclude that ($\tilde{\mathbf{R}}$,+,·) is a field and also that the mapping $\varphi: \mathbf{R} \to \tilde{\mathbf{R}}$ defined by

$$\varphi(a) = (a,0), \quad (a \in \mathbf{R})$$

is an isomorphism of the real number system (**R**,+,·) onto ($\tilde{\mathbf{R}}$,+,·). Thus, even though real numbers are not complex numbers in the strict sense of the word, the isomorphism φ which maps a real number a onto the complex number $(a,0) \in \tilde{\mathbf{R}}$ enables us to regard each real number as a special type of complex number for all arithmetic purposes. For this reason, we shall ordinarily use the notation x for the element $(x,0) \in \tilde{\mathbf{R}}$ corresponding to the real number $x \in \mathbf{R}$. Thus, for example, the unit element $(1,0)$ of **C** will be written as 1 while the zero element $(0,0)$ of **C** will be written as 0.

In view of the preceding notational conventions, it is now possible to relate the traditional notation $z = x + iy$ to the ordered pair notation $z = (x,y)$ in terms of the arithmetic operations $+$ and $\cdot$ on **C**. For if $z = (x,y) \in \mathbf{C}$, then the definitions of addition and multiplication for **C** make it clear that

$$z = (x,y) = (x,0) + (0,y) = (x,0) + (0,1) \cdot (y,0).$$

Hence, if we identify $(x,0)$ with the real number x, $(y,0)$ with the real number y, and use the notation i for the complex number $(0,1)$, the preceding equation can be written as

$$z = x + i \cdot y,$$

where the symbols $+$ and $\cdot$ denote the binary operations of addition and multiplication on **C**. Thus the traditional notation $z = x + iy$ and the ordered pair notation $z = (x,y)$ are consistent even if the former is interpreted as the sum of x with the product $i \cdot y$.

EXERCISES

2.47. Carry out the indicated operations:
  (a) $(-1,2) + (1,-1)$.
  (b) $(-2,3) \cdot (2,-1)$.
  (c) $(1 + i) \cdot (1 - i)$.

2.48. If $z_1 = (x_1,y_1)$ and $z_2 = (x_2,y_2)$ are complex numbers, then:
  (a) Find the ordered pair representing $z_1 - z_2$ (that is, $z_1 + (-z_2)$).
  (b) If z_2 is nonzero, find the ordered pair representation of z_1/z_2 (that is, $z_1 \cdot z_2^{-1}$).

2.49. Suppose z_1 and z_2 are complex numbers. Prove that:
  (a) $\overline{z_1 + z_2} = \bar{z}_1 + \bar{z}_2$.
  (b) $\overline{z_1 \cdot z_2} = \bar{z}_1 \cdot \bar{z}_2$.
  (c) $\overline{-z_1} = -\bar{z}_1$.
  (d) $\overline{z_1^{-1}} = \bar{z}_1^{-1}$, if z_1 is nonzero.
  (e) $z_1\bar{z}_1 = |z_1|^2$.
  (f) $z_1 + \bar{z}_1 = 2\,\mathrm{Re}(z_1)$.
  (g) $|z_1 z_2| = |z_1|\,|z_2|$.
  (h) $|\mathrm{Re}(z_1)| \leq |z_1|$, $|\mathrm{Im}(z_1)| \leq |z_1|$.

2.50. Suppose that $P(z) = a_0 + a_1 z + \cdots + a_n z^n$ is a polynomial and that $a_0, \cdots, a_n$ are real numbers. Prove that if z_0 is a root of $P(z) = 0$, then $\bar{z}_0$ is also a root of this equation.

2.51. Suppose that the binary operation $\cdot$ is defined on the set **C** of complex numbers by
$$z_1 \cdot z_2 = (x_1 x_2,\ y_1 y_2)$$
for all $z_1 = (x_1, y_1) \in \mathbf{C}$, $z_2 = (x_2, y_2) \in \mathbf{C}$. Is the algebraic system $(\mathbf{C}, +, \cdot)$ a commutative ring with unit element? Is it a field?

2.52. Consider the set $\mathcal{C}$ of all polynomials $p(t) = a_0 + a_1 t + \cdots + a_n t^n$ with real number coefficients $a_0, a_1, \cdots, a_n$ and variable t. If $p_1, p_2 \in \mathcal{C}$, define p_1 to be *equivalent* to p_2 (written $p_1 \sim p_2$) if division of p_1 by the polynomial $q(t) = 1 + t^2$ leaves the same remainder as the division of p_2 by q. For example, the remainders upon division of $p_1(t) = 1 + 3t + 2t^3$ and of $p_2(t) = 1 + 2t + t^3$ by $q(t) = 1 + t^2$ are both $1 + t$, so that $p_1 \sim p_2$.

(a) Prove that $\sim$ is an equivalence relation on $\mathcal{C}$ and that each equivalence class with respect to $\sim$ contains one and only one polynomial with degree at most 1.

(b) If $\mathcal{C}'$ is the set of equivalence classes with respect to $\sim$ and if $[p]$ denotes the equivalence class containing the polynomial p, prove that the equations
$$[p] + [q] = [p + q]$$
$$[p] \cdot [q] = [p \cdot q] \qquad [p] \in \mathcal{C}',\ [q] \in \mathcal{C}'$$
unambiguously define binary operations $+$ and $\cdot$ on $\mathcal{C}'$. (In other words, prove that the definitions of $[p] + [q]$ and $[p] \cdot [q]$ do not depend on the particular choice of p in $[p]$ and q in $[q]$.)

(c) Prove that the algebraic system $(\mathcal{C}', +, \cdot)$ is isomorphic to the complex number system $(\mathbf{C}, +, \cdot)$.

(The definition of complex number outlined in this exercise is due to A. Cauchy (1789–1857). The ordered pair definition of complex number was first used by W. R. Hamilton (1805–1865).)

APPENDIX: DECIMAL REPRESENTATION

In this section, we shall examine carefully the process of associating decimal representations with real numbers and show how this process is connected with the notions of Archimedean ordered field and complete ordered field. As we shall see, the full strength of the completeness axiom is not essential in order to simply establish the existence of a decimal representation for a given real number; the Archimedean property is adequate for this purpose. However, the role of completeness in the real number system is illuminated when we reverse the ideas and ask whether or not a given decimal determines some particular real number. The Completeness Property will play a vital role in answering this question.

We first define some basic terms.

APPENDIX: DECIMAL REPRESENTATION

Definition. A *decimal fraction* (or finite decimal) is a rational number of the form

$$N + \frac{x_1}{10} + \frac{x_2}{10^2} + \cdots + \frac{x_n}{10^n},$$

where $N, x_1, x_2, \cdots, x_n$ are nonnegative integers and $0 \leq x_i \leq 9$ for each $i = 1, 2, \cdots, n$. Numbers of this type will be denoted by $N.x_1 x_2 \cdots x_n$.

For example, $64/25$ is a decimal fraction since

$$\frac{64}{25} = 2 + \frac{5}{10} + \frac{6}{10^2} = 2.56.$$

Note that each decimal fraction can be written in the form $a/10^n$ for some natural numbers a and n, and each number of this type is a decimal fraction.

Definition. A *decimal* is an infinite sequence $\{N, x_1, x_2, x_3, \cdots\}$ of integers such that $0 \leq x_n \leq 9$ for each $n = 1, 2, 3, \cdots$. N is called the *integral part* of the decimal and x_n is called the nth *term* of the decimal. We shall employ the notation

$$N.x_1 x_2 x_3 \cdots$$

to denote the decimal with integral part N and the nth term x_n for $n \in \mathbf{N}$.

The decimal $N.x_1 x_2 x_3 \cdots$ is said to *represent* the nonnegative real number x if $N \leq x < N+1$ and

$$N.x_1 x_2 \cdots x_n \leq x < N.x_1 x_2 \cdots x_n + \frac{1}{10^n}$$

for each natural number n. In this case, we write $x = N.x_1 x_2 x_3 \cdots$ and refer to $N.x_1 x_2 x_3 \cdots$ as the *decimal representation* of x. For negative x we write $x = -N.x_1 x_2 x_3 \cdots$ where $N.x_1 x_2 x_3 \cdots$ is the decimal representing the positive number $-x$.

We can geometrically interpret the representation of x by a decimal $N.x_1 x_2 x_3 \cdots$ as follows. For each $n \in \mathbf{N}$, set $a_n = N.x_1 x_2 \cdots x_n$ and let A_n denote the interval

$$A_n = \left[a_n, a_n + \frac{1}{10^n}\right)$$
$$= \left\{y \in \mathbf{R} : a_n \leq y < a_n + \frac{1}{10^n}\right\}.$$

Note that A_n contains its left-hand endpoint a_n but not its right-hand endpoint $a_n + 1/10^n$. The statement that the decimal represents x is

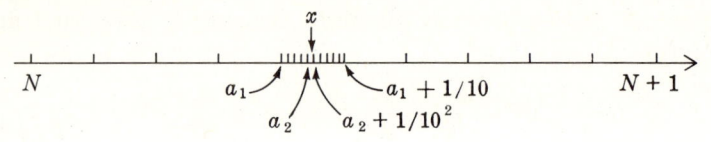

Figure 2.7

equivalent to saying that $x \in A_n$ for all n. (See Figure 2.7.) The sequence of intervals $A_1, A_2, \cdots$ is nested, that is, $A_n \supset A_{n+1}$ for all n. The fact that $x \in A_n$ for all n means that

$$x \in \bigcap_{n \in \mathbf{N}} A_n.$$

Furthermore, since the length of A_n is $1/10^n$, which tends to zero as n becomes large, x is the only real number that is contained in this intersection. For if $y \in \bigcap_{n \in \mathbf{N}} A_n$, then $|x - y| < 1/10^n$ for all n, and hence $y = x$. Thus, *a decimal can represent at most one real number.*

We now show how a unique decimal representation can be constructed for a given real number x. The reader should observe that the following construction depends only on the Archimedean property of the real number system and the Well-Ordering Property of the natural number system. Therefore, the construction can be performed in any Archimedean ordered field as a consequence of the axioms for such a structure. That is, the proof of the next proposition actually shows that *any element of an Archimedean ordered field possesses a decimal representation*. Although the Archimedean property is utilized only once in the following proof, it occurs at a very crucial point, namely, in the very first step.

Proposition 2.15. For each real number x, there exists a unique decimal representing x.

PROOF. It suffices to consider the case $x > 0$. The Archimedean property of the real number system and the Well-Ordering Property of the natural number system imply that there exists a unique integer N such that

$$N \leq x < N + 1.$$

(See Proposition 2.12.) N will be the integral part of the decimal representation of x.

Now set $d_1 = x - N$. Then $0 \leq d_1 < 1$ so that $0 \leq 10d_1 < 10$. It follows from the Well-Ordering Property of the natural number system that there is one and only one nonnegative integer x_1 such that

$$x_1 \leq 10d_1 < x_1 + 1. \tag{1}$$

Since $d_1 = x - N$, the inequality (1) is equivalent to the inequality

$$N + \frac{x_1}{10} \leq x < N + \frac{x_1}{10} + \frac{1}{10}. \tag{1'}$$

Also, since $0 \leq 10d_1 < 10$, x_1 must satisfy $0 \leq x_1 \leq 9$. x_1 will be the first term in the decimal representation of x.

Now set $d_2 = 10d_1 - x_1$. Then $0 \leq d_2 < 1$ and consequently $0 \leq 10d_2 < 10$. Again, an application of the Well-Ordering Principle produces a unique nonnegative integer x_2 such that

$$x_2 \leq 10d_2 < x_2 + 1 \tag{2}$$

or equivalently,

$$N + \frac{x_1}{10} + \frac{x_2}{10^2} \leq x < N + \frac{x_1}{10} + \frac{x_2}{10^2} + \frac{1}{10^2}. \tag{2'}$$

Also, $0 \leq x_2 \leq 9$ since $0 \leq 10d_2 < 10$. The digit x_2 will be the second term of the decimal representation of x.

We proceed inductively to obtain the nth term of the decimal representation of x for all $n \in \mathbf{N}$ as follows. Once d_{n-1} and x_{n-1} have been obtained, we define $d_n = 10d_{n-1} - x_{n-1}$. Then $0 \leq 10d_n < 10$ and x_n is the unique integer satisfying

$$x_n \leq 10d_n < x_n + 1 \tag{n}$$

or equivalently

$$N.x_1 x_2 \cdots x_n \leq x < N.x_1 x_2 \cdots x_n + \frac{1}{10^n}. \tag{n'}$$

Again, $0 \leq x_n \leq 9$ and x_n is the uniquely determined nth term of the decimal representation of x. Thus, we see that the decimal $N.x_1 x_2 x_3 \cdots$ defined in this way represents x and is the unique decimal accomplishing this task.

Example 2.17. The terms of the decimal representation of a number x are uniquely determined by the defining inequalities. If $x = \sqrt{5}$, for example, we have

$$2 \leq \sqrt{5} < 3 \quad \text{since } 4 \leq 5 < 9$$
$$2.2 \leq \sqrt{5} < 2.3 \quad \text{since } (2.2)^2 = 4.84 \leq 5 < 5.29 = (2.3)^2$$
$$2.23 \leq \sqrt{5} < 2.24 \quad \text{since } (2.23)^2 = 4.9729 \leq 5 < 5.0176 = (2.24)^2.$$

Therefore, $N = 2$, $x_1 = 2$, $x_2 = 3$, and $\sqrt{5} = 2.23 \cdots$.

There is an alternate way to define the notion of a decimal representing a real number x. This is done by relaxing the strictness of the right-hand inequalities $x < N.x_1x_2 \cdots x_n + 1/10^n$ for $n \in \mathbf{N}$ in the definition of decimal representation to allow possible equality. Each approach has its advantages and its drawbacks.

The main advantage in defining representations by decimals as we have done lies in the resulting uniqueness of the representing decimal for each real number. In the alternate approach, certain numbers would have two possible decimal representations. For example, the pair of decimals $0.99 \cdots 9 \cdots$ and $1.00 \cdots 0 \cdots$ would both represent the number 1.

In contrast to this, we note that according to the definition of representing decimal stated above, the decimal $0.99 \cdots 9 \cdots$ cannot represent any real number. Indeed, if x were such a number, then we would have

$$\frac{9}{10} + \cdots + \frac{9}{10^n} \leq x < \frac{9}{10} + \cdots + \frac{9}{10^n} + \frac{1}{10^n} = 1$$

or equivalently

$$0 < 1 - x \leq \frac{1}{10^n} \tag{*}$$

for every natural number n. But this is impossible, for if we choose a natural number k such that $(1-x)^{-1} < k$, then

$$0 < \frac{1}{1-x} < k < 10^k$$

and hence

$$0 < \frac{1}{10^k} < 1 - x,$$

which contradicts (*). Thus, the decimal $0.99 \cdots 9 \cdots$ cannot represent x.

EXERCISES

2.53. Find the integral part and the first two terms of the decimals representing $\sqrt{3}$ and $\sqrt{7}$.

2.54. Suppose that $x = 1.213 \cdots$ and $y = 2.351 \cdots$ where decimal terms past the third term are arbitrary. Prove that $x + y = 3.56 \cdots$ and $xy = 2.85 \cdots$. Can anything be said about the third terms in the decimal representations of $x + y$ and xy?

2.55. Prove that a decimal fraction $x = N.x_1x_2 \cdots x_n$ has decimal representation $N.x_1x_2 \cdots x_n 0 \cdots 0 \cdots$ (that is, $x_k = 0$ for $k > n$).

2.56. Prove that if $x = N.x_1x_2x_3 \cdots$ and if k is a natural number, then

$10^k x$ has decimal representation $M.y_1 y_2 y_3 \cdots$ where $M = 10^k N + 10^{k-1} x_1 + \cdots 10 x_{k-1} + x_k$ and $y_n = x_{n+k}$ for $n = 1, 2, 3, \cdots$. (This is expressed more casually by saying that if x is multiplied by 10^k, then the decimal point is moved k places to the right.)

2.57. Suppose x and y are positive real numbers and $y < x$. Show that if $x = N.x_1 x_2 x_3 \cdots$ and $y = M.x_1 x_2 x_3 \cdots$, then $x - y$ has decimal representation $L.00 \cdots 0 \cdots$ where $L = N - M$.

When we first encountered decimal representation in school, we learned to compute the decimal representation of rational numbers by the "long division" process. For example, if x is the rational number 30/4, the decimal representation of x was obtained as follows:

$$
\begin{array}{r}
7.5000 \\
4 \overline{\smash{)}30.0000} \\
\underline{28} \\
20 \\
\underline{20} \\
0
\end{array}
$$

so that the decimal representation of $x = 30/4$ was given as $7.5000 \cdots$. On the other hand, if $x = 1/7$, then

$$
\begin{array}{r}
0.14285714 \cdots \\
7 \overline{\smash{)}1.00000000 \cdots} \\
\underline{7} \\
30 \\
\underline{28} \\
20 \\
\underline{14} \\
60 \\
\underline{56} \\
40 \\
\underline{35} \\
50 \\
\underline{49} \\
10 \\
\underline{7} \\
30
\end{array}
$$

100 NUMBER SYSTEMS AND THEIR PROPERTIES

Although the preceding long division of 7 into 1 does not "terminate" with remainder 0 after a finite number of steps as in the case of long division of 4 into 30, we observe that the remainder obtained in each step of the long division by 7 must always be an integer from 0 to 6 since 7 is the divisor. Therefore, in the long division of any integer by 7, some remainder must be repeated after at most 7 steps, and so we conclude that the digits in the quotient must repeat themselves in cycles. For these reasons, we would conclude that the decimal representation of $x = 1/7$ is 0.142857142857 $\cdots$ where the block of digits 142857 is repeated.

Let us now analyze the familiar long division procedure in terms of our construction of the decimal representation for the rational number $x = a/b$, where a and b are integers, $b \neq 0$, and verify that the long division process does in fact yield the decimal representation of a/b. For simplicity, we shall assume that $0 < a < b$, so that the integral part of a/b is 0. Let us compare the first term obtained by long division with the first term of the decimal representation. When we divide b into $10a$, we obtain a quotient q_1 and a remainder r_1 such that $10a = q_1 b + r_1$ and $0 \leq r_1 < b$. As pointed out in the discussion of the division algorithm at the end of Section 2.2, q_1 is the unique integer such that

$$q_1 b \leq 10a < (q_1 + 1)b. \qquad (1)$$

Also, it follows from the construction in the proof of Proposition 2.15 that the first term x_1 in the decimal representation of a/b is the unique integer such that

$$x_1 \leq 10\frac{a}{b} < (x_1 + 1). \qquad (2)$$

Comparing (1) and (2), we conclude that $x_1 = q_1$. Also note that

$$r_1 = 10a - bx_1 = b\left(10\frac{a}{b} - x_1\right) = bd_2,$$

where d_2 is the integer used in the construction of the decimal representation to obtain the second term x_2.

The next step in the long division algorithm is the division of $10r_1$ by b. A quotient q_2 and remainder r_2 are obtained where $10r_1 = q_2 b + r_2$, $0 \leq r_2 < b$. q_2 is the unique integer satisfying

$$q_2 b \leq 10 r_1 < (q_2 + 1)b.$$

But $r_1 = bd_2$, so that this inequality can be rewritten as

$$q_2 \leq 10 d_2 < (q_2 + 1).$$

This is precisely the defining inequality for the second term x_2 of the decimal representation of a/b. (See the proof of Proposition 2.15.) Hence,

$x_2 = q_2$. We also see that $r_2 = bd_3$ where $d_3 = 10d_2 - x_1$. A similar analysis shows that the next decimal term x_3 is the quotient q_3 obtained in the next step of the long division process. It may now be verified in general by mathematical induction that the long division of b into a does in fact lead to the decimal representation of a/b.

Now that we know that the familiar long division procedure for finding the decimal representation of a rational number x does in fact produce the decimal representation of x, we can draw some interesting conclusions concerning the nature of decimal representations for rational numbers. Our first observation is that the decimal representation $N.x_1x_2x_3 \cdots$ of a rational number x must be one of the following two sorts:

(a) $N.x_1x_2x_3 \cdots$ is a *terminating* decimal. That is, there is a natural number k such that $x_n = 0$ for all $n \geq k$.

(b) $N.x_1x_2x_3 \cdots$ is a *periodic* (or repeating) decimal. That is, there exist natural numbers k and m such that $x_n = x_{n+m}$ for all $n \geq k$. In this case, the block of digits $x_k x_{k+1} \cdots x_{k+m-1}$ repeats once the kth term is reached. The smallest number m with this property is called the *period* of $N.x_1x_2x_3 \cdots$.

Example 2.18. If $x = \frac{30}{4}$, then $x_n = 0$ for $n \geq 2$ and $x = 7.5$, so that x has a terminating decimal representation.

Example 2.19. If $x = \frac{1}{7}$, then the decimal representation of x is periodic with $k = 1$ and period $m = 6$ since $x_1 = x_7 = x_{13} = \cdots$, $x_2 = x_8 = x_{14} = \cdots$, and so on. The block of digits 142857 repeats and we have $x = 0.142857 \cdots 142857 \cdots$.

Example 2.20. The decimal representation of $\frac{19}{88}$ is $0.2159090 \cdots 90 \cdots$. The block 90 repeats, but the repeating block does not appear until the fourth term is reached. Thus, in the definition of periodic decimal, we have $k = 4$ and $m = 2$.

Example 2.21. Sometimes the period of a decimal is fairly large and the fact that it is actually a periodic decimal is not immediately apparent. For example,
$$\tfrac{1}{19} = 0.052631578947368421 \cdots$$
where the given block of 18 terms repeats. Similarly, the period of the decimal expansion of $\frac{1}{29}$ is $m = 28$.

That one of the two alternatives mentioned above must hold for the decimal representation of a rational number $x = a/b$, where a and b are integers, $b > 0$, is a consequence of the fact that the long division algorithm applied to divide b into a always produces remainders at the end of each step that are integers from 0 to $b - 1$. (See the discussion of the decimal representation of $\frac{30}{4}$ and $\frac{1}{7}$ given previously.) We also see from this, that in the case a/b is represented by a periodic decimal, the period m of the decimal must satisfy $m \leq b - 1$.

The converse situation also holds. That is, if the decimal representation of a real number x is either terminating or periodic, then x must be a rational number. Consider first the case in which x is represented by a terminating decimal. Then $x = N.x_1x_2x_3 \cdots$ and there is a natural number k such that $x_n = 0$ for all $n \geq k$. Then $y = 10^{k-1}x$ has decimal representation $M.y_1y_2y_3 \cdots$ where $M = 10^{k-1}N + 10^{k-2}x_1 + \cdots + x_{k-1}$ and $y_n = x_{n+k-1} = 0$ for all $n = 1, 2, 3, \cdots$. (See Exercise 2.56 above.) We conclude that $10^{k-1}x = M$, so that $x = M/10^{k-1}$ which is a rational number. For example, if $x = 32.14700 \cdots$, then $1000x = 32,147.000 \cdots = 32,147$, so that $x = 32,147/1000$.

Next suppose that the decimal representation $N.x_1x_2x_3 \cdots$ of x is periodic with period m and that k is such that $x_n = x_{n+m}$ for $n \geq k$. Then the decimal representation $M.y_1y_2y_3 \cdots$ of $y = 10^{k-1}x$ is also periodic with period m. The same is true of the decimal representation $L.z_1z_2z_3 \cdots$ of $z = 10^m y = 10^{k-1+m}x$. Also, $z_n = y_n$ for all n. It follows that the decimal representation of $z - y$ is $(L - M).000 \cdots$. (See Exercise 2.57.) Thus, we have

$$10^{k-1+m}x - 10^{k-1}x = z - y = L - M$$

or

$$x = \frac{L - M}{10^{k-1+m} - 10^{k-1}}$$

and x is therefore a rational number.

For example, if $x = 7.31414 \cdots 14 \cdots$, then $k = 2$ and $m = 2$. Following the procedure of the previous paragraph, we have

$$y = 10x = 73.1414 \cdots$$
$$z = 1000x = 7314.1414 \cdots$$
$$z - y = 1000x - 10x = 7314 - 73 = 7241$$

and hence,

$$x = \frac{7241}{990}.$$

Collecting the results of the above discussion together, we have the following.

Proposition 2.16. A real number is rational if and only if its decimal representation is either terminating or periodic.

EXERCISES

2.58. Find the decimal representations of 1/8, 2/11, 40/13, 15/56.
2.59. Find the rational numbers having decimal representations:
(a) $0.729729 \cdots 729 \cdots$,
(b) $2.79191 \cdots 91 \cdots$,
(c) $52.12350615061 \cdots 5061 \cdots$.
2.60. Show that multiplication of the number $x = 142{,}857$ by any one of the numbers 2, 3, 4, 5, 6 results in a number whose digits are a cyclic arrangement of the digits of x. [*Hint:* Consider the decimal representaton of $\frac{1}{7}$ obtained by long division.]

We have seen that each real number is represented by a unique decimal, and also that there exist decimals that are not the decimal representation of any real number (for example, $0.999 \cdots$). However, it is possible to associate a unique real number with each decimal by utilizing the Completeness Property of the real number system. This is done as follows.

If $N.x_1x_2x_3 \cdots$ is a decimal, set $a_n = N.x_1x_2x_3 \cdots x_n$. a_n is the decimal fraction obtained by truncating the given decimal at the nth term. The set $\{a_n : n \in \mathbf{N}\}$ is bounded above (by $N + 1$, for example) and hence, its least upper bound

$$x = \text{lub } \{a_n : n \in \mathbf{N}\}$$

exists since the system of real numbers is a complete ordered field. In this way, we may associate a real number with the given decimal.

This number x determined by the decimal $N.x_1x_2x_3 \cdots$ can also be obtained as follows. If we define the interval B_n by

$$B_n = \left[a_n,\, a_n + \frac{1}{10^n} \right]$$
$$= \left\{ y \in \mathbf{R} : a_n \leq y \leq a_n + \frac{1}{10^n} \right\},$$

then it is easily verified that $B_{n+1} \subset B_n$ for all n. That is, the intervals $B_1, B_2, \cdots$ are nested. Hence, by the Nested Interval Property (see Proposition 2.13) $\bigcap_{n \in \mathbf{N}} B_n$ is not empty. Since the length of B_n is $1/10^n$, and since $1/10^n$ tends to zero as n becomes large, it follows that $\bigcap_{n \in \mathbf{N}} B_n$ contains exactly one point. Hence, $\bigcap_{n \in \mathbf{N}} B_n = \{x\}$ where $x = \text{lub } \{a_n : n \in \mathbf{N}\}$.

For example, the interval B_n corresponding to the decimal $0.999 \cdots$

is $B_n = [1 - 1/10^n, 1]$ and we have $\bigcap_{n \in \mathbb{N}} B_n = \{1\}$ in this case. Note that if $A_n = [1 - 1/10^n, 1)$, then $\bigcap_{n \in \mathbb{N}} A_n = \emptyset$. This happens because $0.999 \cdots 99 \cdots$ is not the decimal representation of 1 according to our definition of decimal representation.

Thus, we see that while a given decimal uniquely determines a real number x, the decimal representation of x need not coincide with that given decimal. However, the situation is not as complicated as it might first appear, for only two possibilities can occur as we now show. Suppose we are given the decimal $N.x_1x_2x_3 \cdots$. Set $B_n = [a_n, a_n + 1/10^n]$ and $A_n = [a_n, a_n + 1/10^n)$ where $a_n = N.x_1x_2 \cdots x_n$. The interval A_n is simply B_n with the right-hand endpoint removed. If $x \in \bigcap_{n \in \mathbb{N}} B_n$, then two cases arise:

(1) $x \in \bigcap_{n \in \mathbb{N}} A_n$ (2) $\bigcap_{n \in \mathbb{N}} A_n = \emptyset$.

In case (1), $x \in A_n$ for all n means that x satisfies the inequalities

$$N.x_1x_2 \cdots x_n \leq x < N.x_1x_2 \cdots x_n + \frac{1}{10^n}$$

for all n, and hence the decimal representation of x is just the given decimal $N.x_1x_2x_3 \cdots$.

The case (2) can occur if and only if $x \notin A_m$ for some m. Choose m to be the smallest nonnegative integer for which this happens. This means that $x \notin A_m$, but $x \in B_m$, so that x must be the right-hand endpoint of the interval B_m. That is,

$$x = N.x_1x_2 \cdots x_m + \frac{1}{10^m}.$$

Because the intervals are nested, that is, $B_{n+1} \subset B_n$ for all n, it must be the case that x is the right-hand endpoint of B_n for every n such that $n \geq m$. (See Figure 2.8.) This implies that $x_{m+k} = 9$ for all $k = 1, 2, \cdots$.

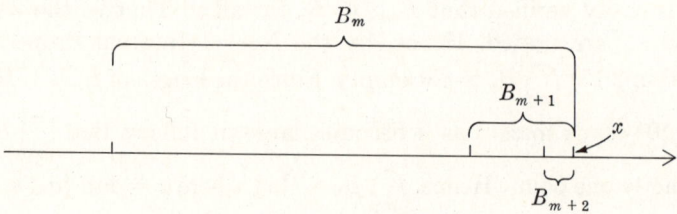

Figure 2.8

For example, since

$$N + \frac{x_1}{10} + \cdots + \frac{x_m}{10^m} + \frac{1}{10^m} = x$$
$$= N + \frac{x_1}{10} + \cdots + \frac{x_m}{10^m} + \frac{x_{m+1}}{10^{m+1}} + \frac{1}{10^{m+1}}$$

we see that

$$\frac{1}{10^m} = \frac{x_{m+1}}{10^{m+1}} + \frac{1}{10^{m+1}}$$

and multiplying both sides of this equality by 10^{m+1} yields the fact that $x_{m+1} = 9$. Proceeding inductively, we obtain that $x_{m+k} = 9$ for $k = 1, 2, \cdots$. Also note that $x_m \leq 8$ since m was chosen to be the smallest nonnegative integer with the stated property. It is now readily seen that the decimal representation of x must be the terminating decimal $N.x_1 x_2 \cdots x_{m-1}(x_m + 1)00 \cdots$ for x is nothing more than the rational number

$$x = N + \frac{x_1}{10} + \cdots + \frac{x_m + 1}{10^m}.$$

Conversely, suppose $x = N.x_1 x_2 \cdots x_n$ is a decimal fraction. We can assume that $x_n \neq 0$. Then it is easy to verify that the intervals B_n defined as above by the decimal $N.x_1 x_2 \cdots (x_n - 1)99 \cdots 9 \cdots$ have the property that $x \in \bigcap_{n \in \mathbf{N}} B_n$. Thus, we conclude that the only numbers that can be determined in this way by decimals different from their decimal representations are the decimal fractions, that is, rational numbers of the form $a/10^n$.

We mentioned earlier that there is an alternate way to view the idea of a decimal representing a real number x. In this alternate approach, the decimal $N.x_1 x_2 x_3 \cdots$ is said to represent x if the inequalities

$$N.x_1 x_2 \cdots x_n \leq x \leq N.x_1 x_2 \cdots x_n + \frac{1}{10^n}$$

are satisfied for each natural number n. This is equivalent to saying that $x \in \bigcap_{n \in \mathbf{N}} B_n$ where the intervals B_n for $n \in \mathbf{N}$ are defined as above. With this viewpoint, each decimal uniquely determines a real number, but certain numbers will have two decimal representations. As the preceding discussion shows, the numbers having two decimal representations in this sense are precisely the decimal fractions where one decimal representation terminates while the other is constantly 9 from some term on. For example, $5.71300 \cdots$ and $5.71299 \cdots$ both would be regarded as decimal representations of 5713/1000. Although this viewpoint is quite acceptable and is in common use, the definition given at the outset of this section has the desirable feature of uniqueness of representation and the property that it coincides with the decimal representation of rational numbers obtained through long division.

We conclude this discussion by observing that the notion of decimal representation can be used to show that each Archimedean ordered field $(A, +, \cdot, \leq)$ can be identified with a subset of the real number system. In the remarks preceding Proposition 2.15, we noted that the proof of Proposition 2.15 is valid in Archimedean ordered fields. Thus, each element $x \in A$ can be associated with a unique decimal $N.x_1x_2x_3 \cdots$ and this decimal in turn determines a unique real number. If we denote this real number by $\varphi(x)$, then we obtain a one-to-one mapping $\varphi \colon A \to \mathbf{R}$ and A corresponds to the subset $\varphi(A)$ of $\mathbf{R}$. Furthermore, if $(A, +, \cdot, \leq)$ is complete, it then follows that φ is a mapping from A onto $\mathbf{R}$ and from this one can conclude that each complete ordered field can be identified with the real number system.

EXERCISE

2.61. Show that if x and y are real numbers, $0 < y < x$, then there exists a rational number r such that $y < r < x$. [*Hint:* Choose n so that $1/10^n < x - y$ and consider the decimal fraction $r = N.x_1x_2 \cdots x_n$ determined by the decimal representation $N.x_1x_2x_3 \cdots$ of x.]

REMARKS AND REFERENCES

All modern secondary school algebra programs place considerable emphasis on the study of the various number systems and on developing skill in working with numerical expressions. Although these two points of emphasis are obviously related, it must be realized that neither is a consequence of the other. That is, the study of number systems as algebraic systems does not necessarily lead to the acquisition of the essential skills required to work with fractions, radicals, inequalities, decimals, and so on. Conversely, emphasis on such skills may not lead to an appreciation of the structural properties of the various number systems.

The situation for the secondary school teacher is further complicated by the fact that it does not seem to be advisable to approach each of these worthwhile aspects of the algebra program simultaneously and with equal vigor on a day-to-day and topic-by-topic basis. Too much emphasis on the deductive justification of each step in the manipulation of an algebraic expression in terms of definitions and axiomatic properties of number systems will interfere with, rather than assist in the development of arithmetic skills, even for the beginning algebra student. Some skill with arithmetic manipulation is required before an appreciation of the deductive character of the number systems can be gained. In other works, although the logical textbook organization of mathematics ordinarily presents "theory" first and "skills" or "applications" second, it is often necessary to reverse the roles of these two aspects of the subject in the classroom.

In developing skills with algebraic expressions, it is important to proceed one step at a time when a new procedure or problem type is encountered. However, as the student begins to acquire some facility and understanding of the technique involved, he should be encouraged to combine two or more simple steps so that computations are not unnecessarily belabored. Also, the teacher should begin to evolve certain code phrases such as "cross multiply in the equation" and "transpose the constant terms to the right-hand side of the equation," and so on. Such phrases are abbreviated descriptions of certain sequences of algebraic steps. Their use is both legitimate and convenient provided that their basis is explained at the outset and that these explanations are reinforced from time to time by analyzing a particular example in terms of definitions and axiomatic properties, that is, by "doing the problem the hard way" as some students so aptly express it. Constant appeal to axioms and continually drawn out computations can interfere with concentration on the main point under discussion.

Students often find it difficult to appreciate the basic axiomatic properties of the various number systems because of their familiarity with these systems attained through their training in arithmetic. For this reason, it is quite instructive to analyze other binary operations such as

$$(x,y) \rightarrow \frac{x}{y},$$

$$(x,y) \rightarrow \frac{x+y}{2},$$

$$(x,y) \rightarrow \text{maximum of } x \text{ and } y,$$

on suitable domains of numbers. For example, it can be determined whether these operations are commutative or associative, whether a distributive property holds for two such operations, and so on.

The commutative ring with unit $(\mathbf{I}_m, +, \cdot)$ of residue classes modulo an integer $m > 1$ can be discussed at a considerably lower level than it is in this book, and therefore can be used in the classroom in connection with the discussion of number systems. One very effective and appealing way to do this is in the context of "clock arithmetic." The students are already familiar with the fact that times on the clock are added modulo 12, although, of course, this terminology is not familiar to them. For example, they will readily agree that if the classroom clock now reads 11 o'clock, then 5 hours from now the clock will read 4 o'clock. That is,

$$11 + 5 = 4 \text{ (on the clock).}$$

After considering a number of such examples, the class will ordinarily experience little difficulty in filling in a complete addition table for the clock. It is useful to observe at this point that 12 o'clock behaves like the zero element and to change the twelves in the table to zeros.

The class can then be led to the notion of multiplication modulo 12 with problems of the following sort. Suppose you begin a train trip at noon and by 4 o'clock you have completed one-fifth of your trip time. If you never change time zones, what time can you expect to arrive at your destination? Students should have little trouble in concluding that the arrival time will be 8 o'clock. That is,

$$4 \cdot 5 = 8 \text{ (on the clock).}$$

Once the multiplication table for the clock is completed, it is a simple matter to lead the class to the proper conclusions concerning negatives, inverses, and so on, in $(\mathbf{I}_{12}, +, \cdot)$. At this point, the class is usually ready for, in fact, anticipating the change from a 12-hour clock to a clock with a different number of hours. The addition and multiplication tables for $(\mathbf{I}_5, +, \cdot)$ can be constructed and the fact that each nonzero element of $\mathbf{I}_5$ has an inverse can be contrasted with the absence of that feature in $\mathbf{I}_{12}$. If the teacher chooses, clock arithmetic can also be used to introduce congruence problems and to illuminate concepts related to equivalence relations.

Mathematical induction is generally regarded to be one of the most difficult topics in the algebra program. Algebra textbooks often slight the discussion of the meaning of the hypothesis and conclusion of the Principle of Mathematical Induction and their examples and exercises often lack variety. The development of this topic in Section 2.1 should help the teacher to supplement the textbook treatment in both of these respects.

It is important that the secondary school students, particularly those who intend to study calculus, gain an appreciation for the distinction and relation between the systems of rational and real numbers. Most algebra texts identify the set of real numbers with the set of points on the real number line. The rational numbers can be located on the number line by subdividing intervals between successive integers. For example, to locate the rational number $\frac{15}{8}$ on the number line, we divide the interval between 1 and 2 on the number line into eight equal subintervals and then identify

$\frac{15}{8}$ with the left-hand endpoint of the first subinterval on the right. (See Figure 2.9.)

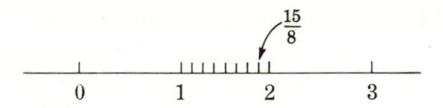

Figure 2.9

After locating rational numbers on the number line in this way, it comes as a surprise to many students that not all real numbers are rational numbers. This is ordinarily demonstrated by proving that $\sqrt{2}$ is not a rational number. The proof of this result "by contradiction" is a classic example of a proof of this type and certainly deserves to be included in any discussion of rational and irrational numbers. Briefly, the proof proceeds as follows. It is supposed that $\sqrt{2}$ is a rational number so that $\sqrt{2} = a/b$ where a and b are positive integers with no common factor other than 1. Then $2b^2 = a^2$, which implies that a^2 is an even integer. But the square of an integer is even if and only if the integer itself is even and hence, a is an even integer. Since a and b have no common factors, this implies that b must be an odd integer. However, if $a = 2c$ for some positive integer c, then $2b^2 = a^2 = 4c^2$, so that $b^2 = 2c^2$. As before, it follows that b is an even integer, and we have arrived at a contradiction to the fact that no integer is both even and odd. Therefore, $\sqrt{2}$ cannot be a rational number.

The treatment of decimals found in the Appendix is substantially more detailed than the treatment found in most secondary school mathematics programs. However, these programs ordinarily do include the long division procedure for finding the decimal representation of rational numbers and the procedure for determining the fraction corresponding to a repeating decimal. The subject is usually treated rather informally in terms of examples, which is quite appropriate in view of the background of the students at this level. However, it is less understandable why some texts choose to neglect the geometric basis for the decimal representation of points on the real number line. The discussion of this aspect of the subject in terms of nested intervals corresponding to specific real numbers not only serves to clarify the meaning of the digits in the decimal representation, but also provides an excellent opportunity to develop some understanding of the role of the completeness property for the real number system. Some fourth-year texts treat decimals from the point of view of infinite series. Although this approach is quick and clean, it suffers from a lack of geometric content and it ties the students' understanding of decimals to his understanding of infinite series, which is ordinarily meager, even among good students at this level.

Complex numbers are ordinarily first encountered in algebra in connection with the solutions of quadratic equations. The arithmetic properties of the complex number system are interesting and instructive, but it is the geometric study of complex numbers that provides the most insight and removes the air of mystery that sometimes surrounds these numbers. Chapter 8 is devoted to a rather detailed account of the geometry of complex numbers and further remarks on this subject will be made at the end of that chapter.

The basic approach to the number systems given in this chapter is axiomatic in character. Existence of the various number systems is taken for granted and emphasis is placed upon the deductive examination of these number systems from a basic list of assumed properties. It is necessary for the prospective teacher to have some acquaintance with this type of development since the current trend is toward axiomatic treatments of school algebra. In fact, several curriculum designers have promoted axiomatic developments of school mathematics to an extreme degree, some going so far as to suggest that calculus be presented axiomatically in the fourth-year program. The teacher must be wary of such proposals and keep in mind that axiomatic structures, while important, are far from encompassing the extent and spirit of mathematics. It is necessary to maintain balance, and overemphasis of this one aspect of mathematics is misleading at best.

It is possible to construct models of the various number systems on the basis of set theory. This constructive approach uses the theory of sets to first construct a model that satisfies the Peano postulates for the natural number system. Then models are constructed in turn for the systems of integers, rational numbers, and real numbers. In this chapter, the complex number system was constructed on the basis of the real number system. Frequently, the rational numbers are defined to be equivalence classes of ordered pairs of integers. Specifically, the equivalence relation is defined on $\mathbf{I} \times \mathbf{I}_0$ (where $\mathbf{I}_0$ denotes the set of nonzero integers) by $(a,b) \sim (c,d)$ if and only if $ad = bc$. The binary operations of addition and multiplication are then defined on the resulting collection of equivalence classes and are shown to satisfy the axioms for a field. The details of verification involved in this and other constructions are quite laborious. We have not presented this type of development of the number systems because it does not appear to us to be as germane to the secondary school algebra program as the axiomatic development. These constructions are frequently misinterpreted as explaining what numbers "really are," but such is not the case. The main issue involved is one of logic, namely, the verification of the consistency of the axioms in question. Students have no qualms about accepting the existence of numbers and classroom time can be spent to greater advantage on other topics.

We shall now list some references for further reading on the subject matter of this chapter. The book by Beaumont and Pierce [1] contains a discussion of the algebraic features of the number systems in the context of rings, integral domains, and fields as well as applications to the theory of equations and number theory. A careful treatment of the construction of the number systems on the basis of set theory is given in the book by Hamilton and Landin [4]. A very detailed account of the axiomatic theory of number systems together with their constructions can be found in the book by Henkin, Smith, Varineau, and Walsh [5]. For a very nice discussion of the role of axiom structures in mathematics in general, see the book by Wilder [6]. Much information concerning the historical evolution of the number systems, a subject that is a source of enjoyment to many students, can be found in the books by Dantzig [2] and Groza [3].

1. Beaumont, R. A. and R. S. Pierce, *The Algebraic Foundations of Mathematics*, Reading, Mass.: Addison-Wesley Publishing Company, Inc., 1963.
2. Dantzig, T., *Number, The Language of Science*, Garden City, N.Y.: Doubleday and Company, Inc., 1954.
3. Groza, V., *A Survey of Mathematics*, New York: Holt, Rinehart and Winston, Inc., 1968.
4. Hamilton, N. and J. Landin, *Set Theory, the Structure of Arithmetic*, Boston: Allyn and Bacon, Inc., 1961.
5. Henkin, L., W. N. Smith, V. J. Varineau, and M. J. Walsh, *Retracing Elementary Mathematics*, New York: The Macmillan Company, 1962.
6. Wilder, R. L., *Introduction to the Foundation of Mathematics*, Second Edition, New York: John Wiley & Sons, Inc., 1965.

Chapter 3

THE THEORY OF NUMBERS

The theory of numbers is a branch of mathematics that requires relatively little formal background to be enjoyed. Number games and puzzles provide a popular pastime and frequently these puzzles are related to ideas and theorems found in the theory of numbers. Although the objects of study are the familiar numbers of everyday life and the basic definitions are quite simple, research in number theory has flourished since ancient times and still continues to be an important field of mathematical activity. Part of the fascination of this subject lies in the fact that there are a number of simply stated problems in the theory that have remained unsolved for centuries.

The history of number theory has its beginnings shrouded in the mysticism of ancient numerology. Indeed, many superstitions concerning numbers persist even today. The most famous of the ancient number theoreticians is probably Pythagoras (circa 500 B.C.). Pythagoras and his

Pythagoras
David Smith Collection

followers regarded numbers as possessing human, musical, and even divine characteristics. For example, they regarded even numbers as feminine and odd numbers other than 1 as masculine. (The number 1, being the source of all numbers, was neither feminine nor masculine.) As a result, they viewed the number five, which is the sum of the first feminine number and the first masculine number, as a symbol of marriage! Many results of a more mathematical nature were developed in the context of the Pythagorean philosophy.

By the time of Euclid (circa 300 B.C.) the study of numbers had become more scientific in character. The basic theory of numbers was presented by Euclid in the ninth book of his *Elements*, a work usually associated with geometry even though it also contains a fairly lengthy discussion of number theory.

The modern theory of numbers begins with the work of Pierre de Fermat (1601–1665). He could be termed an amateur mathematician

Pierre de Fermat
David Smith Collection

since he was professionally a lawyer and published almost nothing in scientific journals. But his notebooks and his correspondence with mathematicians of his day contain many brilliant contributions to geometry, calculus, and probability, as well as to the theory of numbers. Fermat's "Last Theorem," discussed in Section 3.6, is one of the most famous unsolved problems of mathematics.

Many of the great mathematicians of the last few centuries were attracted to the theory of numbers at some point in their careers, most notably the versatile geniuses Leonard Euler (1707–1783) and Karl Friederick Gauss (1777–1855). Gauss, probably the greatest mathemati-

Karl F. Gauss
Library of Congress

cian of the modern era, referred to the theory of numbers as the Queen of Mathematics, a fitting description for this charming and fascinating subject.

3.1. PRIME NUMBERS

Recall that an integer b is said to divide an integer a if there exists an integer c such that $a = bc$. We shall write $b|a$ to signify that b divides a and refer to the integers b and c as *divisors* or *factors* of a. We shall also describe the fact that $b|a$ by saying that a is a *multiple* of b. Since $a = 1 \cdot a = (-1) \cdot (-a)$, we see that ± 1 and $\pm a$ are always divisors of a.

Proposition 3.1. Let a, b, c be integers. Then:

(a) If $a|b$ and $a|c$, then $a|bm + cn$ for any integers m and n.
(b) If $a|b$ and $b \neq 0$, then $|a| \leq |b|$.

PROOF. (a) Since $a|b$ and $a|c$ there exist integers s and t such that $b = as$ and $c = at$. Then $bm + cn = (as)m + (at)n = a(sm + tn)$. Hence, $a|bm + cn$.

(b) Since $a|b$, there exists an integer c such that $b = ac$. Also, $c \neq 0$ since $b \neq 0$, hence, $|c| \geq 1$. Thus, $|b| = |ac| = |a| \, |c| \geq |a|$.

If a is a nonzero integer, the set of positive divisors of a will be denoted by $D(a)$. For example, $D(12) = \{1,2,3,4,6,12\}$ and $D(-15) = \{1,3,5,15\}$. Clearly, 1 and $|a|$ are always members of $D(a)$ and any $b \in D(a)$ satisfies $1 \leq b \leq |a|$.

3.1. PRIME NUMBERS

One of the most basic concepts in the theory of numbers is that of prime number. A natural number p is called *prime* if it cannot be written as the product of two natural numbers each of which is different from p. That is, a natural number p is prime if $p \neq 1$ and $D(p) = \{1,p\}$. The first few prime numbers are 2, 3, 5, 7, 11, 13, 17, 19, $\cdots$. A natural number n is called *composite* if $n \neq 1$ and n is not prime. The numbers 4 and 6 are composite since $4 = 2 \cdot 2$ and $6 = 2 \cdot 3$. Thus, a natural number n is prime if and only if $D(n)$ has exactly two members and it is composite if and only if $D(n)$ has more than two members.

Suppose we wished to find all the primes between 1 and a given number, say 100. We can do this by screening out all the composite numbers in this range by a method known as the Sieve of Eratosthenes. First note that no even number greater than 2 is prime since it would have 2 as a factor. This leaves 2 and the odd numbers as candidates for primality. Next we discard all multiples of 3 except 3 itself. This amounts to crossing out every third number starting at 9 as in the following list.

2	3	5	7	~~9~~	11	13	~~15~~	17	19
~~21~~	23	25	~~27~~	29	31	~~33~~	35	37	~~39~~
41	43	~~45~~	47	49	~~51~~	53	55	~~57~~	59
61	~~63~~	65	67	~~69~~	71	73	~~75~~	77	79
~~81~~	83	85	~~87~~	89	91	~~93~~	95	97	~~99~~

Next we discard multiples of 5 by crossing out every fifth number in the above list starting at 15 (or since this has already been crossed out, we can start at 25). Crossing out every seventh number, starting with 21 or 35, we then eliminate all multiples of 7. It is unnecessary to consider multiples of 4, 6, or 8 since these numbers are even and have already been discarded, and multiples of 9 have fallen by the wayside since they are multiples of 3. Also, there is no need to continue this deletion procedure past multiples of 9 since if n is a number less than 100 such that $n = ab$, then either a or b must be smaller than 10. (See Exercise 3.3.) Any number that has a factor smaller than 10 has already been discarded, hence the numbers remaining must all be primes. These are the 25 primes smaller than 100:

2, 3, 5, 7, 11, 13, 17, 19, 23, 29, 31, 37, 41, 43, 47, 53, 59, 61, 67, 71, 73, 79, 83, 89, 97.

For large numbers, the Sieve of Eratosthenes is somewhat tedious and not too practical, and sieves of a more sophisticated nature can be developed to handle this situation. Present-day methods involving computers have made it possible to list all the primes between 1 and 10,000,000.

EXERCISES

3.1. Are the following statements true or false? Justify your assertions.
   (a) If $a|b + c$, then either $a|b$ or $a|c$.
   (b) If $a|c$ and $a + b = c$, then $a|b$.
   (c) If $a|c$ and $b|c$, then $ab|c$.
   (d) If $a|bc$, then either $a|b$ or $a|c$.

3.2. Let n be a composite number. Show that if n is a square (that is, $n = a^2$ for some natural number a), then $D(n)$ has an odd number of elements, and if n is not a square, then $D(n)$ has an even number of elements.

3.3. Show that if $n = ab$, then either $|a| \leq \sqrt{n}$ or $|b| \leq \sqrt{n}$.

3.4. Find all the primes between 1 and 200.

The statement of the next proposition may seem somewhat obvious because of our computational familiarity with numbers, but the proof is not completely trivial and makes use of the Well-Ordering Property for the natural numbers.

Proposition 3.2. Each composite number can be written as the product of primes.

PROOF. Suppose the assertion is false. Then the set S of all composite numbers that cannot be written as the product of primes is nonempty, and hence, S has a least element n. Since n is composite, $n = ab$, where a, b are natural numbers such that $1 < a < n$ and $1 < b < n$. Now a is either prime or can be written as a product of primes since $a \notin S$. Similarly, b is either prime or can be written as a product of primes. It follows that $n = ab$ can be written as the product of primes, which is a contradiction. Therefore, S must be empty and the proof is complete.

The previous result is an illustration of the importance of primes in number theory. They are the building blocks from which all natural numbers can be manufactured and many results about numbers can be obtained by analyzing their prime factorizations. So far, we have established only the existence of a prime factorization for any natural number. This leaves unanswered the question of how many different factorizations into primes a given natural number can have. As the next proposition states, if the order in which the prime factors are written is ignored, then there is only one way in which a composite number can be expressed as a product of primes. Together, Propositions 3.2 and 3.3 constitute what is known as the *Fundamental Theorem of Arithmetic*, which states that the factorization of each composite number into primes exists and is unique up to a rearrangement of the factors.

In the proof of Proposition 3.3, we shall use the fact that if a prime p divides a product $a_1 a_2 \cdots a_n$, then p divides some particular term a_i in the product. This assertion will be proved in the next section as a consequence of the Euclidean algorithm.

Proposition 3.3. The factorization of a natural number into the product of primes is unique except for rearrangement of factors.

PROOF. Suppose, contrary to the assertion, that there exists a composite number having prime factorizations that are different. Then the set S of such numbers is not empty and must contain a least element n. Then

$$n = p_1 \cdots p_r = q_1 \cdots q_s,$$

where each p_i and q_j is prime. Since $p_1 | n$, we know that p_1 divides the product $q_1 \cdots q_s$. Using the divisibility assertion given just before the statement of the proposition, we obtain the fact that $p_1 | q_j$ for some j. Since q_j is prime and $p_1 \neq 1$, we conclude that $p_1 = q_j$. Using the cancellation property of the integers, we obtain

$$n' = p_2 \cdots p_r = q_1 \cdots q_{j-1} q_{j+1} \cdots q_s.$$

Since $n' < n$, and since n was the least number having more than one prime factorization, we see that the primes occurring in the two factorizations of n' displayed above must coincide. Therefore, $r = s$ and $(q_1, \cdots, q_s)$ must be a rearrangement of $(p_1, \cdots, p_r)$. Hence, n itself has a unique factorization into primes which contradicts the assumption that $n \in S$. Thus, the set of natural numbers having nonunique prime factorizations is empty and the proposition is proved.

The factorization of a natural number into primes can be written systematically as follows. Given a natural number n, let $p_1, \cdots, p_k$ be the *distinct* prime divisors of n written in increasing order, that is, $p_1 < p_2 < \cdots < p_k$. It may happen that a prime divisor occurs several times in the factorization of n. For example, since $24 = 2 \cdot 2 \cdot 2 \cdot 3$, the factor 2 occurs three times. The number of times a prime factor is used in the factorization will be exhibited by employing exponents. We write

$$n = p_1^{a_1} p_2^{a_2} \cdots p_k^{a_k},$$

where $a_1, a_2, \cdots, a_k$ are natural numbers indicating the number of times the corresponding prime is repeated. Thus, for example, we write $24 = 2^3 \cdot 3$. Written in this way, the prime factorization of n is unique and is called the *canonical prime factorization* of n.

118 THE THEORY OF NUMBERS

The essential part of the proof of Proposition 3.2, which asserts the existence of prime factorizations, was the use of the Well-Ordering Property of the natural numbers. However, in proving uniqueness, an additional property involving divisibility was utilized. In other number systems, it is possible to have existence of prime factorizations without having the accompanying uniqueness property as is illustrated in the following example.

Example 3.1. Consider the set M of natural numbers of the form $4k + 1$ where k is a nonnegative integer. The first few elements of M are 1, 5, 9, 13, 17, 21, 25, $\cdots$. The product of two numbers in M is again in M, and M possesses many properties in common with the set **N** of natural numbers. Let us define an element $p \neq 1$ in M to be *M-prime* if it is not the product of any elements *of M* except 1 and p itself. The notion of M-prime is the natural analogue of prime in the usual sense. For example, the numbers 5 and 9 are both M-prime. In fact, the first number in M that is not M-prime is 25 since $25 = 5 \cdot 5$. It can be shown that each element of M can be written as the product of M-primes. However, the factorization of an element of M into M-primes need not be unique! For example, the number 693 belongs to M, but 693 can be written both as $9 \cdot 77$ and $21 \cdot 33$, and all four numbers 9, 77, 21, 33 are M-prime.

EXERCISES

3.5. Write the canonical prime factorizations of 1235 and 5124.
3.6. Show that if $p_1, \cdots, p_k$ are distinct primes and $a_1, \cdots, a_k$ are natural numbers, then the number of positive divisors of $n = p_1^{a_1} \cdots p_k^{a_k}$ is $(a_1 + 1)(a_2 + 1) \cdots (a_k + 1)$. [*Hint:* Note that each p_i will appear 0 or 1 or 2 or $\cdots$ or a_i times as a factor in any positive divisor of n.]
3.7. The set $E = \{2,4,6,8,\cdots\}$ of even natural numbers has the property that the product of two members of E is also in E. Define what is meant by E-prime and find the first five E-prime numbers. Show that factorization of a member of E into E-primes is not unique by exhibiting two such factorizations for 60.

The fact that the set of natural numbers is infinite generates a feeling that the set of prime numbers is also unending, that is, that there is no *last* prime. Of course a feeling is not a proof. The first recorded proof that there are infinitely many prime numbers is found in Book IX of Euclid's *Elements* and is essentially the proof given below.

Proposition 3.4. There are infinitely many prime numbers.

PROOF. Suppose to the contrary there are only finitely many primes. Then the set of all primes is given by $P = \{p_1, p_2, \cdots, p_n\}$ where p_n is the supposed last prime. Consider the number $q = p_1 \cdots p_n + 1$ obtained by multiplying all the primes together and adding 1. Clearly $q \notin P$, so that q must be a composite number. Since q can be written as a product of primes by Proposition 3.2, some $p_i \in P$ must be a divisor of q. That is, $p_1 \cdots p_n + 1 = a \cdot p_i$ for some natural number a. But then

$$1 = ap_i - p_1 \cdots p_n = (a - p_1 \cdots p_{i-1} p_{i+1} \cdots p_n) p_i$$

which implies that p_i is a divisor of 1. This is impossible. We conclude that the set of prime numbers must be infinite.

Note that the number q used in the above proof need not itself be prime. In fact, $2 \cdot 3 \cdot 5 \cdot 7 \cdot 11 \cdot 13 + 1 = 30{,}031 = 59 \cdot 509$ is not prime.

EXERCISES

3.8. Show that each odd prime is of the form $4k + 1$ or $4k - 1$ for some $k \in \mathbf{N}$.

3.9. Modify the proof of Proposition 3.4 to show that there are infinitely many primes of the form $4n - 1$. [*Hint:* Consider $4p_1 \cdots p_n - 1$ and use Exercise 3.8.]

The fact that there are infinitely many primes can be proved in many other ways. For example, it is an immediate consequence of the next proposition.

Proposition 3.5. If n is a natural number greater than 2, then there exists a prime p such that $n < p < n!$.

PROOF. Let $q = n! - 1$. Then $q > 1$ since $n > 2$ and hence q is either prime or has a prime divisor. If q is prime, then $p = q$ satisfies the conclusion of the proposition. If q is not prime, then q has a prime divisor p. Clearly $p < q$. It remains to prove that $n < p$. To this end, suppose that $p \leq n$. Then p would be a divisor of $n!$. But p is also a divisor of $q = n! - 1$. Thus, p must be a prime divisor of the difference $n! - q = 1$, which is impossible. Thus we have $n < p < n!$.

The problem of determining how the primes are distributed throughout the natural numbers is a continuing area of study in the theory of numbers. A look at a long list of primes shows that primes do not seem to appear with any regularity or in any pattern. A sharper version of Proposition 3.5 states that if $n > 3$, then there exists a prime between n and $2n - 2$. The proof of this result is lengthy and will not be discussed here.

The most famous result concerning the distribution of primes is known as the *Prime Number Theorem*. Let $\pi(n)$ denote the number of primes not greater than n and let $\ln n$ denote the natural logarithm of n. Then the Prime Number Theorem states that

$$\lim_{n \to \infty} \frac{\pi(n)}{n/\ln n} = 1.$$

That is, for large n, the number $\pi(n)$ is approximately equal to $n/\ln n$. This is not an exact measure of the primes not greater than n for any given n, but the approximation improves as n gets larger. This is illustrated by the following table.

n	$\pi(n)$	$n/\ln n$
10^3	168	144
10^6	78,498	72,382
10^7	664,579	620,417
10^8	5,761,455	5,428,646
10^9	50,847,478	48,254,630

This theorem was conjectured around 1800 by Gauss on the basis of direct calculations using tables of primes. However, it was not proved until 1896 using analytic machinery, such as modern integration theory, which was not available in Gauss' time.

Even though we know that arbitrarily large prime numbers exist, they are quite difficult to find in practice. Checking possible factors of a number having a thousand digits is not an easy task. Even with the aid of electronic computers, the search for primes is quite difficult. The largest prime number discovered to date is $2^{11,213} - 1$. This number, which has 3376 digits, was shown to be prime by Donald Gillies on the computer ILLIAC II at the University of Illinois in 1963.

Prime numbers that have the form $2^n - 1$ for some natural number n are called *Mersenne primes*, named after the Franciscan friar Marin Mersenne (1588–1648). It is not difficult to show that if $2^n - 1$ is prime, then n must also be prime (see Exercise 3.13). However, $2^n - 1$ is not prime whenever n is prime. For example, if $n = 2, 3, 5, 7$, then $2^n - 1$ is prime, but $2^{11} - 1 = 2047 = 23 \cdot 89$ is composite. For large values of n, it is very difficult to determine whether or not $2^n - 1$ is in fact a prime number and research in this area continues.

Interest in Mersenne primes stems to a large degree from their relationship with perfect numbers. A natural number n is called *perfect* if it equals the sum of its positive divisors, excluding itself. For example,

3.1. PRIME NUMBERS

6 and 28 are perfect since $6 = 1 + 2 + 3$ and $28 = 1 + 2 + 4 + 7 + 14$. (The appellation "perfect" was attached to these numbers by the ancient numerologists. Note that heaven and earth were created in 6 days and the moon takes 28 days to travel around the earth.) Euclid proved in the ninth book of his *Elements* that if $2^n - 1$ is prime, then $2^{n-1}(2^n - 1)$ is perfect. For example, the perfect numbers 496 and 8128 correspond to the choices $n = 5$ and $n = 7$, respectively. Approximately 2000 years later, Euler proved a partial converse: Each even perfect number is of the form $2^{n-1}(2^n - 1)$ where $2^n - 1$ is prime. This left open the question of whether or not there are any odd perfect numbers. The question is still unanswered although it was shown by Bryant Tuckerman in 1968 that if an odd perfect number exists, it must be larger than 10^{36}.

Another outstanding unsolved problem in number theory is known as the twin prime problem. It is a natural impulse to conjecture that certain patterns are generally valid on the basis of studying a handful of special cases. A look at the list of primes between 1 and 100 reveals that there are several pairs of primes that are consecutive odd numbers, for example, 3 and 5, 5 and 7, 11 and 13, 17 and 19, and so on. These are called *twin primes* or *prime pairs* and the question was raised as to whether or not there are infinitely many twin primes. The answer to this question is not known.

Another problem concerning primes that can be raised is the possibility of finding a simple algebraic formula that will yield nothing but prime numbers. Fermat thought he had such a formula when he conjectured that $2^{2^n} + 1$ was prime for all nonnegative integers n. It is true that numbers of this type are prime when $n = 0, 1, 2, 3, 4$. However, Euler proved in 1739, almost a century after Fermat made his conjecture, that $2^{2^5} + 1$ is composite. (In fact, $2^{32} + 1 = 641 \cdot 6{,}700{,}417$.) No values of n for $n > 5$ have ever been found for which $2^{2^n} + 1$ is prime, but neither has it been proved that none exist.

Another attempt to obtain a simple prime-producing formula is provided by the function $f(n) = n^2 - n + 41$. It can be shown that for $n = 1, 2, 3, \cdots, 40$, the number $f(n)$ is indeed prime. On the basis of such strong empirical evidence, one might be tempted to quit checking and have faith that the formula will produce primes indefinitely. Unfortunately, $f(41) = 41 \cdot 41$, which is not prime. As an indication of the futility in searching for a simple formula that generates only primes, we next show that no nonconstant polynomial can have only prime values on the natural numbers.

Proposition 3.6. Let $f(x) = a_n x^n + \cdots + a_1 x + a_0$ be a polynomial with integer coefficients where $a_n > 0$ and $n \geq 1$. Then there exists a natural number k such that $f(k)$ is composite.

122   THE THEORY OF NUMBERS

PROOF. Since $a_n > 0$, there exists a natural number x_0 such that $m = f(x_0) > 1$ and such that $f(x) > f(x_0)$ for $x > x_0$. We shall now show that $f(x_0 + m)$ is composite. For each natural number i, the number $(x_0 + m)^i - x_0^i$ is divisible by m. Hence, the number $f(x_0 + m) - f(x_0)$ is divisible by m, so that $f(x_0 + m) - m = bm$ for some integer b. Therefore, $f(x_0 + m) = (b + 1)m$. Since $f(x_0 + m) > f(x_0) = m$, we see that $b + 1 > 1$, and hence the number $f(x_0 + m)$ is composite. Thus, we may take k in the statement of the proposition to be $k = x_0 + m$.

EXERCISES

3.10. Show that there is only one prime triplet (that is, three consecutive odd numbers that are all prime).

3.11. Find all primes p and q that satisfy $p - q = 3$.

3.12. Show that for any natural number n, there are n consecutive numbers that are all composite. [*Hint:* Note that the numbers $2 + 50!$, $3 + 50!$, $\cdots$, $50 + 50!$ are all composite.]

3.13. (a) Verify that $a^{n+1} - 1 = (a - 1)(a^n + a^{n-1} + \cdots + a + 1)$ for each natural number n by mathematical induction.
(b) Show that if $a^n - 1$ is prime for some $n \geq 2$, then $a = 2$.
(c) Show that if n is composite, then $2^n - 1$ is not prime.

3.14. Show that if $2^n + 1$ is prime, then n must be a power of two, that is, $n = 2^k$ for some nonnegative integer k. [*Hint:* Show that if m is an odd natural number, then $a + 1$ is a factor of $a^m + 1$ for any integer a.]

3.2. GREATEST COMMON DIVISORS AND THE EUCLIDEAN ALGORITHM

If a and b are integers, then the set $D(a) \cap D(b)$ consists of all positive divisors common to both a and b. If either a or b is not zero, then $D(a) \cap D(b)$ is bounded above (for example, $|a| + |b|$ is an upper bound) and hence this set contains a greatest element that is referred to as the *greatest common divisor* of a and b. Thus, the greatest common divisor of two integers a and b, not both zero, is the unique natural number d with the following properties:

(1) d divides both a and b.
(2) If d' divides both a and b, then $d' \leq d$.

The greatest common divisor of a and b is denoted by (a,b) and sometimes by $\gcd(a,b)$ when there is a danger of notational confusion. Note that $(a,0) = |a|$ for any nonzero integer a, and that $(0,0)$ is not defined. The

3.2. GREATEST COMMON DIVISORS AND THE EUCLIDEAN ALGORITHM

case in which $(a,b) = 1$ is of particular interest in many situations, and we say that a and b are *relatively prime* if $(a,b) = 1$.

Example 3.2. Since $D(12) = \{1,2,3,4,6,12\}$ and $D(-28) = \{1,2,4,7,14,28\}$, we have $D(12) \cap D(-28) = \{1,2,4\}$. Hence, $(12,-28) = 4$.

If the numbers a and b are large, then it can become quite tedious to list all of their divisors, and the above procedure for calculating the greatest common divisor is not very efficient. The Euclidean algorithm is a systematic method for finding the greatest common divisor of two integers that is based on repeated use of the division algorithm. Recall that the division algorithm states that if a and b are given integers, then there exist unique integers q and r such that $a = bq + r$ and $0 \leq r < b$. (See Proposition 2.8.) The key relation between greatest common divisors and the division algorithm is given in the next proposition.

Proposition 3.7. If a, b, q, r are integers such that $a = bq + r$, then $(a,b) = (b,r)$.

PROOF. If k is a common divisor of b and r, then k divides $a = bq + r$ by Proposition 3.1. Also, if k divides both a and b, then k divides $r = a - bq$. Hence, the sets $D(a) \cap D(b)$ and $D(b) \cap D(r)$ coincide and therefore $(a,b) = (b,r)$.

Before discussing the procedure of the Euclidean algorithm in general, let us consider a concrete example.

Example 3.3. Suppose we wish to find the greatest common divisor of 1804 and 328. Applying the division algorithm, we obtain

$$1804 = 328 \cdot 5 + 164.$$

Then, by Proposition 3.7, we have $(1804,328) = (328,164)$. A second application of the division algorithm yields

$$328 = 164 \cdot 2 + 0.$$

Hence, $(1804,328) = (328,164) = (164,0) = 164$.

The Euclidean algorithm formalizes the techniques used in the preceding example and proceeds in general as follows. Let a and b be given nonzero integers. If b divides a, then $(a,b) = |b|$ and there is nothing more to do. If not, then there exist integers q_1 and r_1 such that $a = bq_1 + r_1$ and $0 < r_1 < b$. If r_1 divides b, then Proposition 3.7 implies that $(a,b) =$

$(b,r_1) = r_1$ and we are done. Otherwise, we apply the division algorithm to b and r_1 to obtain q_2 and r_2 such that $b = r_1 q_2 + r_2$ and $0 < r_2 < r_1$. If r_2 divides r_1, then $(a,b) = (b,r_1) = (r_1,r_2) = r_2$ and we are done. If r_2 does not divide r_1, then we apply the division algorithm to r_1 and r_2. We thus obtain the following sequence of steps that must eventually terminate with remainder zero since the remainder at each step is a nonnegative integer strictly less than the remainder of the preceding step.

$$
\begin{aligned}
(1) \quad & a = bq_1 + r_1, & 0 < r_1 < b \\
(2) \quad & b = r_1 q_2 + r_2, & 0 < r_2 < r_1 \\
(3) \quad & r_1 = r_2 q_3 + r_3, & 0 < r_3 < r_2 \\
& \quad \cdots \cdots \cdots \\
(n-1) \quad & r_{n-3} = r_{n-2} q_{n-1} + r_{n-1}, & 0 < r_{n-1} < r_{n-2} \\
(n) \quad & r_{n-2} = r_{n-1} q_n + r_n, & 0 < r_n < r_{n-1} \\
(n+1) \quad & r_{n-1} = r_n q_{n+1} + 0.
\end{aligned}
$$

We now see that the last nonzero remainder r_n is the greatest common divisor of a and b since repeated application of Proposition 3.7 yields

$$
\begin{aligned}
(a,b) &= (b,r_1) = (r_1,r_2) = \cdots = (r_{n-1},r_n) \\
&= (r_n, 0) = r_n.
\end{aligned}
$$

For example, in order to find $(315,66)$, we apply the Euclidean algorithm as follows:

$$
\begin{aligned}
315 &= 66 \cdot 4 + 51 \\
66 &= 51 \cdot 1 + 15 \\
51 &= 15 \cdot 3 + 6 \\
15 &= 6 \cdot 2 + 3 \\
6 &= 3 \cdot 2.
\end{aligned}
$$

Hence, $(315,66) = 3$.

EXERCISES

3.15. Find the greatest common divisor of:
(a) 24 and -82.
(b) 963 and 657.
(c) 3997 and 2947.

3.16. Suppose that a, b are nonzero integers and k is a natural number. Show that $(ka,kb) = k(a,b)$.

3.17. The greatest common divisor of n integers $a_1, a_2, \cdots, a_n$ is the largest natural number that divides each a_i, $1 \leq i \leq n$, and is denoted by $(a_1, a_2, \cdots, a_n)$. Use mathematical induction to show that $(a_1, a_2, \cdots, a_n) = ((a_1, \cdots, a_{n-1}), a_n)$ for any $n \geq 3$.

3.18. Find the greatest common divisor of 280, 192, and 120.

3.2. GREATEST COMMON DIVISORS AND THE EUCLIDEAN ALGORITHM 125

While the Euclidean algorithm provides a simple method of computing the greatest common divisor of two integers, its primary importance lies in its theoretical ramifications rather than its computational utility. One consequence of particular importance is the following.

Proposition 3.8. If a and b are nonzero integers, then there exist integers x and y such that
$$ax + by = (a,b).$$

PROOF. The proof proceeds by reversing the steps of the Euclidean algorithm. From the step (n) of the algorithm, we have
$$(a,b) = r_n = r_{n-2} - r_{n-1}q_n.$$
The preceding step $(n-1)$ of the algorithm can then be used to replace r_{n-1} by $r_{n-3} - r_{n-2}q_{n-1}$. We get
$$\begin{aligned}(a,b) &= r_{n-2} - (r_{n-3} - r_{n-2}q_{n-1})q_n \\ &= (1 + q_{n-1}q_n)r_{n-2} - q_n r_{n-3}.\end{aligned}$$

We can now back up one more step in the algorithm to eliminate r_{n-2}. This yields an expression involving r_{n-3} and r_{n-4}, and another step up the algorithm eliminates r_{n-3}. This process can be continued until the first step of the algorithm is reached. At this point, we shall have expressed (a,b) as the sum of a multiple of a and a multiple of b.

Example 3.4. The procedure of the above proof can best be seen by means of an example. We had previously calculated $(315,66) = 3$. Let us write down the steps used in finding $(315,66)$ in reverse order and with the remainders on the left-hand side of the equality sign as follows:
$$\begin{aligned}3 &= 15 - 6 \cdot 2, \\ 6 &= 51 - 15 \cdot 3, \\ 15 &= 66 - 51 \cdot 1, \\ 51 &= 315 - 66 \cdot 4.\end{aligned}$$
Replacing the 6 in the first line by the expression in the second line yields
$$\begin{aligned}3 &= 15 - (51 - 15 \cdot 3) \cdot 2 \\ &= 7 \cdot 15 - 2 \cdot 51.\end{aligned}$$
Now replacing 15 by the expression of the third line gives us
$$\begin{aligned}3 &= 7 \cdot (66 - 51 \cdot 1) - 2 \cdot 51 \\ &= 7 \cdot 66 - 9 \cdot 51.\end{aligned}$$

Finally, substituting for 51 by using the fourth line results in the desired expression
$$3 = 7 \cdot 66 - 9(315 - 66 \cdot 4)$$
$$= 43 \cdot 66 - 9 \cdot 315.$$

Thus, if $a = 66$ and $b = 315$, then the x and y of Proposition 3.8 are $x = 43$ and $y = -9$.

Although this procedure leads to a particular pair of integers x, y satisfying $ax + by = (a,b)$ for a given pair of integers a, b, these numbers are not unique. There are, in fact, infinitely many possible values for x and y that satisfy the relation of Proposition 3.8. Frequently, it is possible to find appropriate values by simple inspection. For example, if $a = 5$ and $b = 3$, then $(a,b) = 1$ and we seek integers x, y satisfying $5x + 3y = 1$. A moment's observation might very well lead us to choose $x = 2$ and $y = -3$ since $10 - 9 = 1$, or we might select $x = 8$ and $y = -13$ since $40 - 39 = 1$. The method of the Euclidean algorithm would give us the solution $x = -1$, $y = 2$.

One of the most important consequences of Proposition 3.8 is the following result which is frequently referred to as Euclid's lemma.

Proposition 3.9. Suppose a, b, c are nonzero integers. If $a|bc$ and $(a,b) = 1$ then $a|c$.

PROOF. By Proposition 3.8, there exist integers x, y such that
$$ax + by = (a,b) = 1.$$
Multiplying by c gives us
$$acx + bcy = c.$$
Since a obviously divides acx and since a divides bcy by hypothesis, it follows from Proposition 3.1 that a divides $acx + bcy = c$.

We now prove as a corollary the divisibility result used earlier in the proof of the Fundamental Theorem of Arithmetic. (See Proposition 3.3.) Note that if p is a prime number and b is an integer, then either $(p,b) = 1$ or $p|b$.

Corollary. If p is prime and p divides the product $a_1 a_2 \cdots a_n$ of the integers $a_1, a_2, \cdots, a_n$, then p divides a_i for some i, $1 \leq i \leq n$.

PROOF. The result will be established by mathematical induction. If $n = 2$, the result follows from Proposition 3.9. Suppose the result holds for a given $n > 2$, and suppose $p|a_1 \cdots a_n a_{n+1}$. Then, by Proposition 3.9, either $p|(a_1 \cdots a_n)$ or $p|a_{n+1}$. In the latter case, we are done. In the

3.2. GREATEST COMMON DIVISORS AND THE EUCLIDEAN ALGORITHM

former case, we have by assumption that $p|a_i$ for some i, $1 \leq i \leq n$. Thus, the result follows for any $n \geq 2$ by mathematical induction.

Proposition 3.9 has many varied and interesting consequences. As an illustration of its utility, we prove a very useful result known as the Rational Root Test for polynomials with integer coefficients.

Proposition 3.10. Let $P(x) = a_n x^n + \cdots + a_1 x + a_0$ be a polynomial with integer coefficients where $a_n \neq 0$. If c and d are integers such that $(c,d) = 1$ and c/d is a root of the equation $P(x) = 0$, then $c|a_0$ and $d|a_n$.

PROOF. We are given that $P(c/d) = 0$. Multiplying both sides of this equality by d^n, we obtain

$$a_n c^n + a_{n-1} c^{n-1} d + \cdots + a_1 c d^{n-1} = -a_0 d^n.$$

Since c obviously divides the left-hand side of this equation, we see that $c|a_0 d^n$. Since $(c,d) = 1$, we obtain from Proposition 3.9 that $c|a_0$. The fact that $d|a_n$ can be obtained in a similar manner by noting that

$$a_{n-1} c^{n-1} d + \cdots + a_0 d^n = -a_n c^n.$$

Example 3.5. Let us find all of the rational roots of the equation $3x^3 - x^2 - 6x + 2 = 0$. The Rational Root Test tells us that the only possible rational roots are the numbers $1, 2, \frac{1}{3}, \frac{2}{3}$ and their negatives. Checking these eight possibilities by substitution shows us that $\frac{1}{3}$ is a root of the equation while the other numbers are not. Hence, there is only one rational root of the equation, namely, $x = \frac{1}{3}$.

EXERCISES

3.19. Find integers x, y such that $8x - 6y = 2$.
3.20. Find $(195,210)$ and then find integers x, y such that $195x + 210y = (195,210)$.
3.21. Find $(109,89)$ and then find integers x, y such that $109x + 89y = (109,89)$.
3.22. Use Proposition 3.8 to show that if k divides both a and b, then k divides (a,b).
3.23. Suppose that a,b are nonzero integers.
   (a) Show that if integers x, y exist such that $ax + by = 1$, then $(a,b) = 1$.
   (b) Show that if $a' = a/(a,b)$ and $b' = b/(a,b)$, then $(a',b') = 1$.
3.24. Show that if $(a,b) = 1$, then $(a + b, a - b)$ is either 1 or 2.
3.25. Show that if $(a,b) = 1$, then $(a^2,b^2) = 1$.
3.26. Show that if $a|c$ and $b|c$ and $(a,b) = 1$, then $ab|c$.

128   THE THEORY OF NUMBERS

3.27. Show that for given integers $a_1, a_2, \cdots, a_n$, there exist integers $x_1, x_2, \cdots, x_n$ such that $a_1x_1 + a_2x_2 + \cdots + a_nx_n = g$ where $g = (a_1, a_2, \cdots, a_n)$. (See Exercise 3.17.)

3.28. Find the rational roots of the following equations:
   (a) $x^5 - 2x + 1 = 0$.
   (b) $2x^3 + 6x^2 + 4x + 1 = 0$.
   (c) $x^3 - 3x - 1 = 0$.
   (d) $2x^3 - 11x^2 + 17x - 6 = 0$.

A notion closely related to the greatest common divisor of two integers is their least common multiple. If a and b are natural numbers, then the set of positive common multiples of a and b is not empty (for example, the number ab is a multiple of both a and b). The least element of this set of common multiples is denoted by $[a,b]$ and is called the *least common multiple* of a and b.

Recall that when we add fractions n/a and m/b, the first step is usually the determination of the lowest common denominator, and this is nothing more than the least common multiple $[a,b]$ of the denominators of the fractions.

Proposition 3.11. If m is a common multiple of a and b, then m is a multiple of $[a,b]$.

PROOF. Let $s = [a,b]$. Then $s \leq m$ by definition. Applying the division algorithm, we obtain integers q and r such that $m = sq + r$, $0 \leq r < s$. We must show $r = 0$. If we assume $r \neq 0$, then it follows that $a|r$, since $a|m$ and $a|s$. Also, $b|r$ since $b|m$ and $b|s$. Hence, r is a common multiple of a and b, which implies that $s \leq r$ by definition of least common multiple. This contradiction shows that $r = 0$ and the proof is complete.

The relation between the greatest common divisor and the least common multiple of two integers is given in the next proposition.

Proposition 3.12. If a and b are natural numbers, then $(a,b) \cdot [a,b] = ab$.

PROOF. Let $g = (a,b)$ and $s = [a,b]$. Note that ab/g is a multiple of both a and b and hence $ab/g \geq s$. Thus, $ab \geq gs$. To obtain the opposite inequality, we use the fact that any common multiple of a and b must be a multiple of s. Since ab is a common multiple of a and b, it is therefore a multiple of s and hence ab/s is a natural number. Letting $s = Aa$, we see that

$$A\frac{ab}{s} = \frac{Aab}{Aa} = b$$

3.2. GREATEST COMMON DIVISORS AND THE EUCLIDEAN ALGORITHM

so that ab/s is a divisor of b. Similarly, ab/s divides a. Thus ab/s is a common divisor of a and b which implies that $ab/s \leq g$. Hence, we have $ab \leq gs$, which together with the previous inequality proves the proposition.

Suppose a and b are given natural numbers and suppose that $p_1, p_2, \cdots, p_n$ are the distinct prime numbers that occur in the prime factorization of either a or b. We may then write

$$a = p_1^{\alpha_1} p_2^{\alpha_2} \cdots p_n^{\alpha_n},$$
$$b = p_1^{\beta_1} p_2^{\beta_2} \cdots p_n^{\beta_n},$$

where some of the exponents may be zero. (For example, if the prime number p_i is a factor of a but not of b, then $\beta_i = 0$.) The greatest common divisor and least common multiple of a and b can be described in terms of these primes as follows:

$$(a,b) = p_1^{m_1} p_2^{m_2} \cdots p_n^{m_n}; \tag{1}$$
$$[a,b] = p_1^{M_1} p_2^{M_2} \cdots p_n^{M_n}, \tag{2}$$

where $m_i = \min(\alpha_i, \beta_i)$ and $M_i = \max(\alpha_i, \beta_i)$; that is, m_i is the smaller and M_i is the larger of the two exponents α_i and β_i for $i = 1, 2, \cdots, n$.

For example, if $a = 12$ and $b = 63$, then the primes occurring in the prime factorization of either a or b are 2, 3, 7 and we can write

$$12 = 2^2 \cdot 3^1 \cdot 7^0,$$
$$63 = 2^0 \cdot 3^2 \cdot 7^1.$$

Consequently, we obtain:

$$(12,63) = 2^0 \cdot 3^1 \cdot 7^0 = 3,$$
$$[12,63] = 2^2 \cdot 3^2 \cdot 7^1 = 252.$$

Observe from (1) that the integers a and b are relatively prime (that is, $(a,b) = 1$) if and only if $m_i = \min(\alpha_i, \beta_i) = 0$ for $i = 1, \cdots, n$. In other words, two integers a and b are relatively prime if and only if they have no prime factors in common.

The following result establishes the distributive properties for the greatest common divisor and least common multiple.

Proposition 3.13. If a, b, c are integers, then

$$(a,[b,c]) = [(a,b), (a,c)],$$
$$[a,(b,c)] = ([a,b], [a,c]).$$

This proposition can be proved by using the prime factor representations (1) and (2) given above for the greatest common divisor and least common multiple. We leave the details to the reader as an exercise. (See Exercise 3.30.)

EXERCISES

3.29. Find (a,b) and $[a,b]$ when
(a) $a = 84$, $b = 990$,
(b) $a = 851$, $b = 989$,
(c) $a = 3927$, $b = 20{,}020$.

3.30. (a) Suppose that α, β, γ are nonnegative integers. Prove that:
$\min(\alpha,\max(\beta,\gamma)) = \max(\min(\alpha,\beta),\min(\alpha,\gamma))$
$\max(\alpha,\min(\beta,\gamma)) = \min(\max(\alpha,\beta),\max(\alpha,\gamma))$.
(b) Prove Proposition 3.13.

3.3. LINEAR DIOPHANTINE EQUATIONS

In Proposition 3.8, it was stated that if a and b are given nonzero integers, then there exist integers x and y such that $ax + by = (a,b)$. It was also noted that the integers x and y are not unique. In fact, there are infinitely many values of x and y that satisfy this relationship. Indeed, if $x = x_0$, $y = y_0$ is a particular solution of $ax + by = (a,b)$, then since

$$a(x_0 - bt) + b(y_0 + at) = ax_0 + by_0 = (a,b)$$

we see that $x = x_0 - bt$, $y = y_0 + at$ is also a solution for any integer t.

Let us now consider a general linear equation $ax + by = c$ where a, b, c are given integers such that either $a \neq 0$ or $b \neq 0$. If x and y are allowed to be arbitrary real numbers, then solutions to this equation always exist. In fact, the solution set is simply a straight line in the Cartesian plane. However, if we impose the restriction that the solutions we seek must be integers, then the character of the problem changes radically. Geometrically speaking, the insistence that solutions be integers means that we wish to find points on the straight line that have integer coordinates. (See Figure 3.1.)

It is not always the case that integer solutions will exist. For example, the equation $6x - 9y = 2$ can have no integer solutions, since if integers x_0 and y_0 existed such that $6x_0 - 9y_0 = 2$, then the fact that 3 divides $6x_0 - 9y_0$ would imply that 3 divides 2.

When only integer solutions are sought for an equation with integer coefficients in one or more variables, the equation is referred to as a *Diophantine equation*. The term "Diophantine equation" is used in honor of the mathematician Diophantus who first studied such problems in a systematic way. His work on this and related topics under the title *Arithmetics* was a 13-volume treatise, but only 6 volumes still exist. Almost nothing is known concerning the life of Diophantus except that he

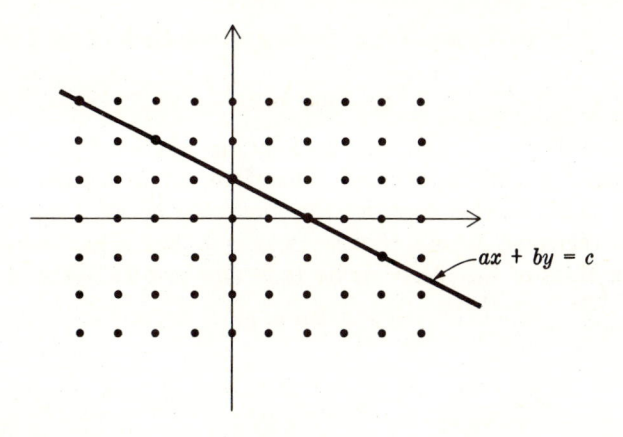

Figure 3.1

lived in Alexandria in the third century A.D. Legend has it that his tomb carried the following inscription:

> Here you see the tomb containing the remains of Diophantus. One sixth of his life God granted him his youth. After a twelfth more his cheeks were bearded. After an additional seventh he kindled the light of marriage, and in the fifth year he fathered a son. Alas the unfortunate son's life span was only half that of his father who consoled his grief in the remaining four years of his life. By this device of numbers, tell us the extent of his life.

If this inscription is fact, then we have some meager knowledge of Diophantus' life, including the fact that he lived to be 84.

We now proceed to a systematic analysis of Diophantine equations of the form $ax + by = c$. In the ensuing discussion, it is always assumed that a, b, c are integers and that at least one of the numbers a, b is different from zero. The question of existence of solutions must be settled first, and this is done in the next proposition.

Proposition 3.14. The Diophantine equation $ax + by = c$ has a solution if and only if (a,b) is a divisor of c.

PROOF. If an integer solution x_0, y_0 exists, then (a,b) must divide $ax_0 + by_0$ since it divides both a and b. Hence, (a,b) is a divisor of c. Conversely, suppose $c = k(a,b)$ for some integer k. We have already seen in Proposition 3.8 that integers x_0, y_0 exist such that $ax_0 + by_0 = (a,b)$. Then $a(kx_0) + b(ky_0) = k(a,b) = c$, which shows that the given equation has the integer solution kx_0, ky_0.

We now turn to the problem of finding *all* solutions of the Diophantine equation
$$ax + by = c. \qquad (1)$$
If c is not a multiple of (a,b), then no solutions of (1) exist and our search for solutions is quickly concluded. Therefore, let us assume that c is a multiple of (a,b). In this case, we can make the added assumption that a and b are relatively prime, that is, $(a,b) = 1$. For if $(a,b) \neq 1$, we may divide both sides of Equation (1) by (a,b) and rewrite (1) in the form
$$a'x + b'y = c', \qquad (2)$$
where
$$a' = \frac{a}{(a,b)}, \qquad b' = \frac{b}{(a,b)}, \qquad c' = \frac{c}{(a,b)},$$
Note that a', b', c' are integers and that $(a',b') = 1$. Moreover, the Diophantine equations (1) and (2) are equivalent in the sense that they have the same solutions. Therefore, when considering the Diophantine equation (1) in which c is a multiple of (a,b), we shall assume that a and b are relatively prime.

In practice, a particular solution x_0, y_0 of $ax + by = c$ can be obtained by inspection or by applying the Euclidean algorithm. In the latter case, we use the Euclidean algorithm as described in the previous section to obtain a solution x_1, y_1 of $ax + by = (a,b) = 1$ and then note that cx_1, cy_1 is a solution of $ax + by = c$. Another method of finding solutions will be presented in Section 3.4.

In any case, once a particular solution x_0, y_0 of $ax + by = c$ is obtained, then an infinite number of solutions is determined, namely,
$$x = x_0 + bt,$$
$$y = y_0 - at,$$
where t is any integer. This is verified by direct substitution into the equation. We now show that all solutions of the equation are obtained in this way.

Proposition 3.15. If x_0, y_0 is a solution of the Diophantine equation $ax + by = c$ and $(a,b) = 1$, then every solution is given by
$$x = x_0 + bt,$$
$$y = y_0 - at,$$
where $t \in \mathbf{I}$, the set of all integers.

PROOF. Suppose that x_1, y_1 is any other solution of $ax + by = c$. Then subtracting
$$ax_0 + by_0 = c$$

from
$$ax_1 + by_1 = c$$
we obtain
$$a(x_1 - x_0) + b(y_1 - y_0) = 0$$
or equivalently
$$a(x_1 - x_0) = -b(y_1 - y_0).$$

Thus, we have the situation that $a(x_1 - x_0)$ is a multiple of b and at the same time a and b are relatively prime. We conclude from Proposition 3.9 that $x_1 - x_0$ must be a multiple of b. Hence, there exists an integer t such that $x_1 - x_0 = bt$, that is,
$$x_1 = x_0 + bt.$$
Substituting this into the last equation, we obtain $abt = -b(y_1 - y_0)$, that is,
$$y_1 = y_0 - at.$$
Therefore, the solution x_1, y_1 has the desired form.

Thus, we have found a criterion for the existence of solutions to the Diophantine equation $ax + by = c$, and we have also found a description of all solutions of this equation when solutions exist.

It is important to note that if a and b are not relatively prime, then the conclusion of Proposition 3.15 is not true. That is, if $(a,b) \neq 1$, then there may be solutions of $ax + by = c$ that are not of the form $x_0 + bt$, $y_0 - at$ for any integer t. For example, consider the equation $4x - 2y = 6$. We may take $x_0 = 1$, $y_0 = -1$ as a particular solution. But then the solution $x = 2$, $y = 1$ cannot be written in the form $1 + 2t$, $-1 - 4t$ for any integer t.

The above description of the solution set of the linear Diophantine equation $ax + by = c$ should bring to mind the parametric equations for a straight line. If L denotes the straight line in the Cartesian plane that has equation $ax + by = c$, where x and y now vary over the set $\mathbf{R}$ of all real numbers, and if (x_0, y_0) lies on L, then

$$\begin{aligned} x &= x_0 + bt \\ y &= y_0 - at \end{aligned} \qquad (t \in \mathbf{R}) \qquad (*)$$

are parametric equations of L. If x_0, y_0 are integers, then each integer value of t yields a point on L whose coordinates are integers. If a and b are relatively prime, then Proposition 3.15 states that the converse is true—each point on L with integer coordinates is given by (*) for some integer value of t. Thus, the solutions of the linear Diophantine equation $ax + by = c$ correspond to the points on the line L with integer coordinates and this correspondence is given by (*) when t is restricted to have only integer values.

Frequently, one is only interested in finding the positive solutions of a linear Diophantine equation $ax + by = c$. If the equation has integer solutions and if a and b are nonzero integers with opposite sign, then infinitely many positive solutions will exist. On the other hand, if a and b have the same sign, then at most a finite number of positive solutions can exist. Geometrically speaking, this is because in the first case, the slope of the corresponding line in the plane is positive, while in the second case, the slope is negative. (See Figure 3.2.)

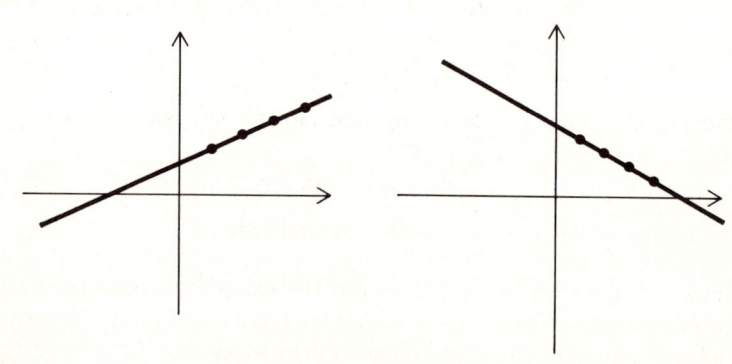

Figure 3.2

Example 3.6. Find all positive solutions of $6x + 9y = 75$. We first note that since $(6,9) = 3$ divides 75, the equation does possess solutions. Dividing both sides of the equation by 3 yields the equivalent equation

$$2x + 3y = 25$$

in which the coefficients are relatively prime. By inspection, we see that $x_0 = -1$, $y_0 = 9$ is a particular solution. Hence, all solutions are given by

$$x = -1 + 3t,$$
$$y = 9 - 2t,$$

where $t \in I$. Since we desire only positive solutions, we want those values of t that satisfy

$$-1 + 3t > 0,$$
$$9 - 2t > 0.$$

The first of these inequalities imposes the condition $t > \frac{1}{3}$ on t and the second gives us $t < \frac{9}{2}$. Putting these together and realizing that t must be an integer, we get the restriction $1 \leq t \leq 4$. Thus, there are four positive solutions (2,7), (5,5), (8,3), (11,1) corresponding to the values $t = 1, 2, 3, 4$, respectively.

3.3. LINEAR DIOPHANTINE EQUATIONS

Diophantine equations frequently arise in solving certain types of traditional "word problems." In fact, one way to broach the subject of Diophantine equations in the classroom is by means of word problems. In developing a systematic method of attacking such problems, the ideas and techniques of the preceding discussion can be introduced in a natural way. The following is an example of a word problem that leads to a linear Diophantine equation. Further examples are provided in the exercises.

Example 3.7. 70 silver dollars are to be divided among 50 men, women, and children in such a way that each man receives \$6, each woman receives \$3, and each child receives \$1. Can this be done, and if so, how many men, women, and children are there?

To solve this, we let x, y, and z denote the number of men, women, and children, respectively. The conditions of the problem give rise to two equations:
$$x + y + z = 50,$$
$$6x + 3y + z = 70.$$

Elimination of z results in the equation
$$5x + 2y = 20.$$

By inspection, we see that one solution of the equation is $x = 0$, $y = 10$. Thus, since $(5,2) = 1$, Proposition 3.15 tells us that the general solution of the equation is given by
$$x = 2t,$$
$$y = 10 - 5t,$$

where t is an integer. Of course, solutions to the stated problem must all be *positive* integers. This forces t to satisfy $2t > 0$ and $10 - 5t > 0$, that is, $0 < t < 2$. Thus, $t = 1$, which yields the solution $x = 2$, $y = 5$, $z = 43$.

In the above problem the solution turned out to be unique. In general, this need not be the case. For example, if there had been 40 people instead of 50, then two solutions would exist, namely, $x = 4$, $y = 5$, $z = 31$, and $x = 2$, $y = 10$, $z = 28$. On the other hand, if the dollar distribution had been \$7 for men, \$4 for women, and \$1 for children, then no solution would be possible.

EXERCISES

3.31. Determine which of the following equations have integer solutions. Find solutions for those equations that have them.
(a) $147x + 21y = 36$.
(b) $5x - 3y = 7$.

136 THE THEORY OF NUMBERS

      (c) $24x - 51y = 9$.
      (d) $33x + 19y = 250$.
      (e) $51x + 85y = 1037$.

3.32. Solve the Problem in Example 3.7 if there were (a) 30 people, (b) 60 people.

3.33. (A fowl problem.) A farmer can buy a turkey for $5, a duck for $7, and a chicken for $3. He wants to buy 30 birds for $100. How many of each kind can he buy?

3.34. Farmer Ferguson owns cows and chickens and in his last count discovered that there were 78 legs and 35 heads. However, he also counted the legs of one or more 3-legged milking stools by mistake. How many cows and chickens does Farmer Ferguson own?

3.35. Suppose 1 diamond costs 5 gold coins, 1 sapphire costs 3 gold coins, and 1 gold coin buys 3 rubies. A merchant has 100 gold coins and wishes to purchase 100 gems, at least one of each kind. How many of each kind can he purchase? (Find all possible solutions.)

3.36. Suppose that we have a balance as shown in Figure 3.3 and a supply of 3- and 5-ounce weights. A given object is to be weighed by placing it on one pan and then achieving the balance position by placing 3- and 5-ounce weights on the two pans. For example, Figure 3.3

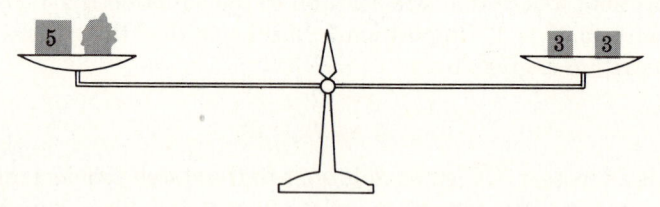

Figure 3.3

illustrates how a 1-ounce object can be weighed by using one 5-ounce and two 3-ounce weights. Show how the problem of weighing a given object of c ounces, where c is a positive integer, can be solved by means of a Diophantine equation such that each solution of the equation provides a method of weighing the given object.

3.37. Suppose we have a balance and a supply of 4- and 6-ounce weights. Is it possible to weigh a 3-ounce object? What objects can be weighed?

3.38. Suppose we are given containers with which we can measure 3 pints and 5 pints of water. Given an unlimited quantity of water and these two containers, how can we obtain exactly 1 pint of water? Show how this problem leads to a Diophantine equation.

3.39. Use the methods of Diophantine equations to determine the number of ways one can obtain 85¢ using only dimes and quarters.

3.40. Let $a_1, a_2, \cdots, a_n, b$ be given integers. Show that the Diophantine equation
$$a_1 x_1 + \cdots + a_n x_n = b$$
has a solution if and only if the greatest common divisor $(a_1, \cdots, a_n)$ divides b. (See Exercise 3.27.)

3.4. CONGRUENCES

Because of the importance of divisibility in number-theoretic considerations, the relation of congruence between integers is a fruitful object of study. This section is devoted to the discussion of some applications of this relation in the theory of numbers.

Although congruence relations were discussed in Sections 1.3, 1.5, and 2.2, the objectives of those sections differ from those of this section, and it is certainly not necessary for the reader to be familiar with the entire content of those sections in order to understand and appreciate the material to be discussed here. For this reason, we shall develop those basic definitions and elementary properties of congruence that are relevant to the present section.

If m is a fixed positive integer, then two integers a and b are *congruent modulo m*, written $a \equiv b \pmod{m}$, if $a - b$ is a multiple of m. For example, $9 \equiv -1 \pmod 5$ since $9 - (-1) = 2 \cdot 5$. Congruence modulo m is easily seen to define an equivalence relation on the set **I** of integers. Moreover, if $a \equiv b \pmod{m}$ and $c \equiv d \pmod{m}$, then

$$a + c \equiv b + d \pmod{m}, \tag{1}$$
$$ac \equiv bd \pmod{m}. \tag{2}$$

For example, since $13 \equiv 1 \pmod 4$ and $2 \equiv -6 \pmod 4$, it follows that $(13 + 2) \equiv (1 - 6) \pmod 4$ and $13 \cdot 2 \equiv 1 \cdot (-6) \pmod 4$. Mathematical induction together with (2) can be used to prove that if $a \equiv b \pmod{m}$, then

$$a^n \equiv b^n \pmod{m} \tag{3}$$

for every natural number n.

An alternate description of the meaning of the congruence relation is provided by the following simple result.

Proposition 3.16. Two integers a and b are congruent modulo m if and only if a and b have the same remainder r when divided by m.

PROOF. Suppose $a \equiv b \pmod{m}$ and $b = qm + r$ where $0 \leq r < m$. Then, since $a = b + tm$ for some $t \in \mathbf{I}$, we have $a = (q + t)m + r$. Thus both a and b yield the same remainder when divided by m.

Conversely, if $a = q_1 m + r$ and $b = q_2 m + r$ where $0 \leq r < m$, then $a - b = (q_1 - q_2)m$, which shows that $a \equiv b \pmod{m}$.

138   THE THEORY OF NUMBERS

Corollary. Each integer is congruent modulo m to exactly one of the numbers $0, 1, 2, \cdots, m - 1$.

Even with the preceding very modest list of results concerning congruences, it is possible to derive some useful divisibility tests. These results are all based on the fact that each positive integer N can be written in the form

$$N = a_n 10^n + \cdots + a_1 10^1 + a_0,$$

where each a_k is an integer satisfying $0 \leq a_k \leq 9$. As a matter of fact, $a_n, \cdots, a_1, a_0$ are simply the successive digits of N. For example, if $N = 30{,}279$, then $a_0 = 9$, $a_1 = 7$, $a_2 = 2$, $a_3 = 0$, $a_4 = 3$.

We shall first derive a test to determine whether or not a given positive integer N is divisible by 9. Since $10 \equiv 1 \pmod{9}$, it follows from property (3) above that $10^k \equiv 1 \pmod{9}$ for any positive integer k. Consequently, by virtue of properties (1) and (2) above, we conclude that

$$N \equiv (a_n + a_{n-1} + \cdots + a_1 + a_0) \pmod{9}.$$

That is, N is congruent to the sum of its digits modulo 9. In view of Proposition 3.16, this in turn implies the following divisibility test for 9: *An integer N is divisible by 9 if and only if the sum of its digits is divisible by 9.* For example, $2{,}005{,}407{,}963$ is divisible by 9 since the sum of its digits is 36, while $24{,}758{,}143$ is not divisible by 9 since the sum of its digits is 34; moreover, we see that the remainder upon division of $24{,}758{,}143$ by 9 must be $3 + 4 = 7$.

This procedure of replacing a number by the sum of its digits is commonly referred to as "casting out nines." The reason for this appellation is that whenever several digits of the number N have a sum of 9, these digits can be deleted or "cast out." For example,

$$\begin{aligned} 24{,}758{,}123 &\equiv (2 + 4 + 7 + 5 + 8 + 1 + 2 + 3) \pmod{9} \\ &\equiv (4 + 5 + 8 + 1 + 2 + 3) \pmod{9} \\ &\equiv (8 + 1 + 2 + 3) \pmod{9} \\ &\equiv 5 \pmod{9}. \end{aligned}$$

Thus the remainder obtained when $24{,}758{,}123$ is divided by 9 is 5. Of course, if certain digits have a sum greater than 9, then that sum can be replaced by the difference between the sum and 9. For example, $8 + 7$ can be replaced by 6 and $6 + 5$ can be replaced by 2 to yield the fact that

$$\begin{aligned} 875 &\equiv (8 + 7 + 5) \pmod{9} \\ &\equiv (6 + 5) \pmod{9} \\ &\equiv 2 \pmod{9}. \end{aligned}$$

A simple test can also be derived to determine whether or not a given integer N is divisible by 11. To do this, we first note that since $10 \equiv$

$-1 \pmod{11}$, it follows from property (3) above that
$$10^{2k+1} \equiv -1 \pmod{11},$$
$$10^{2k} \equiv 1 \pmod{11},$$
for each positive integer k. Therefore, if
$$N = a_n 10^n + a_{n-1} 10^{n-1} + \cdots + a_1 10 + a_0$$
is a given positive integer, then properties (1) and (2) above show that
$$N \equiv ((-1)^n a_n + \cdots + a_2 - a_1 + a_0) \pmod{11}.$$
In other words, N is congruent modulo 11 to the sum of the digits in the even places minus the sum of the digits in the odd places. Therefore, by Proposition 3.16, we have the following divisibility test for 11: *An integer N is divisible by 11 if and only if the sum of the digits of N in the even places minus the sum of the digits of N in the odd places is divisible by 11.* For example, 9,486,344 is not divisible by 11 since $(4 + 3 + 8 + 9) - (4 + 6 + 4) = 10$ is not divisible by 11. However, 940,082 is divisible by 11 since $(2 + 0 + 4) - (8 + 0 + 9) = -11$ is divisible by 11.

Similar rules for divisibility and calculation of remainders can be obtained for other numbers. In theory, a rule can be derived for any m, but in practice, the resulting relation is frequently complicated and the law of diminishing returns takes effect.

One use of these tests is to check the accuracy of certain arithmetic operations. For example, if large numbers N and M are multiplied, then the many calculations needed to perform the multiplication allow for possible errors to be committed. A partial check on the accuracy of the computation can be performed by using congruences. If $L = M \cdot N$, then for any m, it is necessary (but not sufficient!) that $L \equiv M \cdot N \pmod{m}$. A particularly simple choice for m in this check is $m = 9$. Let us consider an example. Suppose $N = 5271$ and $M = 4372$ are multiplied rather quickly to obtain $L = 23,044,812$ and we desire some assurance that the answer is correct. By virtue of our conclusions concerning divisibility by 9, we see that
$$5271 \equiv 6 \pmod{9},$$
$$4372 \equiv 7 \pmod{9}.$$
Hence, we see that $N \cdot M \equiv 42 \pmod{9} \equiv 6 \pmod{9}$. Now if the multiplication is correct, then we necessarily must have $L \equiv 6 \pmod{9}$. A direct computation shows that this is the case. Although this gives us some evidence that the answer is correct, it is not a guarantee. For example, $L' = 23,404,812$ (two digits of L have been interchanged) is also congruent to 6 modulo 9 since an interchange of digits does not affect their sum. However, both L and L' cannot be correct answers. Thus, while the

above procedure is a quick check, it is not foolproof. If it had turned out that L was *not* congruent to 6 modulo 9, then we would have known an error had been committed and a second multiplication would have been in order. A double check on the computation can be performed by using a second value of m, say $m = 11$. Since $N \equiv 2 \pmod{11}$, $M \equiv 5 \pmod{11}$, and $L \equiv 10 \pmod{11}$, we have additional evidence that the multiplication is correct. Note that $L' \equiv 2 \pmod{11}$ shows that L' cannot be the product of N and M.

A similar check can be used in the case of addition. In fact, suppose that $N_1, \cdots, N_p$ are large numbers whose sum has been computed to be K. Then, if K is indeed the correct sum, it must be true that

$$(N_1 + \cdots + N_p) \equiv K \pmod{m}$$

for any positive integer m. In particular, if $m = 9$, if n_i is the sum of the digits of N_i for $i = 1, \cdots, p$, and if k is the sum of the digits of K, then we must have

$$(n_1 + \cdots + n_p) \equiv k \pmod{9}$$

in order for K to be the correct sum. For example, to check the sum

38,179	N_1
432,981	N_2
78,752	N_3
$+\,141,278$	N_4
691,190	K

we simply note that $n_1 = 28$, $n_2 = 27$, $n_3 = 29$, $n_4 = 23$, and $k = 26$. Since $(28 + 27 + 29 + 23) = 107$ and $107 \equiv 8 \pmod{9} \equiv 26 \pmod{9}$, we can have some confidence that K is the correct sum. Of course, a double check could be made by using $m = 11$.

As a final illustration of the use of congruence in connection with divisibility of large numbers, let us consider the problem of determining the remainder upon division by 7 of the number $N = 2^{55} + 11^2$. First note that $2^6 = 64 \equiv 1 \pmod{7}$ so that $2^{54} = (2^6)^9 \equiv 1 \pmod{7}$. Therefore, $2^{55} = 2 \cdot 2^{54} \equiv 2 \pmod{7}$. Also, $11 \equiv 4 \pmod{7}$ so that $11^2 \equiv 16 \pmod{7} \equiv 2 \pmod{7}$. We conclude that $(2^{55} + 11^2) \equiv (2 + 2) \pmod{7} \equiv 4 \pmod{7}$; that is, the remainder upon division by 7 of $2^{55} + 11^2$ is 4.

EXERCISES

3.41. Is 176,521,221 divisible by 9? by 11?

3.42. Check the following computations by casting out nines and double check with the 11-test.

(a) $(52{,}817) \cdot (3{,}212{,}146) = 169{,}655{,}915{,}282$.

(b)   6,814,279
     13,182,147
     16,145,152
    + 7,018,143
    ───────────
     43,159,721

3.43. A natural number is called a *palindrome* if reversal of the digits results in the same number. For example, 21,312 and 473,374 are palindromes. Prove that palindromes with an even number of digits are divisible by 11.

3.44. Find the missing digit in the product

$$(211{,}476) \cdot (31{,}574) = 6{,}67\_\_, 143{,}224.$$

3.45. Find divisibility tests for 3 and 4.

3.46. Show that if the positive integer N_2 is obtained from the positive integer N_1 by rearranging the digits of N_1, then $N_1 - N_2$ is divisible by 9.

3.47. Find the remainder when:
 (a) 15^{37} is divided by 13.
 (b) $10^{49} + 5^3$ is divided by 7.
 (c) $2^{37} - 1$ is divided by 223.

3.48. Prove that if $a \equiv b \pmod{m}$, then $(a,m) = (b,m)$.

The cancellation law for the system of integers states that if a, b, c are integers such that $ab = ac$ and $a \neq 0$, then $b = c$. This is not always valid when dealing with congruences. For example, it is true that $3 \cdot 2 \equiv 3 \cdot 4 \pmod 6$, but 2 is not congruent to 4 modulo 6. However, cancellation is valid for congruences in certain circumstances as the next proposition shows.

Proposition 3.17. If a and m are relatively prime, then $ab \equiv ac \pmod{m}$ implies that $b \equiv c \pmod{m}$.

PROOF. We are given that $a(b - c) = km$ for some integer k. Since a and m are relatively prime, a must divide k so that $k = ta$ for some integer t. This implies that $a(b - c) = atm$, and since $a \neq 0$, we have $b - c = tm$. Hence, $b \equiv c \pmod{m}$.

Corollary. If p is prime and does not divide a, then $ab \equiv ac \pmod{p}$ implies that $b \equiv c \pmod{p}$.

If a is an integer other than 1 or -1, then no integer b can be found such that $ba = 1$. That is, except for 1 and -1, no integer has a reciprocal,

142 THE THEORY OF NUMBERS

or inverse, in the system of integers. In contrast to this, many integers do have inverses in the context of congruences. More precisely, it frequently happens that for a given integer a, there exists an integer b such that $ba \equiv 1 \pmod{m}$. For example, if $m = 9$ and $a = 4$, then $b = 7$ can play the role of the inverse of a modulo 9 since $7 \cdot 4 \equiv 1 \pmod{9}$. However, note that $a = 3$ can have no inverse modulo 9 in this sense since $3b - 1$ is not divisible by 9 for any integer b. The next very useful result will settle the existence of inverses modulo m when m is prime.

Proposition 3.18. *(Fermat's Theorem)* If p is prime and does not divide a, then

$$a^{p-1} \equiv 1 \pmod{p}.$$

PROOF. Consider the set $A = \{a, 2a, \cdots, (p-1)a\}$. We assert that no two distinct elements of A can be congruent modulo p. For suppose $sa \equiv ta \pmod{p}$ where $1 \leq s \leq p - 1$ and $1 \leq t \leq p - 1$. Since a and p are relatively prime, cancellation is permissible and we have $s \equiv t \pmod{p}$. However, since both s and t lie between 1 and $p - 1$, this is possible only if $s - t = 0$. Thus, we have proved that A consists of mutually incongruent elements.

We next claim that each element of A is congruent to some element of the set $B = \{1, 2, \cdots, p - 1\}$. To see this, note that every integer must be congruent modulo p to one of the numbers $0, 1, 2, \cdots, p - 1$, but no element of A can be congruent to 0 since p does not divide a.

Now since A and B have the same number of elements and each set consists of mutually incongruent elements, each element of A is congruent to some element of B, and conversely. It follows that the product of the elements of A is congruent to the product of the elements of B. That is,

$$a \cdot 2a \cdots (p-1)a \equiv 1 \cdot 2 \cdots (p-1) \pmod{p}.$$

Since the numbers p and $1 \cdot 2 \cdots (p-1)$ are relatively prime, cancellation is legitimate and we obtain

$$a^{p-1} \equiv 1 \pmod{p}.$$

Corollary. If p is prime and does not divide a, then there exists an integer b such that $ba \equiv 1 \pmod{p}$.

PROOF. If $p = 2$, then $a \equiv 1 \pmod{p}$, since otherwise a would be a multiple of p, and we may take $b = 1$. If p is a prime greater than 2, then $b = a^{p-2}$ has the desired property by Fermat's Theorem.

Note that Fermat's Theorem can also be formulated as follows: If p is prime, then $a^p \equiv a \pmod{p}$ for any integer a. In the case that a is not a multiple of p, then $(a,p) = 1$ and cancellation of a returns us to the original statement of the theorem.

Example 3.8. If n is a natural number, we can show that the units digit of the fifth power n^5 of n must be the same as the units digits of n as follows. First, we have by Fermat's Theorem that $n^5 \equiv n \pmod{5}$. Also, since $n^2 \equiv n \pmod{2}$, we have $n^5 \equiv n \pmod{2}$. (In fact, $n^k \equiv n \pmod{2}$ for any natural number k.) Thus, 5 and 2 are each divisors of $n^5 - n$, and since 5 and 2 are relatively prime, it follows that $10 = 5 \cdot 2$ is a divisor of $n^5 - n$. Hence, we have $n^5 \equiv n \pmod{10}$, which shows that the units digits of n^5 and n are the same.

EXERCISES

3.49. Show that if $ab \equiv ac \pmod{m}$ and $d = (a,m)$, then $b \equiv c \pmod{m'}$ where $m' = m/d$.

3.50. Find an integer b such that $3b \equiv 1 \pmod{7}$. Find such an integer b satisfying $0 \leq b \leq 7$.

3.51. Is $5^{36} - 1$ divisible by 13?

3.52. Find the remainder when 4^{444} is divided by 17.

3.53. Prove that if p is prime, then $(a + b)^p \equiv a^p + b^p \pmod{p}$ for all integers a and b. Is this true if p is not prime?

3.54. Show that $a^7 - a$ is divisible by 21 for all integers a.

3.55. Fermat's Theorem can also be proved by using the binomial theorem:

(a) Show that if p is prime, then the binomial coefficient
$$\binom{p}{k} = \frac{p(p-1)\cdots(p-k+1)}{k!}$$
is divisible by p for $k = 1, 2, \cdots, p - 1$.

(b) Use mathematical induction to show that $a^p \equiv a \pmod{p}$ for any integer a. [*Hint:* The binomial theorem states that
$$(x + y)^p = \sum_{k=0}^{p} \binom{p}{k} x^{p-k} y^k.$$
Consider $a^p = [(a - 1) + 1]^p$.]

3.56. Show that if $a \equiv b \pmod{p}$ and p is prime, then $a^p \equiv b^p \pmod{p^2}$. [*Hint:* Choose t so that $a = b + tp$ and consider $(b + tp)^p$ as in Exercise 3.55.]

3.57. Show that if p is prime, then the algebraic system $(\mathbf{I}_p, +, \cdot)$ is an integral domain. (See Section 2.2.)

We now turn to a discussion of the problem of solving congruences involving a variable. Consider the following examples:

(a) $3x \equiv 6 \pmod{7}$,
(b) $12x \equiv 7 \pmod{6}$,
(c) $x^2 - x \equiv 2 \pmod{4}$.

Since 3 and 7 are relatively prime, the congruence (a) can be solved by cancelling the 3 which yields $x \equiv 2 \pmod 7$. Thus, x is a solution of (a) if and only if it is congruent to 2 modulo 7. Congruence (b) has no solutions because $12x$ is divisible by 6 for any integer x while 7 is not divisible by 6. Example (c) can be solved by directly checking possible values of x. It is only necessary to check the values of 0, 1, 2, 3 because every integer is congruent modulo 4 to one of these numbers, and an integer will be a solution if and only if it is congruent to a solution that lies among these numbers. Direct substitution shows us that $x = 0$ and $x = 1$ are not solutions of (c) while $x = 2$ and $x = 3$ are solutions. Thus, x is a solution of (c) if and only if it is congruent to either 2 or 3 modulo 4.

In what follows, we shall deal only with congruences that can be put in the form $f(x) \equiv 0 \pmod{m}$ where $f(x)$ is a polynomial in x with integer coefficients. It is readily seen that if x_0 is a solution to such a congruence, then every integer congruent to x_0 modulo m is also a solution. Furthermore, each solution must be congruent to a solution that lies among the numbers $0, 1, \cdots, m - 1$. For these reasons, we shall refer to those solutions x_0 of the congruence $f(x) \equiv 0 \pmod{m}$ satisfying $0 \leq x_0 \leq m - 1$ as *basic solutions*. Note that distinct basic solutions are not congruent to one another modulo m, and that every solution of the congruence is congruent modulo m to exactly one basic solution. Thus, in solving congruences of this type, it suffices to find all basic solutions since all other solutions are then completely determined.

A congruence of the form $ax \equiv b \pmod{m}$ is called a *linear congruence*. The theory of linear congruences can be developed by utilizing the closely related theory of linear Diophantine equations presented in the previous section. An integer x_0 is a solution of $ax \equiv b \pmod{m}$ if and only if $ax_0 - b$ is a multiple of m, that is, $ax_0 - b = my_0$ for some integer y_0. Thus, the linear congruence

$$ax \equiv b \pmod{m}$$

has a solution if and only if the linear Diophantine equation

$$ax - my = b$$

has a solution. Since it follows from Proposition 3.14 that this Diophantine equation has a solution if and only if (a,m) is a divisor of b, we have the following corresponding result for linear congruences.

Proposition 3.19. The linear congruence $ax \equiv b \pmod{m}$ has a solution if and only if (a,m) is a divisor of b.

For example, the congruence $42x \equiv 14 \pmod{45}$ has no solutions since $(42,45) = 3$ is not a divisor of 14.

This settles the question of when the congruence $ax \equiv b \pmod{m}$ possesses solutions. We now determine the number of basic solutions such a congruence has.

Proposition 3.20. If the linear congruence $ax \equiv b \pmod{m}$ is such that $d = (a,m)$ is a divisor of b, then this congruence has exactly d basic solutions.

PROOF. Since $d|b$, we know that there is at least one basic solution of $ax \equiv b \pmod{m}$. If x_0 is a solution, then there exists an integer y_0 such that x_0, y_0 is a solution of the Diophantine equation $ax - my = b$. This Diophantine equation is equivalent to the equation

$$\frac{a}{d}x - \frac{m}{d}y = \frac{b}{d}$$

in which the coefficients are relatively prime. Therefore, by Proposition 3.15, the solutions to this Diophantine equation are given by

$$x = x_0 + \frac{m}{d}t,$$

$$y = y_0 + \frac{a}{d}t,$$

where $t \in \mathbf{I}$. Therefore, the solutions of the congruence $ax \equiv b \pmod{m}$ are all of the form $x = x_0 + (m/d)t$ where $t \in \mathbf{I}$. While some of these solutions will be congruent to one another modulo m, it is easily seen that none of the solutions in the set

$$B = \left\{ x_0, x_0 + \frac{m}{d}, x_0 + 2\frac{m}{d}, \cdots, x_0 + (d-1)\frac{m}{d} \right\}$$

obtained by taking $t = 0, 1, 2, \cdots, d-1$ are congruent to one another modulo m.

We next show that any solution not in B must be congruent modulo m to some solution in B. Consider a solution $x_0 + t(m/d)$ where t is not in the set $\{0,1,2, \cdots, d-1\}$. Then $t = dq + r$ where $q \neq 0$ and $0 \leq r \leq d - 1$. Therefore,

$$\left(x_0 + t\frac{m}{d}\right) - \left(x_0 + r\frac{m}{d}\right) = (t - r)\frac{m}{d} = qm$$

so that $x_0 + t(m/d)$ is congruent modulo m to the solution $x_0 + r(m/d)$ which lies in B since $0 \leq r \leq d - 1$.

Thus, we see that there exist d mutually incongruent solutions of the congruence with the property that any other solution is congruent modulo m to one of these solutions. We conclude that there must be exactly d basic solutions.

Suppose that $ax \equiv b \pmod{m}$ is a linear congruence such that (a,m) divides b. If we set $d = (a,m)$, then the congruence

$$\frac{a}{d}x \equiv \frac{b}{d}\left(\text{mod } \frac{m}{d}\right)$$

will be called the *reduced* congruence associated with the congruence $ax \equiv b \pmod{m}$. It is readily seen that the original congruence and the associated reduced congruence have the same solution sets; that is, an integer will be a solution of $ax \equiv b \pmod{m}$ if and only if it is a solution of the reduced congruence. However, the number of basic solutions of the two congruences need not be the same. In fact, since a/d and m/d are relatively prime, the preceding proposition shows that the reduced congruence has one basic solution, while the original congruence has d basic solutions. If x_0 denotes the basic solution of the reduced congruence, then the basic solutions of the original congruence are given by

$$x = x_0 + t\frac{m}{d}, \quad t = 0, 1, 2, \cdots, d - 1.$$

Now that the theory of the existence of solutions of linear congruences has been settled, we come to the practical problem of actually finding solutions. In view of the relationship between congruences and Diophantine equations, the technique involving the Euclidean algorithm does provide one method. However, there are frequently simpler ways of solving linear congruences, which in turn yield solutions of linear Diophantine equations. The following examples illustrate some of the techniques that can be employed in solving linear congruences.

Example 3.9.

(a) $2x \equiv 3 \pmod{5}$. Since $(2,5) = 1$, we see from Proposition 3.20 that there is one basic solution. One approach to solving this congruence is to note that $2^4 \equiv 1 \pmod{5}$ by Fermat's Theorem, so that multiplication of both sides of the congruence by $2^3 = 8$ yields $x \equiv 24 \pmod{5}$. Thus, the basic solution is $x = 4$.

A simpler approach is to note that cancellation would be permissible since 2 and 5 are relatively prime, but unfortunately 3 is not divisible by 2. However, we may replace the 3 appearing in the given congruence by 8 since $3 \equiv 8 \pmod 5$, and the resulting congruence $2x \equiv 8 \pmod 5$ is easily solved by cancelling the 2.

(b) $6x \equiv 14 \pmod{20}$. The fact that $(6,20) = 2$ tells us there are two basic solutions. The reduced congruence in this case is $3x \equiv 7 \pmod{10}$, which is equivalent to $-7x \equiv 7 \pmod{10}$ since $-7x$ and $3x$ are congruent modulo 10 for any integer x. Cancelling the -7 gives us $x \equiv -1 \pmod{10}$ and we see that $x = 9$ is the basic solution of the reduced congruence. Since the basic solutions of the original congruence are those solutions of the reduced congruence that lie between 0 and 19 (inclusive), we see that the two basic solutions of $6x \equiv 14 \pmod{20}$ are $x = 9$ and $x = 19$.

(c) $7x \equiv 11 \pmod{19}$. Our approach here is to multiply both sides by some number that will simplify the coefficient of x. Since 21 is close to 19, we multiply both sides by 3 and get $21x \equiv 33 \pmod{19}$. This is equivalent to $2x \equiv 14 \pmod{19}$, and hence, the basic solution is $x = 7$.

(d) $3x \equiv 4 \pmod 8$. When using the device of multiplying both sides of a congruence by a number, one must be sure that the number is relatively prime to the modulus. For example, suppose we multiplied both sides of $3x \equiv 4 \pmod 8$ by 2. This would yield $6x \equiv 8 \pmod 8$, or equivalently, $6x \equiv 0 \pmod 8$. While $x = 0$ is a solution of this congruence, it is *not* a solution of the original one. However, multiplication by 3 is legitimate and the resulting congruence is $9x \equiv 12 \pmod 8$. Replacing $9x$ by x, we see that the basic solution is $x = 4$.

(e) $256x \equiv 188 \pmod{524}$. We first find $(256, 524) = 4$. Then the reduced congruence is $64x \equiv 47 \pmod{131}$. Multiplying both sides of the reduced congruence by 2 and replacing $128x$ by $-3x$, we get $-3x \equiv 94 \pmod{131}$. Now that the coefficient of x has been simplified, we try to replace 94 by a number divisible by 3. $94 + 131 = 225$ is such a number. Thus we obtain $-x \equiv 75 \pmod{131}$, or equivalently, $x \equiv 56 \pmod{131}$. The four basic solutions of the original congruence are therefore $x = 56$, 187, 318, 449.

The relationship between linear congruences and linear Diophantine equations that was mentioned previously gives us a convenient method for solving linear Diophantine equations by means of congruences. This method is illustrated in the following examples.

148   THE THEORY OF NUMBERS

Example 3.10. Solve $12x - 53y = 17$. This equation can be converted into either of the two congruences

$$12x \equiv 17 \pmod{53},$$

or

$$53y \equiv -17 \pmod{12}.$$

Since 12 is the smaller modulus, we shall use the latter congruence. Replacing 53 by 5 and -17 by -5, we obtain $5y \equiv -5 \pmod{12}$. The solutions of this congruence are given by

$$y = -1 + 12t$$

for $t \in \mathbf{I}$. We now substitute this into the Diophantine equation in order to get the corresponding values of x. Thus,

$$12x = 17 + 53(-1 + 12t),$$

so that

$$x = -3 + 53t$$

for $t \in \mathbf{I}$.

Example 3.11. (The Coconut Problem) Three sailors are shipwrecked on a desert island and gather a pile of coconuts for food. They decide to divide the coconuts in the morning and retire for the night. One awakens and distrustfully decides to take his share during the night. He attempts to divide the pile of coconuts into three equal shares, but finds there is one coconut left over. He discards the extra one, takes his share to a hiding place, and goes back to sleep. Later, another of the sailors awakens and upon dividing the pile of remaining coconuts into three equal shares, finds there is one left over. He also discards the extra one, hides his share and goes back to sleep. Later, the third sailor repeats the process once again. He finds that by discarding one, the remaining pile of coconuts can be divided into thirds, and he hides his share before returning to sleep. We now ask: How many coconuts were in the original pile?

Let x denote the number of coconuts in the original pile and let y denote the number of coconuts left in the morning. The first sailor leaves $\frac{2}{3}(x - 1)$ coconuts after taking his share, the second sailor leaves

$$\tfrac{2}{3}[\tfrac{2}{3}(x - 1) - 1].$$

coconuts, and the number of coconuts remaining after the third sailor takes his share is

$$y = \tfrac{2}{3}[\tfrac{2}{3}[\tfrac{2}{3}(x - 1) - 1] - 1].$$

This simplifies to

$$8x - 27y = 38.$$

Since coconuts come only in whole numbers, we seek positive solutions to this linear Diophantine equation. To solve this equation, we use the congruence $-27y \equiv 38 \pmod 8$. Replacing $-27y$ by $5y$ and 38 by 30, we obtain the equivalent congruence $5y \equiv 30 \pmod 8$ which has solutions $y = 6 + 8t$ for $t \in \mathbf{I}$. Substitution into the Diophantine equation gives us

$$x = 25 + 27t$$

for $t \in \mathbf{I}$. Each positive value of t yields a solution to the coconut problem. The least number of coconuts possible is $x = 25$.

If the problem had involved four sailors instead of three, then we would have been led to the Diophantine equation

$$81x - 256y = 525$$

which has solutions

$$x = -3 + 256t,$$
$$y = -3 + 81t.$$

The minimum number of coconuts needed in this case is $x = 253$.

EXERCISES

3.58. Solve the congruences:
   (a) $x^2 - 2x + 5 \equiv 0 \pmod 4$,
   (b) $x^3 - 3x^2 + 2 \equiv 0 \pmod 5$.

3.59. Find all basic solutions of the following congruences:
   (a) $5x \equiv 7 \pmod{12}$,
   (b) $6x \equiv 19 \pmod{24}$,
   (c) $13x \equiv 2 \pmod{40}$,
   (d) $9x \equiv 15 \pmod{60}$,
   (e) $24x \equiv 18 \pmod{64}$,
   (f) $187x \equiv 69 \pmod{211}$.

3.60. Solve the following Diophantine equations using congruences:
   (a) $3x - 5y = 6$,
   (b) $43x + 71y = 9$,
   (c) $27x - 8y = 14$,
   (d) $51x + 17y = 10$.

3.61. Solve the coconut problem for:
   (a) four sailors,
   (b) five sailors.

3.62. Verify that if $d = (a,m)$ divides b, then the congruence $ax \equiv b \pmod m$ has the same solutions as the reduced congruence

$$\frac{a}{d} x \equiv \frac{b}{d} \left(\bmod \frac{m}{d} \right).$$

3.5. THE CHINESE REMAINDER THEOREM

We now turn to the problem of solving a system of linear congruences. A solution of the system:

$$\begin{cases} a_1 x \equiv b_1 (\mathrm{mod}\ m_1) \\ a_2 x \equiv b_2 (\mathrm{mod}\ m_2) \\ \quad \cdot \quad \cdot \quad \cdot \\ \quad \cdot \quad \cdot \quad \cdot \\ a_n x \equiv b_n (\mathrm{mod}\ m_n) \end{cases} \quad (1)$$

is an integer that satisfies each of the congruences in the system. Thus, the solution set of (1) is the intersection of the solution sets of the individual congruences in the system. The method of substitution illustrated in the following example provides a straightforward procedure for solving systems of linear congruences.

Example 3.12. Solve

$$\begin{cases} 2x \equiv 1 (\mathrm{mod}\ 5) \\ 3x \equiv 7 (\mathrm{mod}\ 8). \end{cases}$$

We begin by solving the first congruence $2x \equiv 1 (\mathrm{mod}\ 5)$ to obtain

(a) $\qquad\qquad x = 3 + 5t \qquad (t \in \mathbf{I}).$

Thus, the solution set for the first congruence consists of all integers congruent to 3 modulo 5. Next we proceed to find which (if any) of these solutions also satisfy the second congruence. To do this, we substitute $x = 3 + 5t$ into the second congruence to obtain

$$3(3 + 5t) \equiv 7 (\mathrm{mod}\ 8)$$

or equivalently

$$15t \equiv -2 (\mathrm{mod}\ 8).$$

Since this latter congruence has the solutions

(b) $\qquad\qquad t = 2 + 8u \qquad (u \in \mathbf{I})$

we can find all values of the variable x that satisfy both congruences by substituting (b) into (a) to obtain

$$x = 3 + 5(2 + 8u) = 13 + 40u \qquad (u \in \mathbf{I}).$$

Thus, the set of all solutions of this pair of congruences consists of all integers congruent to 13 modulo 40.

If the system consists of more than two congruences, the substitution procedure illustrated in the above example is first applied to obtain the solutions of the first two congruences, then again to obtain the solutions of the first three congruences, and so on, until the last congruence is reached.

Of course, if any of the congruences in a system do not have solutions, then the system of congruences cannot have solutions. For this reason, *we shall assume in the sequel that each congruence $a_i x \equiv b_i (\text{mod } m_i)$ in the system* (1) *has solutions*. It should be carefully noted that this assumption does not imply that there are solutions to the system (1). For example, each congruence in the system:

$$\begin{cases} x \equiv 1 (\text{mod } 4) \\ x \equiv 2 (\text{mod } 6) \end{cases}$$

satisfies this assumption; yet this system has no solution since any solution to the first congruence must be an odd integer while any solution to the second congruence must be an even integer. Note that if the method of substitution is used in an attempt to solve this system, one is led to the congruence $4t \equiv 1 (\text{mod } 6)$ which has no solutions since $(4,6) = 2$ is not a divisor of 1.

Since we are assuming that each congruence $a_i x \equiv b_i (\text{mod } m_i)$ in the system (1) has solutions, we know from our work in the preceding section that the solution set for each congruence $a_i x \equiv b_i (\text{mod } m_i)$ consists of all integers congruent to some integer c_i modulo m'_i where m'_i may be distinct from m_i. For this reason, we can always replace the given system (1) by the simpler system:

$$\begin{cases} x \equiv c_1 (\text{mod } m'_1) \\ x \equiv c_2 (\text{mod } m'_2) \\ \quad \cdot \quad \cdot \quad \cdot \\ \quad \cdot \quad \cdot \quad \cdot \\ \quad \cdot \quad \cdot \quad \cdot \\ x \equiv c_n (\text{mod } m'_n). \end{cases} \quad (2)$$

The systems (1) and (2) are equivalent in the sense that they have the same solution set. In fact, each congruence in (2) describes the solution set of the corresponding congruence in (1); consequently, every solution of the system (2) is certainly a solution of the system (1). On the other hand, if x_0 is a solution of the system (1), it must be congruent to c_i modulo m'_i for $i = 1, \cdots, n$. That is, every solution of the system (1) is also a solution of the system (2).

Thus, if we are given a system (1) of congruences for which the individual congruences in the system have solutions, we can always replace the given system by an equivalent system of the form (2). In case the moduli in (2) are relatively prime in pairs, the solutions are given by the following result having the romantic title of the Chinese Remainder

152   THE THEORY OF NUMBERS

Theorem. This title is apparently due to the fact that many ancient Chinese manuscripts on arithmetic dealt with problems of the following sort: Find a number that leaves remainders of 2, 3, 4 when divided by 3, 5, 7, respectively. When formulated in terms of congruences, a number of the desired sort is provided by any solution of the system of congruences:

$$\begin{cases} x \equiv 2 \pmod{3} \\ x \equiv 3 \pmod{5} \\ x \equiv 4 \pmod{7}. \end{cases}$$

This problem will be solved below as an application of the Chinese Remainder Theorem.

Proposition 3.21. *(Chinese Remainder Theorem)* Let $m_1, m_2, \cdots, m_n$ be positive integers that are relatively prime in pairs, and let M be their product. Then the solutions of the system of congruences

$$\begin{cases} x \equiv a_1 \pmod{m_1} \\ x \equiv a_2 \pmod{m_2} \\ \cdot \cdot \\ \cdot \cdot \\ \cdot \cdot \\ x \equiv a_n \pmod{m_n} \end{cases}$$

are given by

$$x = \sum_{i=1}^{n} a_i b_i \frac{M}{m_i} + Mt, \qquad (t \in \mathbf{I})$$

where b_i is any integer satisfying

$$b_i \frac{M}{m_i} \equiv 1 \pmod{m_i}, \qquad i = 1, 2, \cdots, n.$$

PROOF. Since the positive integers M/m_i and m_i are relatively prime, an integer b_i satisfying the stated condition does exist for each $i = 1, 2, \cdots, n$. Let

$$x_0 = \sum_{i=1}^{n} a_i b_i \frac{M}{m_i}.$$

We must show that $x_0 \equiv a_i \pmod{m_i}$ for $i = 1, 2, \cdots, n$. For $k \neq i$, M/m_k is a multiple of m_i; consequently, we see that $a_k b_k (M/m_k) \equiv$

3.5. THE CHINESE REMAINDER THEOREM

$0 \pmod{m_i}$ for $k \neq i$. Also, since $b_i(M/m_i) \equiv 1 \pmod{m_i}$, we see that $a_i b_i(M/m_i) \equiv a_i \pmod{m_i}$. Hence, it follows that $x_0 \equiv a_i \pmod{m_i}$. Thus, x_0 is a solution of the system. If x is congruent to x_0 modulo M, then x is also a solution of the system since $M \equiv 0 \pmod{m_i}$ for $i = 1, 2, \cdots, n$.

It remains to show that if x_1 is any solution of the system, then $x_1 \equiv x_0 \pmod{M}$. Since $x_1 \equiv a_i \pmod{m_i}$ and $b_i(M/m_i) \equiv 1 \pmod{m_i}$, we have $x_1 \equiv a_i b_i(M/m_i) \pmod{m_i}$ for each i. And since $a_k b_k(M/m_k) \equiv 0 \pmod{m_i}$ for $k \neq i$, we obtain $x_1 \equiv x_0 \pmod{m_i}$. This implies that for each i, m_i divides $x_1 - x_0$. But since the numbers $m_1, m_2, \cdots, m_n$ are relatively prime in pairs, this can occur only if $M = m_1 \cdots m_n$ divides $x_1 - x_0$. That is, $x_1 \equiv x_0 \pmod{M}$.

Let us now apply the Chinese Remainder Theorem to solve the problem mentioned in the remarks preceding the statement of the theorem. We have $m_1 = 3$, $m_2 = 5$, $m_3 = 7$. Thus, $M = 3 \cdot 5 \cdot 7 = 105$ and the b_1, b_2, b_3 of the theorem must satisfy

$$35 b_1 \equiv 1 \pmod{3},$$
$$21 b_2 \equiv 1 \pmod{5},$$
$$15 b_3 \equiv 1 \pmod{7}.$$

We find that $b_1 = 2$, $b_2 = 1$, $b_3 = 1$ are such numbers. Since $a_1 = 2$, $a_2 = 3$, $a_3 = 4$, we see that the integer

$$x_0 = 2 \cdot 2 \cdot 35 + 3 \cdot 1 \cdot 21 + 4 \cdot 1 \cdot 15 = 263$$

leaves remainders 2, 3, 4 when divided by 3, 5, 7, respectively. Moreover, all other integers having this property are given by

$$x = 263 + 105t$$

for $t \in I$. In particular, the smallest positive integer with this property is 53.

Problems of this type can be formulated as a game as follows. Ask someone to think of a number between 1 and 100 and then ask him to state the remainders when that number is divided by 3, 5, 7, respectively. The number can then be determined as follows. Let x_0 denote the unknown number and let a_1, a_2, a_3 be the stated remainders. Then x_0 must be a solution of the system

$$\begin{cases} x \equiv a_1 \pmod{3} \\ x \equiv a_2 \pmod{5} \\ x \equiv a_3 \pmod{7}. \end{cases}$$

Then $M = 3 \cdot 5 \cdot 7 = 105$ and, as we have already noted, $b_1 = 2$, $b_2 = 1$, $b_3 = 1$. Therefore, the Chinese Remainder Theorem tells us that x_0 is the basic solution of the congruence:

$$x \equiv 70a_1 + 21a_2 + 15a_3 \pmod{105}.$$

For example, if the given remainders are $a_1 = 1$, $a_2 = 3$, $a_3 = 4$, then $x \equiv 193 \pmod{105}$ so that the unknown number must be $x_0 = 88$.

EXERCISES

3.63. Solve the following systems of congruences:

(a) $\begin{cases} x \equiv 3 \pmod 5 \\ x \equiv 1 \pmod 3, \end{cases}$ (b) $\begin{cases} 2x \equiv 3 \pmod 6 \\ 5x \equiv 10 \pmod{15}, \end{cases}$

(c) $\begin{cases} x \equiv 3 \pmod 4 \\ 2x \equiv 1 \pmod 5 \\ 3x \equiv 4 \pmod 7, \end{cases}$ (d) $\begin{cases} x \equiv 5 \pmod 7 \\ x \equiv 1 \pmod 5 \\ x \equiv 6 \pmod 9. \end{cases}$

3.64. A certain number between 1 and 1200 has remainders 1, 2, 6 when divided by 9, 11, 13, respectively. What is the number? What is the number if the remainders are 3, 8, 5, respectively?

3.65. A gang of 13 thieves stole a sack of silver dollars. When they tried to divide the booty evenly, there were 3 dollars left over. In a fight over the extra dollars, 2 of the thieves were killed. The money was distributed again, but this time there were 5 dollars left over. Another argument ensued and 1 more thief was killed. The money could now be divided evenly among the remaining thieves. What is the least possible amount of money that could have been stolen?

3.66. (Ancient Hindu Problem.) A woman with a basket of eggs finds that if she removes the eggs from the basket either 2, 3, 4, 5, or 6 at a time, there is always 1 egg left over. However, when she removes them 7 at a time, there are no eggs left over. What is the least number of eggs the woman can have in her basket?

3.67. Prove that the system

$$\begin{cases} x \equiv a \pmod m \\ x \equiv b \pmod n \end{cases}$$

has a solution if and only if $a - b$ is a multiple of (m,n). Show that in this case, if x_0 is one solution of the system, then the general solution of the system is given by $x \equiv x_0 \pmod{[m,n]}$.

We conclude this section with a few examples that illustrate how systems of congruences can be utilized to solve congruences that are complicated either by a large modulus or the occurrence of powers of x higher than one.

Example 3.13. Solve $17x \equiv 5 \pmod{252}$. This can be solved by techniques discussed earlier, but the modulus 252 is not particularly easy to work with. However, since $252 = 4 \cdot 7 \cdot 9$, and since 4, 7, 9 are relatively prime in pairs, the requirement that $17x - 5$ be divisible by 252 is equivalent to the requirement that x satisfy the system of congruences:

$$\begin{cases} 17x \equiv 5 \pmod 4 \\ 17x \equiv 5 \pmod 7 \\ 17x \equiv 5 \pmod 9. \end{cases}$$

Thus, solving the given congruence is equivalent to solving the system

$$\begin{cases} x \equiv 1 \pmod 4 \\ x \equiv 4 \pmod 7 \\ x \equiv 4 \pmod 9, \end{cases}$$

which is a simplified version of the system above. Solving this system, we get $x \equiv 193 \pmod{252}$.

Example 3.14. Solving congruences of degree greater than one is considerably more complicated than solving linear congruences. We will discuss one technique that is sometimes helpful. If the modulus m of a congruence $f(x) \equiv 0 \pmod m$ is small, then a solution can be obtained by direct examination of the cases $x = 0, 1, 2, \cdots, m - 1$. For example, we can solve

$$x^3 - x + 1 \equiv 0 \pmod 5$$

by checking $x = 0, 1, 2, 3, 4$ one at a time. We see that $x = 0, 1, 2, 4$ are not solutions while $x = 3$ is a solution. If the modulus is large, then it becomes tedious to check all of the possible cases. However, it is frequently possible to replace the given congruence by a system of congruences. For example, suppose we wish to solve

$$x^3 - x + 1 \equiv 0 \pmod{35}.$$

Since $35 = 5 \cdot 7$ and 5, 7 are relatively prime, x is a solution of this congruence if and only if it is a solution of the system of congruences:

$$\begin{cases} x^3 - x + 1 \equiv 0 \pmod 7 \\ x^3 - x + 1 \equiv 0 \pmod 5. \end{cases}$$

A direct calculation shows that all solutions of the first congruence are given by

$$x = 3 + 5t \qquad (t \in \mathbf{I})$$

while all solutions of the second congruence are

$$x = 2 + 7t \qquad (t \in \mathbf{I}).$$

Hence, the solutions of the original congruence are precisely the solutions of the system
$$\begin{cases} x \equiv 3 \pmod 5 \\ x \equiv 2 \pmod 7. \end{cases}$$

The solutions of this system are readily determined to be all integers that are congruent to 23 modulo 35.

EXERCISES

3.68. Solve the congruence $23x \equiv 12 \pmod{210}$.

3.69. Solve the following:
   (a) $x^3 - 3x^2 + 27 \equiv 0 \pmod{35}$,
   (b) $x^2 - 7x \equiv 9 \pmod{15}$.

3.6. PYTHAGOREAN TRIPLES AND FERMAT'S LAST THEOREM

Diophantine equations of degree higher than one are generally difficult to solve. There are no systematic procedures that can be applied to solve general Diophantine equations; special equations must be attacked in special ways. We shall discuss an important Diophantine equation of higher degree known as the Pythagorean Equation and the famous unsolved problem commonly referred to as Fermat's Last Theorem.

Probably the most famous Diophantine equation is the equation $x^2 + y^2 = z^2$. This equation is referred to as the *Pythagorean equation* because of its obvious relationship to the Pythagorean Theorem for right triangles. As usual, the adjective Diophantine indicates that we are seeking only integer solutions. Of course, one obvious solution is $x = 0$, $y = 0$, $z = 0$; moreover, for any integer a, it is clear that $x = a$, $y = 0$, $z = \pm a$ as well as $x = 0$, $y = a$, $z = \pm a$ are also solutions. We shall refer to these solutions as *trivial solutions* of the Pythagorean equation. Any solution $x = x_0$, $y = y_0$, $z = z_0$ that is *not* a trivial solution will be called a *Pythagorean triple* and will be denoted by (x_0, y_0, z_0). A Pythagorean triple (x_0, y_0, z_0) is *primitive* if x_0, y_0, z_0 are positive integers that are relatively prime in pairs.

In general, it is not true that three integers are relatively prime in pairs if they have no prime factor in common. For example, the numbers 2, 6, 3 have no prime factor in common, yet $(2,6) = 2$ and $(6,3) = 3$. However, if (x_0, y_0, z_0) is a Pythagorean triple, then this conclusion is valid. For if any two of the integers x_0, y_0, z_0 have a common prime factor p, then the relation $x_0^2 + y_0^2 = z_0^2$ forces the third integer in this triple to have p as a factor. Consequently, it follows that any positive Pythagorean triple (that is, a Pythagorean triple of positive integers) is a multiple (kx_0, ky_0, kz_0)

of a primitive Pythagorean triple (x_0, y_0, x_0). Furthermore, since the square of a negative integer is a positive integer, we see that any Pythagorean triple can be obtained from a positive Pythagorean triple by simply changing the signs of the appropriate terms of the latter. Thus, once the set of primitive solutions of the Pythagorean equation is determined, we can then generate the set of all nontrivial solutions by changing signs and taking multiples of primitive solutions.

We now wish to describe the set of primitive Pythagorean triples. We first note that if (x_0, y_0, z_0) is such a triple, then one of the numbers x_0, y_0 must be odd and the other even. In fact, the assumption that (x_0, y_0, z_0) is a primitive solution rules out the possibility that both x_0 and y_0 are even, for in this case x_0, y_0, and z_0 would all have a factor of 2. Nor can both x_0 and y_0 be odd, for if $x_0 = 2n + 1$ and $y_0 = 2m + 1$, then $z_0^2 = 4(n^2 + m^2 + n + m) + 2$, which implies that z_0^2 must be even but not divisible by 4 and this is impossible. Let us assume that y_0 is even and x_0 is odd. Since the Pythagorean equation is symmetric in x and y, it does not matter which is assumed to be even and which odd, and we take y_0 even and x_0 odd simply to fix the notation. Then clearly, z_0 is odd.

Set $y_0 = 2a$. Then

$$4a^2 = z_0^2 - x^2 = (z_0 - x_0)(z_0 + x_0).$$

Since both x_0 and z_0 are odd, we see that both $z_0 - x_0$ and $z_0 + x_0$ are even. Let us set $z_0 + x_0 = 2u$ and $z_0 - x_0 = 2v$. Then, we have $x_0 = u - v$, $z_0 = u + v$ and $a^2 = uv$. The fact that x_0 and z_0 are relatively prime implies that u and v are relatively prime. For if u and v have a common prime factor p, then $u - v$ and $u + v$ would also each have the factor p. Thus, we have the situation that u and v have no factor in common; yet their product uv is a square a^2. This forces u and v to be squares themselves. Hence, we may set $u = m^2$ and $v = n^2$ where m and n are relatively prime since u and v have this property. If we now substitute into the expressions for x_0, y_0 and z_0 we obtain

$$x_0 = m^2 - n^2, \qquad y_0 = 2mn, \qquad z_0 = m^2 + n^2.$$

Since we desire positive solutions, we require that $m > n$. Also, since both x_0 and z_0 are odd, we have that one of the numbers m, n must be even and the other odd. We summarize our discussion as follows.

Proposition 3.22. If y is an even integer, then (x, y, z) is a primitive Pythagorean triple if and only if

$$x = m^2 - n^2,$$
$$y = 2mn,$$
$$z = m^2 + n^2,$$

where m and n are relatively prime positive integers such that one is even, one is odd, and $m > n$.

A few primitive solutions of $x^2 + y^2 = z^2$ are listed below:

m	n	x	y	z
2	1	3	4	5
3	2	5	12	13
4	3	7	24	25
4	1	15	8	17
5	2	21	20	29

EXERCISES

3.70. Find four more primitive Pythagorean triples.

3.71. Find all right triangles of relatively prime integer sides such that the hypotenuse differs from one leg by 2.

3.72. Show that if (x,y,z) is a Pythagorean triple, then xyz is divisible by 60.

3.73. Show that if a right triangle has integer sides, then the radius r of the inscribed circle is an integer. [*Hint:* Note that in Figure 3.4 the

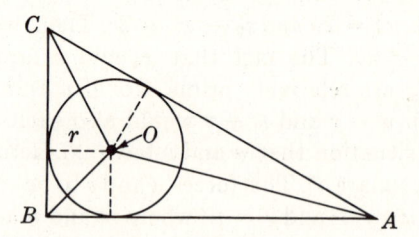

Figure 3.4

area of triangle ABC equals the sum of the areas of triangles OAB, OAC, OBC where O is the center of the inscribed circle.]

The Method of Infinite Descent

We now consider the Diophantine equation $x^4 + y^4 = z^2$. We shall show that this equation has no positive solutions. Of particular interest is the method of proof. It is a method due to Fermat called the method of infinite descent. The idea is as follows. It is assumed that a positive solution (x_0, y_0, z_0) exists. An argument then proceeds to show that another positive solution (x_1, y_1, z_1) must exist such that $z_1 < z_0$. However, this cannot be possible, for we could then produce a sequence (x_n, y_n, z_n) of positive solutions for which we would have $z_1 > z_2 > \cdots > z_n > \cdots$. Such an infinite descent is impossible since there are only finitely many positive integers smaller than any fixed positive integer z_0.

We see the method of infinite descent is a close relative of mathematical induction. The method can be generally described as follows. Suppose $P(n)$ is a statement about the natural number n for each $n \in \mathbf{N}$. If it can be shown that the truth of $P(n_0)$, where n_0 is some fixed natural number, implies the truth of $P(n_1)$ where n_1 is a natural number smaller than n_0, then we may conclude that $P(n)$ cannot be true for any $n \in \mathbf{N}$. For otherwise there would exist infinitely many natural numbers smaller than n_0 for which $P(n)$ would be true, and this is evidently impossible.

Proposition 3.23. The Diophantine equation $x^4 + y^4 = z^2$ has no positive solutions.

PROOF. Suppose a positive solution (x_0, y_0, z_0) exists. By arguing in a manner similar to that used in discussing the Pythagorean equation, we may assume that x_0, y_0, z_0 are relatively prime in pairs and that x_0 is odd and y_0 is even. Then (x_0^2, y_0^2, z_0) is a primitive Pythagorean triple and we may set $x_0^2 = m^2 - n^2$, $y_0^2 = 2mn$, $z_0 = m^2 + n^2$ where m and n are relatively prime positive integers, one even, one odd, and $m > n$. We now proceed to obtain a second positive solution (x_1, y_1, z_1) of $x^4 + y^4 = z^2$ where $z_1 < z_0$. Since $x_0^2 + n^2 = m^2$, we see that (x_0, n, m) is a primitive Pythagorean triple and thus there exist relatively prime positive integers a and b, one even and one odd, where $a > b$, such that

$$x_0^2 = a^2 - b^2, \quad n = 2ab, \quad m = a^2 + b^2.$$

Using these expressions for m and n, we obtain

$$y_0^2 = 2(a^2 + b^2) \cdot 2ab$$
$$= 4ab(a^2 + b^2).$$

The fact that a and b are relatively prime implies that a, b, $a^2 + b^2$ are relatively prime in pairs. Since the product of these numbers is a square $(y_0/2)^2$, we conclude that a, b, $a^2 + b^2$ are all squares. Hence, we may set $a = x_1^2$, $b = y_1^2$, and $a^2 + b^2 = z_1^2$. But now we see that

$$z_1^2 = a^2 + b^2 = x_1^4 + y_1^4$$

so that (x_1, y_1, z_1) is a positive solution of the equation $x^4 + y^4 = z^2$. Furthermore,

$$z_1 \leq a^2 + b^2 = m < m^2 + n^2 = z_0.$$

Hence, the method of infinite descent shows that no positive solutions can exist.

Fermat's Last Theorem

For each positive integer n, we consider the Diophantine equation $x^n + y^n = z^n$ in x, y, and z. For $n = 1$, this is just the linear equation $x + y = z$ which has infinitely many solutions. For $n = 2$, we have the Pythagorean equation which has been discussed above. For general n, we always have the trivial solutions $(0,0,0)$ $(a,0,\pm a)$ and $(0,a,\pm a)$ for any integer a. Moreover, if n is odd, then $(a,-a,0)$ is a solution for every integer a. The problem of solving $x^n + y^n = z^n$ becomes interesting if we insist on nontrivial solutions.

Fermat, while studying the works of Diophantus, wrote in the margin of one of the volumes that he had discovered a truly wonderful proof that $x^n + y^n = z^n$ has no positive solutions whenever $n \geq 3$. He added that the margin was too small to contain his proof. Unfortunately, Fermat never recorded his proof in any location whatever. It is possible that he later discovered an error in his argument, but did not bother to alter his marginal note. In any case, his remark has led mathematicians to hunt unsuccessfully for a proof of "Fermat's Last Theorem" for over 300 years. These efforts should not be regarded as wasted, however, since the search for a proof has stimulated deep and significant research in number theory and related parts of mathematics.

While Fermat's Last Theorem has not been proved in general, sophisticated techniques involving the modern computer have been used to show that the assertion of the theorem is true for all n satisfying $3 \leq n < 25{,}000$. This result was established in 1964 by John Selfridge and Barry Pollack.

We conclude this discussion by stating some simple results related to Fermat's Last Theorem. The proofs are left as exercises.

Proposition 3.24. If n is a multiple of 4, then $x^n + y^n = z^n$ has no positive solutions.

Proposition 3.25. If there exists an $n \geq 3$ such that $x^n + y^n = z^n$ has a positive solution, then there exists a prime $p \geq 3$ such that $x^p + y^p = z^p$ has a positive solution.

EXERCISES

3.74. Prove Proposition 3.24. [*Note:* If (x_0,y_0,z_0) is a solution of $x^4 + y^4 = z^4$, then (x_0,y_0,z_0^2) is a solution of $x^4 + y^4 = z^2$.]

3.75. Prove Proposition 3.25.

3.76. Prove that if $n \equiv 3 \pmod{4}$, then the Diophantine equation $x^2 + y^2 = n$ has no solutions. [*Hint:* Show that x^2 is congruent to either 0 or 1 modulo 4 for any integer x.]

REMARKS AND REFERENCES

Apart from a rather cursory review of least common multiple, greatest common divisor, and prime factorization, most secondary school textbooks, even those intended for the final year, omit consideration of topics from the theory of numbers. This omission is due in large part to the fact that a knowledge of number theory is not required for the study of college level mathematics and that the subject has comparatively few applications to other disciplines. However, despite these rather practical reasons for excluding consideration of this subject at the secondary school level, there are several equally compelling arguments that promote its inclusion, at least as enrichment material.

First of all, many topics in number theory are quite accessible to students with a modest background in algebra. Secondly, the basic ideas involved can be nicely motivated by means of puzzles and problems that invariably arouse a student's interest. This is not to be dismissed lightly since the amount a student learns is often directly related to the degree of his interest in the subject matter. Some such recreational approaches have been mentioned in the text and exercises of this chapter. Finally, the techniques and methods of proof used in number theory are quite varied and instructive. In this subject perhaps more than almost any other, conjectures and proofs can be discovered through the analysis of particular examples, and this discovery process by examination of special cases illustrates very nicely the spirit of mathematical research.

The notion of one integer dividing another is basic and familiar to everyone. A discussion of divisibility tests can be presented without the notational trappings of congruences by using only the definition and fundamental properties of divisibility. However, the notation and language of congruences constitute a convenient way to discuss divisibility systematically. Algebraic manipulations of congruences are quite simple and enable one to discover and prove divisibility results that are not as transparent when viewed directly in terms of the definition. It should be mentioned that the notion of congruence can be very effectively introduced by means of clock arithmetic. (See the Remarks and References for Chapter 2.)

Material concerning prime numbers, divisibility tests, the Euclidean algorithm, Pythagorean triples, and so on, can and should be woven into the regular algebra and geometry program. On the other hand, the study of Diophantine equations and the Chinese Remainder Theorem probably cannot be handled so incidentally. Consequently, these topics are best suited for a special enrichment unit or for independent study projects.

There are many interesting books on the subject of number theory.

We shall limit ourselves to mentioning a few that seem appropriate as sources for further reading. The book by Ore [2] is an excellent reference containing historical material on number theory and discussions of the relation of this subject to word problems. An indication of how puzzles can be used to motivate ideas in number theory can be found in the book by Stein [5]. The paperback book by Beiler [1] deals extensively with the recreational aspects of number theory, and the little book by Sierpinski [4] contains much fascinating information on Pythagorean triples. Shockley [3] and Stewart [6] present readable accounts of a fairly broad range of topics in number theory.

1. Beiler, A. H., *Recreations in the Theory of Numbers—The Queen of Mathematics Entertains*, New York: Dover Publications, Inc., 1964.
2. Ore, Oystein, *Number Theory and Its History*, New York: McGraw-Hill Book Company, Inc., 1948.
3. Shockley, J. E., *Introduction to Number Theory*, New York: Holt, Rinehart and Winston, Inc., 1967.
4. Sierpinski, W., *Pythagorean Triangles*, The Scripta Mathematica Studies, No. 9, New York: Graduate School of Science, Yeshiva University, 1962.
5. Stein, Sherman, *Mathematics—The Man-made Universe*, Second Edition, San Francisco: W. H. Freeman and Company, 1969.
6. Stewart, B. M., *Theory of Numbers*, Second Edition, New York: The Macmillan Company, 1964.

Chapter 4

BINOMIAL COEFFICIENTS AND COUNTING TECHNIQUES

The main purpose of this chapter is the development of the basic counting techniques that are necessary for the study of probability theory. Since binomial coefficients are closely related to many of these counting techniques, the first section is devoted to a discussion of the properties of this

Blaise Pascal
David Smith Collection

set of numbers. Although the so-called Pascal triangle of binomial coefficients appeared first in a Chinese manuscript in 1261, it was Blaise Pascal (1623–1662) who first systematically studied the binomial coefficients and their relation to counting and games of chance. His treatise, *Traité du Triangle Arithmétique*, appeared posthumously in 1665. This work is also historically interesting because it contains one of the first careful formulations of the principle of mathematical induction.

4.1. THE BINOMIAL COEFFICIENTS

If n is a nonnegative integer, then the expression $(1 + x)^n$ can be expanded by multiplying and collecting like powers of x to yield a polynomial in x of degree n. For example,

$$(1 + x)^0 = 1,$$
$$(1 + x)^1 = 1 + x,$$
$$(1 + x)^2 = 1 + 2x + x^2,$$
$$(1 + x)^3 = 1 + 3x + 3x^2 + x^3,$$
$$(1 + x)^4 = 1 + 4x + 6x^2 + 4x^3 + x^4.$$

It is clear that the coefficient of x^n in the expansion of $(1 + x)^n$ is 1 and that the constant term (that is, the coefficient of x^0) is 1. This leaves the problem of determining the coefficients of x^k for $k = 1, 2, \cdots, n - 1$. It is traditional to denote the coefficient of x^k in the expansion of $(1 + x)^n$ by $\binom{n}{k}$ and to regard this as the definition of the symbol $\binom{n}{k}$. The numbers $\binom{n}{k}$ are called the *binomial coefficients*. Adopting this notation, we have

$$(1 + x)^n = \sum_{k=0}^{n} \binom{n}{k} x^k,$$

where the terms are arranged in increasing powers of x. Thus, $\binom{n}{0} = \binom{n}{n} = 1$ since these numbers are the constant term and the coefficient of x^n, respectively. Note that $\binom{0}{0} = 1$.

We shall now examine some of the basic properties of the binomial coefficients. For small values of n, the numbers $\binom{n}{k}$ can be obtained by expanding $(1 + x)^n$ directly and noting the coefficients of x^k. For example, from the expansion of $(1 + x)^3$ given above, we see that

$$\binom{3}{0} = 1, \quad \binom{3}{1} = 3, \quad \binom{3}{2} = 3, \quad \binom{3}{3} = 1.$$

If we arrange the binomial coefficients for the cases $n = 0, 1, 2, 3, 4$ in a triangular array we get the beginning of what is known as Pascal's triangle.

4.1. THE BINOMIAL COEFFICIENTS

$n = 0$:   1
$n = 1$:   1   1
$n = 2$:   1   2   1
$n = 3$:   1   3   3   1
$n = 4$:   1   4   6   4   1

By definition, *Pascal's triangle* is an infinite triangular array of integers such that:

(a) The edge numbers are all 1.
(b) Each interior number is the sum of the two numbers diagonally above it.

According to these defining conditions, the $n = 5$ row of Pascal's triangle is

$n = 5$:   1   5   10   10   5   1

since the edge numbers must be 1 by (a) and since $5 = 1 + 4$, $10 = 4 + 6$ by (b).

We now have two collections of numbers: the set of binomial coefficients and the set of numbers in Pascal's triangle. Although these two sets are defined in different ways, we have seen that they coincide for $n = 0, 1, 2, 3, 4$. Moreover, a direct calculation of the coefficients in the expansion of $(1 + x)^5$ shows that this coincidence persists for $n = 5$. On the basis of these observations, we are led to conjecture that for every nonnegative integer n, the nth row of Pascal's triangle consists of the binomial coefficients $\binom{n}{k}$ for $k = 0, 1, \cdots, n$.

In order to prove this conjecture, we must prove that the coefficient $\binom{n+1}{k}$ of x^k in the expansion of $(1 + x)^{n+1}$ is the sum of the coefficient $\binom{n}{k-1}$ of x^{k-1} and the coefficient $\binom{n}{k}$ of x^k in the expansion of $(1 + x)^n$. That is, we must show that for each positive integer n,

$$\binom{n+1}{k} = \binom{n}{k-1} + \binom{n}{k},$$

for $k = 1, \cdots, n$. Since $(1 + x)^{n+1}$ is the product of $(1 + x)^n$ and $(1 + x)$, let us multiply

$$(1 + x)^n = \sum_{k=0}^{n} \binom{n}{k} x^k$$

by $(1 + x)$ and collect terms to see if this relation is valid.

$$\begin{aligned}(1 + x)^{n+1} &= (1 + x)(1 + x)^n \\ &= (1 + x) \sum_{k=0}^{n} \binom{n}{k} x^k \\ &= \sum_{k=0}^{n} \binom{n}{k} x^k + \sum_{j=0}^{n} \binom{n}{j} x^{j+1} \\ &= \sum_{k=0}^{n} \binom{n}{k} x^k + \sum_{k=1}^{n+1} \binom{n}{k-1} x^k \\ &= 1 + \sum_{k=1}^{n} \left[\binom{n}{k} + \binom{n}{k-1} \right] x^k + x^{n+1}.\end{aligned}$$

By definition of $\binom{n+1}{k}$, we also have

$$(1 + x)^{n+1} = \sum_{k=0}^{n+1} \binom{n+1}{k} x^k.$$

Comparing these expressions, we see that the conjectured relation is valid and we have the following proposition.

Proposition 4.1. The binomial coefficients have the following properties for each nonnegative integer n:

$$\binom{n}{0} = \binom{n}{n} = 1, \quad \text{(Edge condition.)} \tag{1}$$

$$\binom{n+1}{k} = \binom{n}{k-1} + \binom{n}{k}, \quad k = 1, 2, \cdots, n. \tag{2}$$
(Recursion formula.)

Although the recursion formula and edge condition uniquely determine the binomial coefficients and can be used to calculate any $\binom{n}{k}$, it is desirable to have a nonrecursive expression for these numbers. Such an expression, due to Pascal, is given in the next proposition. As we shall see in the next section, this formula arises quite naturally if the calculation of the coefficient of x^k in the expansion of $(1 + x)^n$ is viewed as a counting problem.

Proposition 4.2. For each nonnegative integer n,

$$\binom{n}{k} = \frac{n!}{k!(n-k)!}, \quad \text{for } k = 0, 1, \cdots, n. \tag{3}$$

PROOF. We shall prove this result by showing that the stated expression satisfies the edge condition and recursion formula given in Proposition 4.1. The edge condition (1) is easily verified for any n by direct substitution of $k = 0$ and $k = n$. (As is customary, we define $0! = 1$ for notational convenience.)

Since

$$\frac{n!}{k!(n-k)!} = \frac{n(n-1)\cdots(n-k+1)}{k!}, \quad \text{for } k = 1, 2, \cdots, n, \tag{4}$$

the calculation

$$\frac{n(n-1)\cdots[n-(k-1)+1]}{(k-1)!} + \frac{n(n-1)\cdots(n-k+1)}{k!}$$

$$= \frac{k\cdot n(n-1)\cdots(n-k+2) + n(n-1)\cdots(n-k+1)}{k!}$$

$$= \frac{(k+n-k+1)n(n-1)\cdots(n-k+2)}{k!}$$

$$= \frac{(n+1)n\cdots(n-k+2)}{k!}$$

shows that the recursion formula (2) is satisfied for $n+1$ and $k = 1, 2, \cdots, n$. Hence the formula is proved.

In view of the symmetry of k and $n-k$ in the denominator of (3), we have the following.

Corollary. $\binom{n}{k} = \binom{n}{n-k}$ for $k = 0, 1, \cdots, n$.

The above discussion yields, in a somewhat roundabout way, a proof of the Binomial Theorem. This theorem is frequently stated for a binomial $a + b$ rather than $1 + x$, but the substitution $x = b/a$ yields this version. In fact,

$$\left(1 + \frac{b}{a}\right)^n = \frac{(a+b)^n}{a^n}$$

so that

$$(a+b)^n = a^n \sum_{k=0}^{n} \binom{n}{k} \left(\frac{b}{a}\right)^k$$

$$= \sum_{k=0}^{n} \binom{n}{k} a^{n-k} b^k.$$

In this expression, the powers of a decrease from n to 0 while the powers of b increase from 0 to n and $\binom{n}{k}$ is the coefficient of $a^{n-k}b^k$. In this form, the binomial theorem is stated as follows.

Proposition 4.3. *(The Binomial Theorem)* If a and b are numbers, then

$$(a+b)^n = \sum_{k=0}^{n} \binom{n}{k} a^{n-k} b^k, \tag{5}$$

where

$$\binom{n}{k} = \frac{n!}{k!(n-k)!}, \quad \text{for } k = 0, 1, 2, \cdots, n.$$

There are many relationships between binomial coefficients that are interesting and useful. Two identities that follow immediately from the Binomial Theorem are stated in the following corollary. Others can be found in the exercises.

Corollary. For each positive integer n,

$$\sum_{k=0}^{n} \binom{n}{k} = 2^n, \tag{6}$$

$$\sum_{k=0}^{n} (-1)^k \binom{n}{k} = 0. \tag{7}$$

PROOF. If we set $a = 1$ and $b = 1$ in (5), we obtain

$$2^n = (1+1)^n = \sum_{k=0}^{n} \binom{n}{k} 1^{n-k} 1^k = \sum_{k=0}^{n} \binom{n}{k}.$$

This proves (6). Similarly, if we take $a = 1$ and $b = -1$ in (5), we obtain (7).

4.1. THE BINOMIAL COEFFICIENTS

The binomial coefficients can be interpreted in several different ways and the fact that different beginnings result in the same set of numbers enhances interest in the Binomial Theorem. One approach is in the nature of a geometric puzzle.

Consider the network of Figure 4.1, which can be regarded as the map of a section of a city where the lines represent streets of the city.

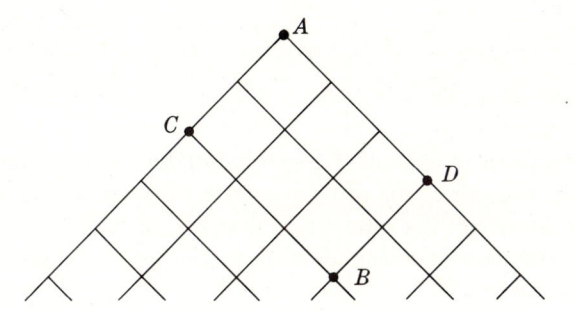

Figure 4.1

Suppose we wish to move from the point A to a point B along the streets in the shortest possible manner. Such a route will be a zig-zag path from A to B that always moves down the map since an upward movement from any point will result in a route that is longer than one that always proceeds in a downward direction. There are several such shortest routes. For example, in Figure 4.1 we could go from A to C and then from C to B along straight paths, or we could go from A to D and then from D to B. Note that any shortest path from A to B must lie within the rectangle determined by A, B, C, and D. Further reflection shows that any downward zig-zag route from A to B must have the same length. In Figure 4.1, each shortest path from A to B has a length of five blocks.

These considerations suggest the following problem: Given a point B of intersection in the network, how many different shortest routes are there from the apex A to B? The answer may not come immediately to mind. In a problem of this sort, it is best to consider easy special cases with the idea of discovering a pattern that may lead to the final solution. Let us regard the points as lying in rows and see what the answer is when we consider a point B lying in the first few rows. (See Figure 4.2.) If B lies in row 1, then we can reach it in one step and there is exactly one shortest path from A for each of the two possible positions for B. If B lies in row 2, there are two cases. If B is on the boundary of the network, there is again just one shortest path to B. However, if B is the middle point of row 2, then there are two shortest paths (for we can either zig and then zag, or first zag and then zig). In row 3, a little checking reveals that there

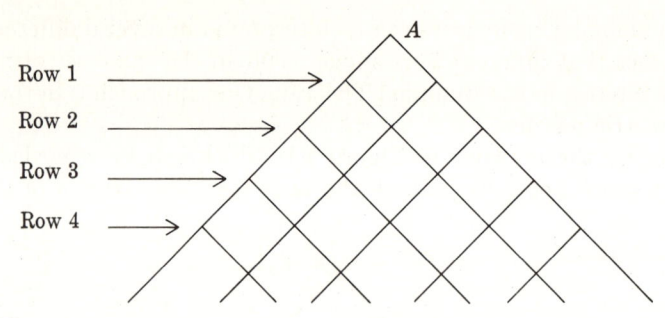

170 BINOMIAL COEFFICIENTS AND COUNTING TECHNIQUES

Row 1 ⟶
Row 2 ⟶
Row 3 ⟶
Row 4 ⟶

Figure 4.2

is one shortest path to each of the two points on the boundary and three shortest paths to each of the interior points. If we write our answers in a triangular array where the number of shortest paths from A to a point B is written in the same relative location as the point B being considered, we get the following.

$$
\begin{array}{ccccccc}
 & & & 1 & & & \\
 & & 1 & & 1 & & \\
 & 1 & & 2 & & 1 & \\
1 & & 3 & & 3 & & 1
\end{array}
$$

The resemblance to the Pascal triangle is striking. We now show that the Pascal triangle of binomial coefficients provides the solution to the shortest path problem.

If the point B lies on the edge of the map, then there is exactly one shortest path from A to B. Thus, the edge condition (1) for the Pascal triangle is satisfied. Now consider a point B at an interior intersection. We must show that the number of shortest paths from A to B satisfies the recursion formula (2) for Pascal's triangle. Any shortest path from A to B must pass through exactly one of the points B_1 or B_2 that lie diagonally above B. (See Figure 4.3.) If we consider a shortest path from A to B_1, then there is only one way to proceed to B while keeping the route of minimal length. Similarly, if we proceed from A to B_2 along a shortest path, there is again only one way to complete the path to B to get a short-

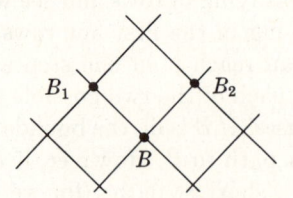

Figure 4.3

est path from A to B. Hence, we see that the number of shortest paths from A to B is equal to the number of shortest paths from A to B_1 plus the number of shortest paths from A to B_2. This is precisely the recursion relation (2) for Pascal's triangle. We therefore conclude that the solution to the shortest path problem is precisely the array of numbers in Pascal's triangle where the numbers and points have corresponding locations.

EXERCISES

4.1. Find the coefficient of x^3y^7 in the expansion of $(2x - y)^{10}$.

4.2. Find the constant term in the expansion of $(3x - 2/x)^8$.

4.3. Which terms in the expansion of $(a + b)^{87}$ have their coefficients equal to the coefficient of the 47th term?

4.4. Approximate the value of $(1.02)^7$ by using the first three terms of the expansion of $(1 + 0.02)^7$.

4.5. Prove the Binomial Theorem directly by using mathematical induction.

4.6. Show that if k is an integer satisfying $0 \leq k \leq \dfrac{n-1}{2}$, then
$$\binom{n}{k} \leq \binom{n}{k+1}.$$

4.7. Prove the identity
$$\binom{k}{k} + \binom{k+1}{k} + \cdots + \binom{n}{k} = \sum_{m=k}^{n} \binom{m}{k} = \binom{n+1}{k+1}$$
for $0 \leq k \leq n$ by mathematical induction. Describe the location of the terms in this identity in Pascal's triangle.

4.8. (a) Use the identity of Exercise 4.7 for $k = 1$ to derive the formula
$$1 + 2 + \cdots + n = \frac{n(n+1)}{2}.$$

(b) Using the fact that $2\binom{m}{2} + m = m^2$ for $m \geq 2$, derive the formula
$$1^2 + 2^2 + \cdots + n^2 = \frac{n(n+1)(2n+1)}{6}.$$

4.9. (a) Show that $k\binom{n}{k} = n\binom{n-1}{k-1}$.

(b) Use (a) to show that $\sum_{k=1}^{n} k \binom{n}{k} = n2^{n-1}$, for $n \geq 1$.

(c) Show that $\sum_{k=1}^{n} (-1)^k k \binom{n}{k} = 0$, for $n > 1$.

4.10. What are the values of the following:

(a) $\binom{n}{0} + \binom{n}{2} + \binom{n}{4} + \binom{n}{6} + \cdots$

(b) $\binom{n}{1} + \binom{n}{3} + \binom{n}{5} + \binom{n}{7} + \cdots$.

The sums terminate when the lower number becomes larger than the upper number. [*Hint:* Use the corollary to Proposition 4.3.]

4.11. In how many ways can the word *ALGEBRA* be spelled by going from the top *A* in the array shown to the bottom *A* by passing from each letter to a letter lying diagonally beneath it? How many ways are there if we further insist that only the middle *E* be used in the spelling?

```
            A
          L   L
        G   G   G
      E   E   E
        B   B   B
          R   R
            A
```

4.2. BINOMIAL COEFFICIENTS AND COUNTING PROBLEMS

An unsatisfying feature of the discussion to this point is the fact that the formula

$$\binom{n}{k} = \frac{n!}{k!(n-k)!}$$

given in Section 4.1 for the binomial coefficients was simply verified to be correct without any indication of how it was obtained. We shall now see how the relation of the binomial coefficients to certain counting problems provides a nice way to derive this formula. It is probably this interpretation that led Pascal to discover the formula in the first place.

The art of counting the number of elements that belong to a given set or that satisfy a prescribed condition has been of interest since ancient times. Today it is an important and flourishing subject of research known as combinatorial analysis. We present here some basic counting techniques that are related to the binomial coefficients. Counting problems will again be encountered in our study of probability theory in Chapter 5 and some of the methods discussed here will be quite useful there. Let us

4.2. BINOMIAL COEFFICIENTS AND COUNTING PROBLEMS

first formalize two elementary facts known as the addition principle and the multiplication principle of counting.

The *addition principle* of counting simply states that if S is a set of m elements, T is a set of n elements, and $S \cap T = \emptyset$, then the set $S \cup T$ has $m + n$ elements. For purposes of counting this is frequently stated in a less precise, but more suggestive manner, as follows. If one thing can be done in m ways and a second thing can be done in n ways, then it is possible to do either the first *or* second thing in $m + n$ ways. For example, if we can travel from a point A to a point B in 7 ways and if we can travel from A to a point C in 15 ways, then we can go from A to either B or C in 22 ways.

The *multiplication principle* of counting states that if S is a set of m elements and T is a set of n elements, then the Cartesian product $S \times T$ has mn elements. That is, if $S = \{a_1, \cdots, a_m\}$ and $T = \{b_1, \cdots, b_n\}$, then there are mn ordered pairs (a_i, b_j) where $i = 1, 2, \cdots, m$ and $j = 1, 2, \cdots, n$. We can phrase this less formally as follows. If we can do one thing in m ways and a second thing in n ways, then we can do the two things together in mn ways.

Of course, these principles extend to any finite number of sets. For example, if $S_1, \cdots, S_k$ are sets having $n_1, \cdots, n_k$ elements respectively, then $S_1 \times \cdots \times S_k$ has $n_1 \cdots n_k$ elements.

Example 4.1. In a certain state, license plates are made with six symbols on them. The first two symbols are letters from the alphabet and the last four symbols are digits from 0 to 9. Applying the multiplication principle of counting, we see that there are $26 \cdot 26 \cdot 10 \cdot 10 \cdot 10 \cdot 10 = 6{,}760{,}000$ possible license plates that can be made in this way. For a more populous state, it might be better if three letters and three digits were used, for it is possible to make 17,576,000 different license plates of this type.

If S is a set of n elements, then a selection of k elements of S in a particular order is called a *k-permutation* of S. In other words, a k-permutation of S is an ordered k-tuple where the terms in the k-tuple are distinct elements of S. For example, if $S = \{1,2,3,4,5\}$, then (4,2,1), (2,1,4) and (5,3,1) are all different 3-permutations of S. An n-permutation of a set S of n elements is simply called a *permutation* of S. A permutation of S can thus be viewed as a particular ordering or arrangement of the elements of S.

Proposition 4.4. If S is a set of n elements, then the number of k-permutations of S is $n(n-1) \cdots (n-k+1)$. In particular, there are $n!$ permutations of S.

PROOF. We must count the number of k-tuples consisting of k distinct elements of S. We count as follows. The first term of such a k-tuple can be chosen in n ways since S has n elements. Once the first term has been selected, there remain $n - 1$ elements from which the second term can be chosen. Similarly, if the first and second terms have been selected, then there are $n - 2$ ways of selecting the third term of the k-tuple. Since k terms are to be chosen, the last term of the k-tuple can be chosen in $n - k + 1$ ways. Thus, the total number of k-permutations of S is $n(n-1) \cdots (n-k+1)$. If $k = n$, this number is just $n(n-1) \cdots 2 \cdot 1 = n!$.

If S is a set of n elements, then a subset of S containing exactly k elements will be called a *k-subset* of S. (Another term frequently used in place of k-subset is that of combination, or combination of n objects taken k at a time, but this terminology will not be used here.) The notions of k-permutation and k-subset are quite different since the order in which elements are selected is ignored when dealing with k-subsets. For example, if $S = \{1,2,3,4,5\}$, then the 3-subsets $\{4,2,1\}$ and $\{2,1,4\}$ are identical while the 3-permutations $(4,2,1)$ and $(2,1,4)$ are distinct.

The number of k-subsets of a set of n elements will be denoted by $C(n,k)$. We may think of $C(n,k)$ as the number of different ways k objects can be chosen from a collection of n distinct objects.

If we would calculate the numbers $C(n,k)$ by direct counting for the cases $n = 0, 1, 2, 3, 4$ and $0 \leq k \leq n$, we would again obtain the beginning of Pascal's triangle. For example, if we listed all of the subsets of a set of three elements, we would find there are one 0-subset (that is, the empty set), three 1-subsets, three 2-subsets, and one 3-subset. This evidence would suggest that $C(n,k)$ is just the binomial coefficient $\binom{n}{k}$. However, to prove this assertion in general, we must show that the numbers $C(n,k)$ satisfy the edge condition and recursion relation determining the binomial coefficients. (See Proposition 4.1.)

Proposition 4.5. For any natural number n and any integer k satisfying $0 \leq k \leq n$,

$$C(n,k) = \binom{n}{k}.$$

PROOF. If S is a set of n elements, then there is only one subset containing no elements, namely the empty set, so that $C(n,0) = 1$. Also, since the only n-subset is S itself, we have $C(n,n) = 1$. Therefore, the edge condition for the binomial coefficients is satisfied by the numbers $C(n,k)$. We

4.2. BINOMIAL COEFFICIENTS AND COUNTING PROBLEMS

now wish to establish the recursion relation

$$C(n + 1, k) = C(n, k - 1) + C(n,k),$$

where $k = 1, 2, \cdots, n$. Consider a set $S = \{x_1, x_2, \cdots, x_{n+1}\}$ having $n + 1$ elements. In how many ways can we select a k-subset of S? If we fix the element x_{n+1}, then a k-subset either will contain x_{n+1} or it will not. Thus, there are two distinct types of k-subsets: (a) those containing x_{n+1} and (b) those not containing x_{n+1}. The number of k-subsets satisfying (a) is $C(n, k - 1)$ since $k - 1$ elements must be chosen from the remaining n elements to get a k-subset of S of this type. The number of k-subsets satisfying (b) is $C(n,k)$ since all k elements must be taken from the n elements different from x_{n+1}. Since a k-subset must satisfy either (a) or (b), we see from the addition principle of counting that the total number $C(n + 1, k)$ of k-subsets of S is $C(n, k - 1) + C(n,k)$. Thus, the recursion relation and edge condition are satisfied by the numbers $C(n,k)$ and we conclude that $C(n,k) = \binom{n}{k}$.

We can now derive the factorial formula for binomial coefficients in a very natural way by simply counting the number of k-permutations of a set of n elements in two different ways. More specifically, we know by Proposition 4.4 that there are $k!$ permutations of each set of k elements. That is, each k-subset of a set S of n elements can be ordered in $k!$ different ways. Therefore, since there are $\binom{n}{k}$ k-subsets of S and since there are $k!$ permutations of each k-subset of S, an application of the multiplication principle of counting shows that there are a total of $k!\binom{n}{k}$ k-permutations of S. On the other hand, we have also established in Proposition 4.4 that the number of k-permutations of S is $n(n - 1) \cdots (n - k + 1)$. Hence, we conclude that $k!\binom{n}{k} = n(n - 1) \cdots (n - k + 1)$, which yields the formula

$$\binom{n}{k} = \frac{n(n - 1) \cdots (n - k + 1)}{k!} = \frac{n!}{k!(n - k)!}.$$

The interpretation of $\binom{n}{k}$ as the number of k-subsets of an n-element set can actually be used to prove the Binomial Theorem as follows. The expression

$$(1 + x)^n = (1 + x)(1 + x) \cdots (1 + x)$$

is the product of n factors, each of which is $1 + x$. Note that the coeffi-

176 BINOMIAL COEFFICIENTS AND COUNTING TECHNIQUES

cient of x^k is nothing more than the number of ways that x can be chosen k times from the n factors. But this is precisely $\binom{n}{k}$, the number of ways k elements can be chosen from n elements. Hence, we have

$$(1 + x)^n = \sum_{k=0}^{n} \binom{n}{k} x^k.$$

We now consider some applications of the counting techniques developed thus far.

Example 4.2. An ordinary deck of playing cards consists of 52 cards, each card bearing a suit (hearts, diamonds, clubs, or spades) and a rank (ace through king). A multitude of counting problems can be raised in this context. We consider a few sample problems.

(a) If 2 cards are drawn from the deck, one after another, the result is an ordered pair of playing cards (that is, a 2-permutation of the 52-element set). There are $52 \cdot 51 = 2652$ different ordered pairs of cards. How many such pairs contain cards of the same suit? To answer this question, we note that the first card can be selected in 52 ways, but then only 12 of the remaining 51 cards will have the same suit as the first card. Hence, there are $52 \cdot 12 = 624$ such ordered pairs.

(b) Suppose now we neglect the order in which the two cards are drawn and simply draw 2 cards from the deck simultaneously (that is, a 2-subset). There are $\binom{52}{2} = 1326$ such unordered pairs of cards. We ask again: How many of these pairs contain cards of the same suit? We may use the result of part (a) by noting that since each 2-subset can be ordered in $2! = 2$ ways, the answer must be $\frac{624}{2} = 312$. Or we could work the problem directly by noting that there are four possible suits and $\binom{13}{2} = 78$ ways of selecting 2 cards from each suit. Hence, there are $4 \cdot 78 = 312$ ways of getting two cards of the same suit.

(c) A 5-subset of the set of 52 playing cards is called a *poker hand*. There are $\binom{52}{5} = 2{,}598{,}960$ different poker hands. The term "full house" is poker parlance for a poker hand that contains 3 cards of one rank and 2 cards of a second rank. Let us compute the number of full houses that are possible. Since there are 13 possible ranks, there are $\binom{13}{2} = 78$ ways of getting two different ranks needed in a full house. There are 4 cards of each rank. Therefore, there are $\binom{4}{3} = 4$ ways of obtaining 3 cards of one rank and $\binom{4}{2} = 6$ ways of getting 2 cards of another rank. Hence, there are $78 \cdot 4 \cdot 6 = 1872$ possible ways of getting a full house.

Example 4.3. The identity $\sum_{k=0}^{n} \binom{n}{k}^2 = \binom{2n}{n}$ can be obtained by a counting argument as follows. Consider a set B of $2n$ balls, n of which are red and n of which are white. We know that the number of n-subsets of B is $\binom{2n}{n}$ by Proposition 4.5. We can also count the number of n-subsets of B in another way. Each n-subset of B contains a certain number of red balls and a certain number of white balls. If an n-subset contains k red balls, then it necessarily contains $n - k$ white balls. The number of n-subsets of B consisting of k red balls and $n - k$ white balls is $\binom{n}{k} \cdot \binom{n}{n-k}$ since there are $\binom{n}{k}$ ways to select k red balls from the n red balls in B and $\binom{n}{n-k}$ ways to select $n - k$ white balls from the n white balls. Now an n-subset can have either 0 red balls, 1 red ball, 2 red balls, and so on. That is, k can vary from 0 to n. Thus, we see that the total number of n-subsets of B is given by $\sum_{k=0}^{n} \binom{n}{k}\binom{n}{n-k}$, which is equal to $\sum_{k=0}^{n} \binom{n}{k}^2$ since $\binom{n}{n-k} = \binom{n}{k}$. Hence, we get $\sum_{k=0}^{n} \binom{n}{k}^2 = \binom{2n}{n}$.

EXERCISES

4.12. A coin is tossed 10 times. How many different ways can exactly 4 heads be obtained? [Answer: 210.]

4.13. (a) A poker hand containing 4 cards of the same rank is called four-of-a-kind. How many such poker hands are there? [Answer: 624.]

(b) A poker hand containing 5 cards of the same suit is called a flush. How many such poker hands are there? [Answer: 5148.]

4.14. Show that if a set S has n elements then $P(S)$, the collection of all subsets of S, has 2^n elements. (See the corollary to Proposition 4.3.)

4.15. Let S be a set of 10 elements. Show that there are more subsets of S having 5 elements than there are subsets having any other given number of elements. Can you generalize? (See Exercise 4.6.)

4.16. A bridge club has 16 members, 9 women and 7 men. In how many ways can a foursome be chosen for a game of bridge? In how many ways can a foursome be chosen if it is to consist of 2 men and 2 women? [Answer: 1820; 756.]

4.17. Eight people are to be seated at a circular table (there is no first or last seat). How many seating arrangements are possible? Generalize to n people.

4.18. Derive the identity of Example 4.3 by computing the coefficient of x^n in the expansion of $(1 + x)^{2n}$.

4.19. Let n and m be natural numbers and let r be an integer satisfying $0 < r < n + m$. Use a counting argument to show that

$$\sum_{k=0}^{r} \binom{n}{k}\binom{m}{r-k} = \binom{n+m}{r}.$$

(We use the convention that $\binom{n}{k} = 0$ if $k > n$.)

4.3. MULTINOMIAL COEFFICIENTS

Let S be a set of n elements and let $n_1, n_2, \cdots, n_k$ be nonnegative integers such that $n_1 + n_2 + \cdots + n_k = n$. Then it is possible to partition S into k disjoint subsets $A_1, A_2, \cdots, A_k$ such that the ith subset A_i has n_i elements. In fact, this can be done in several ways and we shall let

$$\binom{n}{n_1, n_2, \cdots, n_k}$$

denote the total number of ways S can be partitioned into k disjoint subsets having $n_1, n_2, \cdots, n_k$ elements, respectively. Observe if $k = 2$, then this reduces to an ordinary binomial coefficient since the number of ways S can be partitioned into two disjoint subsets having n_1 and $n_2 = n - n_1$ elements respectively is clearly the same as the number of ways of selecting an n_1-subset of S. That is,

$$\binom{n}{n_1, n_2} = \binom{n}{n_1} = \frac{n!}{n_1! n_2!}.$$

Let us now examine the case $k = 3$. Imagine that we have 3 boxes and that we wish to put n_1 elements of S into the first box, n_2 elements into the second box, and n_3 elements into the third box where $n_1 + n_2 + n_3 = n$ is the total number of elements of S. The n_1 elements of S to be placed in the first box can be selected from S in $\binom{n}{n_1}$ ways. Once this is done, the n_2 elements placed in the second box can be chosen from the remaining $n - n_1$ elements of S in $\binom{n - n_1}{n_2}$ ways. The $n_3 = n - n_1 - n_2$ elements left over all go into the third box. Hence, there are

$$\binom{n}{n_1}\binom{n - n_1}{n_2} = \frac{n!}{n_1!(n - n_1)!} \cdot \frac{(n - n_1)!}{n_2!(n - n_1 - n_2)!}$$
$$= \frac{n!}{n_1!n_2!n_3!}$$

ways a set of n elements can be partitioned into three disjoint subsets of n_1, n_2, n_3 elements respectively. That is,

$$\binom{n}{n_1,n_2,n_3} = \frac{n!}{n_1!n_2!n_3!}.$$

The above argument indicates how the general result

$$\binom{n}{n_1,n_2, \cdots ,n_k} = \frac{n!}{n_1!n_2! \cdots n_k!}$$

can be proved and we leave it to the reader to supply the necessary details.

The above numbers are known as *multinomial coefficients* for the following reason. Consider the trinomial $a + b + c$ raised to the nth power; that is,

$$(a + b + c)^n = \underbrace{(a + b + c)(a + b + c) \cdots (a + b + c)}_{n \text{ factors}}.$$

Each term resulting from the multiplication of these n factors will have the form $a^r b^s c^t$ where the sum of the exponents is n, that is, $r + s + t = n$. Such a term is obtained by selecting a from r of the factors, b from s of the factors and c from t of the factors. (Of course, it is possible that one or two of the numbers r, s, t might be zero.) The coefficient of $a^r b^s c^t$ in the expansion of $(a + b + c)^n$ is just the number of different ways this term can be obtained, and this is just the number of ways three disjoint subsets of r, s, t elements respectively can be selected from a set of n elements where $r + s + t = n$. Hence, the coefficient of $a^r b^s c^t$ is $\binom{n}{r,s,t}$. This gives us the trinomial expansion

$$(a + b + c)^n = \sum \frac{n!}{r!s!t!} a^r b^s c^t,$$

where the sum is taken over all triples (r,s,t) of nonnegative integers such that $r + s + t = n$.

180 BINOMIAL COEFFICIENTS AND COUNTING TECHNIQUES

Similarly, it can be shown that the general multinomial expansion has the form

$$(a_1 + a_2 + \cdots + a_k)^n = \sum \frac{n!}{n_1! n_2! \cdots n_k!} a_1^{n_1} a_2^{n_2} \cdots a_k^{n_k},$$

where the sum is taken over all k-tuples $(n_1, n_2, \cdots, n_k)$ of nonnegative integers such that $n_1 + n_2 + \cdots + n_k = n$. For example, the coefficient of $wx^3 y^6 z^2$ in the expression $(w + x + y + z)^{12}$ is

$$\frac{12!}{1!3!6!2!} = 55{,}440.$$

The following examples illustrate some applications of the formula obtained above for the multinomial coefficients.

Example 4.4. Suppose 4 people play poker and each person is dealt a 5-card poker hand. Compute the number of distinct sets of 4 poker hands. A poker hand contains 5 cards and there are 4 hands in each deal. Consequently, since a poker deck consists of 52 cards, there will be $52 - 4 \cdot 5 = 32$ cards remaining after each deal. Therefore, the number of distinct sets of 4 poker hands is

$$\frac{52!}{(5!)^4 32!}$$

which is approximately 148×10^{21}.

Example 4.5. Show that $(n!)!$ is an integer multiple of $(n!)^{(n-1)!}$.

Let S be a set having $n!$ elements. Since $n!$ equals n times $(n-1)!$, it is possible to think of S as being partitioned into $(n-1)!$ disjoint subsets of n elements each. This partitioning can be achieved in

$$\binom{n!}{\underbrace{n, n, \cdots, n}_{(n-1)! \text{ terms}}} = \frac{(n!)!}{(n!)^{(n-1)!}}$$

ways, and this number is necessarily an integer. Hence, $(n!)!$ must be a multiple of $(n!)^{(n-1)!}$.

Up to now we have been concerned with selecting and arranging objects that were always distinguishable from one another; that is, we could tell one object from another during the selection or arrangement process. Let us briefly consider how the situation changes when we deal with objects that are not always distinguishable from one another. As an example, consider the use of Morse code in sending words over a telegraph

4.3. MULTINOMIAL COEFFICIENTS

line. In this code, each letter is represented by an arrangement of dots and dashes in a row. Consider the following question: How many different letters can be sent if each letter is to be represented by exactly 2 dashes and 1 dot? That is, we wish to know how many different arrangements of 2 dashes and 1 dot there are. Evidently, $\cdot - -$, $- \cdot -$, and $- - \cdot$ are the only distinct arrangements of this sort. Note that this is *not* the number of ways the elements of a set having three elements can be arranged (which is 3! = 6), but rather it is the number of ways of selecting two elements from a set of three elements $\left(\text{that is, } \binom{3}{2} = 3\right)$. The reason, of course, is that the two dashes are indistinguishable from one another, so we do not have a set of three distinct symbols. What we do have are three *positions* and what we are concerned with is how to place the three symbols into these three positions to obtain distinct arrangements. Since two of the symbols are indistinguishable, we see the number of distinct arrangements is exactly the number of ways we can choose two positions for these indistinguishable symbols out of the three positions available, that is, $\binom{3}{2} = 3$.

More generally, suppose we have a collection of n objects such that there are n_1 objects of one type, n_2 objects of a second type, $\cdots$, n_k objects of a kth type, where $n_1 + n_2 + \cdots + n_k = n$. It is assumed that all objects of one particular type are indistinguishable from one another. The question posed is: How many different arrangements of these objects are there? The key word here is "different." As when dealing with dots and dashes, if only the objects of some particular type are moved around in a given arrangement, then the resulting new arrangement will have the same appearance as the original arrangement, and thus they cannot be regarded as different. The answer to the question is obtained by considering the n positions to be filled by the n objects. Clearly, two arrangements will be different if and only if there is at least one position such that the objects occupying that position in the two arrangements are of different types. Hence, there are as many different arrangements of the objects as there are ways of partitioning the set of n positions into k different subsets having $n_1, n_2, \cdots, n_k$ elements, respectively. That is, the number of different ways our set of n objects can be arranged is

$$\binom{n}{n_1, n_2, \cdots, n_k} = \frac{n!}{n_1! n_2! \cdots n_k!}.$$

Example 4.6. Let us consider a "word" to be an arrangement of letters from the alphabet. Whether or not a "word" can be located in a dictionary is irrelevant for our purpose. With this agreement, let us determine how many different words can be spelled by using all the letters found in ILLINOIS. We see there are 8 letters, but there are 3 I's, 2 L's, and 1 each of N, O, and S. Thus, there are

$$\binom{8}{3,2,1,1,1} = \frac{8!}{3!2!} = 3360$$

different words that can be spelled using these letters.

EXERCISES

4.20. What is the coefficient of $a^3b^4c^2$ in the expansion of $(a + b + c)^9$?

4.21. What is the coefficient of x^2y^3 in the expansion of $(2x - y + 4)^9$?

4.22. What is the sum of all numbers of the form

$$\frac{9!}{a!b!c!},$$

where a, b, c are nonnegative integers satisfying $a + b + c = 9$?
[Answer: 3^9.]

4.23. Show that $(n^2)!$ is an integer multiple of $(n!)^n$.

4.24. Show that there are $\binom{n+1}{k}$ ways of arranging n dots and k dashes, $n \geq k$, in a row so that no 2 dashes are adjacent.

4.25. How many different words can be spelled using all the letters of MISSISSIPPI? [Answer: 34,650.]

4.26. Suppose there are 4 dashes and 3 dots.
 (a) In how many different ways can the 7 symbols be arranged in a row?
 (b) Suppose we first choose any 4 of the symbols and then arrange them in a row. Show that there are 15 different arrangements obtainable in this way.

4.4. THE PRINCIPLE OF INCLUSION-EXCLUSION

In this section, we shall discuss an interesting counting method frequently referred to as the Principle of Inclusion-Exclusion. In what follows, the number of elements in a finite set S will be denoted by $N(S)$.

As a prelude to our general discussion, let us consider the following simple problem. If a set of playing cards consists of all of the spades and aces in a standard deck, how many cards does this set contain? There are 13 spades and 4 aces; however, the correct answer is not $13 + 4 = 17$ since this would count the ace of spades twice. Of course, the correct answer is $13 + 4 - 1 = 16$.

4.4. THE PRINCIPLE OF INCLUSION-EXCLUSION

More generally, if S_1 and S_2 are finite sets, then the number of elements of $S_1 \cup S_2$ is given by

$$N(S_1 \cup S_2) = N(S_1) + N(S_2) - N(S_1 \cap S_2).$$

The subtraction of the term $N(S_1 \cap S_2)$ is required in order to avoid counting each element of $S_1 \cap S_2$ twice.

For three finite sets S_1, S_2, S_3, the problem of counting the number of elements in the set $S_1 \cup S_2 \cup S_3$ is just slightly more complicated. In this case, an element x of $S_1 \cup S_2 \cup S_3$ may be a member of: (1) exactly one of the sets, (2) exactly two of the three sets, or (3) all three sets. These three possible situations are illustrated in the Venn diagram, Figure 4.4. The correct formula for $N(S_1 \cup S_2 \cup S_3)$ that is suggested by

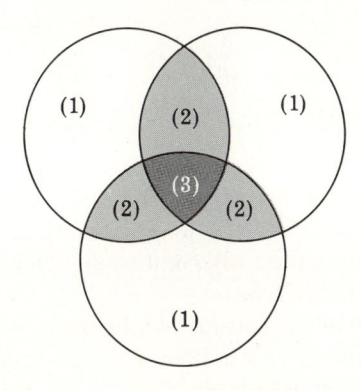

Figure 4.4

this Venn diagram is

$$N(S_1 \cup S_2 \cup S_3) = N(S_1) + N(S_2) + N(S_3) - N(S_1 \cap S_2) \\ - N(S_1 \cap S_3) - N(S_2 \cap S_3) + N(S_1 \cap S_2 \cap S_3).$$

Rather than prove this formula for three sets now, we shall discuss an example and then proceed to establish the general formula for n sets.

Example 4.7. Find the number of integers from 1 to 100 that are multiples of either 2, 3, or 5.

If we denote the sets of integers from 1 to 100 that are multiples of 2, 3, 5 by M_2, M_3, M_5, respectively, then we wish to compute $N(M_2 \cup M_3 \cup M_5)$. Clearly, $N(M_2) = \frac{100}{2} = 50$ and $N(M_5) = \frac{100}{5} = 20$. Also, since $3 \cdot 33 = 99$ is the largest multiple of 3 smaller than 100, we have $N(M_3) = 33$. The set $M_2 \cap M_3$ consists of all integers from 1 to 100 that are multiples of 6 so that $N(M_2 \cap M_3) = 16$. Similarly, we see that $N(M_2 \cap M_5) = 10$, $N(M_3 \cap M_5) = 6$, and $N(M_2 \cap M_3 \cap M_5) = 3$.

184   BINOMIAL COEFFICIENTS AND COUNTING TECHNIQUES

Therefore, applying the above formula, we obtain

$$N(M_2 \cup M_3 \cup M_5) = 50 + 20 + 33 - 16 - 10 - 6 + 3$$
$$= 74.$$

Note that this implies that the number of integers from 1 to 100 that are *not* multiples of either 2, 3, or 5 is $100 - 74 = 26$.

In general, if $S_1, S_2, \cdots, S_n$ are finite sets, then the number of elements belonging to at least one of these sets is given by

$N(S_1 \cup \cdots \cup S_n) =$
(1) $\qquad N(S_1) + N(S_2) + \cdots + N(S_n)$
(2) $\qquad - N(S_1 \cap S_2) - N(S_1 \cap S_3) - \cdots - N(S_{n-1} \cap S_n)$
(3) $\qquad + N(S_1 \cap S_2 \cap S_3) + \cdots + N(S_{n-2} \cap S_{n-1} \cap S_n)$

$\qquad \qquad \qquad \cdots$

(n) $\qquad + (-1)^{n+1} N(S_1 \cap \cdots \cap S_n).$

That is, we *add* the number of elements belonging to the individual sets (line (1)), *subtract* the number of elements belonging to all intersections of pairs of sets (line (2)), *add* the number of elements belonging to all intersections of three sets (line (3)), and so on. This formula is known as the *Principle of Inclusion-Exclusion*.

To verify this formula, consider an element $x \in S_1 \cup \cdots \cup S_n$. There are n possibilities: (1) x can belong to exactly one of the sets, (2) x can belong to exactly two of the sets, $\cdots$, (n) x can belong to all n sets. We must show that in each of these cases, the element x is counted only once when all the numbers in the above formula are added together.

Suppose x is contained in exactly m of the sets where $1 \leq m \leq n$. Then x is not counted at all in lines $(m + 1)$ through (n) above since x is not contained in any intersection involving more than m of the sets. In lines (1) through (m) of the formula, the element x will be counted as follows:

$$\binom{m}{1} = m \quad \text{times in line (1)}$$

$$-\binom{m}{2} \quad \text{times in line (2)}$$

$$+\binom{m}{3} \quad \text{times in line (3)}$$

$$\vdots \qquad \qquad \cdots$$

$$(-1)^{m+1}\binom{m}{m} \quad \text{times in line } (m).$$

4.4. THE PRINCIPLE OF INCLUSION-EXCLUSION

This is because a term in line (k) involving an intersection of k sets will contribute to the count if and only if the k sets selected all are taken from the m sets to which x belongs. Since there are $\binom{m}{k}$ ways of selecting k sets from the m sets containing x, we see that line (k) contributes $(-1)^{k+1}\binom{m}{k}$ to the count. Thus, the total number of times x is counted is

$$\sum_{k=1}^{m}(-1)^{k+1}\binom{m}{k} = \binom{m}{1} - \binom{m}{2} + \binom{m}{3} - \cdots + (-1)^{m+1}\binom{m}{m}.$$

This is easily seen to equal 1 since the Binomial Theorem gives us

$$0 = (1-1)^m = \binom{m}{0} - \binom{m}{1} + \binom{m}{2} - \cdots + (-1)^m\binom{m}{m}$$

and hence

$$\binom{m}{1} - \binom{m}{2} + \cdots + (-1)^{m+1}\binom{m}{m} = \binom{m}{0} = 1.$$

We conclude that each element of $S_1 \cup \cdots \cup S_n$ is counted precisely once by the above formula and the Principle of Inclusion-Exclusion is established.

We conclude this section with an application that will be used later in connection with probability theory (see Example 5.6).

Example 4.8. Suppose that $(x_1, x_2, \cdots, x_n)$ is a permutation of the set $\{1, 2, \cdots, n\}$ of integers from 1 to n. We will say that $(x_1, x_2, \cdots, x_n)$ has a *match* at the ith component if $x_i = i$. If $(x_1, x_2, \cdots, x_n)$ does not have a match at any component, we will call this permutation a *derangement*. For example, if $n = 4$, then $(3,2,1,4)$ has a match at the second and fourth components, while $(3,1,4,2)$ is a derangement.

We know that there are $n!$ permutations of $\{1, 2, \cdots, n\}$. We shall now determine how many of these permutations are derangements. Let S_i denote the set of permutations of the set of integers from 1 to n that have match at the ith component (and perhaps at other components). Then $S_1 \cup \cdots \cup S_n$ is the set of permutations that have at least one match and the number of derangements is $n! - N(S_1 \cup \cdots \cup S_n)$. Thus, we need to compute $N(S_1 \cup \cdots \cup S_n)$.

A permutation $(x_1, x_2, \cdots, x_n)$ is in S_i if and only if $x_i = i$ and $(x_1, \cdots, x_{i-1}, x_{i+1}, \cdots, x_n)$ is a permutation of the remaining $n-1$ integers $1, \cdots, i-1, i+1, \cdots, n$. Since there are $(n-1)!$ permutations of a set of $n-1$ elements, we see that $N(S_i) = (n-1)!$ for each

$i = 1, 2, \cdots, n$. Similarly, permutations in $S_{i_1} \cap S_{i_2}$, where $i_1 \neq i_2$, have two components fixed while the remaining $n - 2$ components are arbitrary. Hence, $N(S_{i_1} \cap S_{i_2}) = (n - 2)!$ In general, if $i_1, \cdots, i_k$ are k distinct integers from 1 to n, then $N(S_{i_1} \cap \cdots \cap S_{i_k}) = (n - k)!$

In line (k) of the above formula for $N(S_1 \cup \cdots \cup S_n)$, there are $\binom{n}{k}$ terms since there are $\binom{n}{k}$ ways of choosing k sets from the n sets $S_1, \cdots, S_n$. Furthermore, in this case, each term in line (k) has the same value as found in the previous paragraph. Therefore, the Principle of Inclusion-Exclusion gives us

$$N(S_1 \cup \cdots \cup S_n) = \binom{n}{1}(n-1)! - \binom{n}{2}(n-2)!$$
$$+ \cdots + (-1)^{n+1}\binom{n}{n}.$$

Since

$$\binom{n}{k}(n-k)! = \frac{n!}{k!},$$

for $k = 1, 2, \cdots, n$, this can be written as

$$N(S_1 \cup \cdots \cup S_n) = n!\left(1 - \frac{1}{2!} + \frac{1}{3!} - \cdots + (-1)^{n+1}\frac{1}{n!}\right).$$

Hence, the number of derangements of the set $\{1, 2, \cdots, n\}$ is

$$n! - N(S_1 \cup \cdots \cup S_n) = n!\left(\frac{1}{2!} - \frac{1}{3!} + \cdots + (-1)^n\frac{1}{n!}\right).$$

EXERCISES

4.27. Find the number of integers from 1 to 105 that are relatively prime to 105 (that is, that have no prime factors in common with $105 = 3 \cdot 5 \cdot 7$). [Answer: 48.]

4.28. How many natural numbers smaller than or equal to 1,000,000 are either perfect squares or perfect cubes? [Answer: 1090.]

4.29. Find the number of integers from 1 to 500 that are multiples of either 2, 3, or 5, but not multiples of 7. [Answer: 314.]

4.30. In a group of 100 students taking three courses, none received a grade lower than C. Also, exactly 60 received one or more A's, exactly 70 received one or more B's, and exactly 80 received one or more C's. What is the least number of students that could have received one A, one B, and one C? [Answer: 10.]

4.31. Suppose that 5 letters are written and 5 envelopes addressed, one for each letter. Suppose that the letters are then placed in the envelopes in a random fashion. In how many of the possible arrangements of letters and envelopes will at least 1 letter be placed in the proper envelope? [Answer: 76.]

4.32. How many permutations of the integers from 1 through n have exactly k matches?

REMARKS AND REFERENCES

The Binomial Theorem is a standard topic in the latter part of most secondary school algebra programs. The use of this result to compute the coefficients in the expansion of a binomial raised to a positive integral exponent is usually found to be easy and enjoyable by most algebra students. However, the relevance of this result to the secondary school mathematics program certainly goes well beyond its utility for expanding such binomial expressions. As we have indicated in this chapter, the binomial coefficients can be considered from the following viewpoints:

(1) As coefficients of the terms $a^k b^{n-k}$ in the expansion of $(a + b)^n$.

(2) As entries in Pascal's triangle; that is, the set of numbers determined by the edge condition and recursion formula of Proposition 4.1.

(3) As the numbers defined by $\binom{n}{k} = \dfrac{n!}{k!(n-k)!}$.

(4) As the number of shortest paths between the apex A and any other point B in the network of Figure 4.1.

(5) As the number of k-subsets of a set of n elements.

(6) As the number of distinct permutations of n objects, k of one type and $n - k$ of another, where the objects of each type are regarded as indistinguishable.

Any one of these viewpoints can be used as a starting point in the classroom discussion of the binomial coefficients. For example, one could begin the discussion with the problem of counting shortest paths, a puzzle that most students find quite interesting. This leads quite naturally to the Pascal triangle with its edge condition and recursion formula. It can then be observed that a second counting problem, namely, counting the number of subsets of a given finite set, leads to the same collection of numbers. The factorial formula can then be derived by contrasting ordered and unordered subsets, and it is but a short step to the proof of the Binomial Theorem by means of a counting argument.

Counting problems are worthy of considerable attention in the secondary school. Discussions of the Principle of Inclusion-Exclusion and the multinomial coefficients are accessible and stimulating topics at this level, but time pressures do not always permit their inclusion. However, they do provide excellent enrichment material and better students might be interested in these and related topics as subjects for special projects. For example, the discovery of a three-dimensional Pascal pyramid for trinomial coefficients analogous to the Pascal triangle for binomial coefficients is an excellent project of this type.

A natural culmination for the discussion of counting techniques is the study of probability measures on finite sample spaces. This subject is discussed in some detail in the next chapter. Most modern text series used in the secondary schools include a unit on probability, and students at this level find the subject to be particularly exciting.

An excellent treatment of counting techniques at an elementary level can be found in the paperback book by Niven [2]. A more sophisticated treatment of this subject including a great variety of applications is given by Liu [1]. The shortest path problem and other related problems are treated very nicely by Polya [3].

1. Liu, C. L., *Introduction to Combinatorial Mathematics*. New York: McGraw-Hill Book Company, Inc., 1968.
2. Niven, Ivan, *Mathematics of Choice: How to Count without Counting*. New York: Random House, 1965.
3. Polya, George, *Mathematical Discovery*. Vol. I, New York: John Wiley & Sons, Inc., 1962.

Chapter 5

PROBABILITY THEORY

Although gamblers and other devotees of games of chance have been interested in odds and probabilities of outcomes since ancient times, mathematical historians usually give 1654 as the year in which the theory of probability was born. This was the year in which a professional gambler, the Chevalier de Méré, posed several problems involving dice games to the mathematician, Blaise Pascal. This led to a correspondence between Pascal and Pierre de Fermat in which they discussed the basic ideas of probability and developed techniques for solving dice problems and problems related to other games of chance.

It should be pointed out that they were not the first to look into the subject. The colorful Gerolamo Cardano, who was a physician, gambler, astrologer, and mathematician among other things, wrote his *Liber de Ludo Aleae* (The Book on Games of Chance) in 1520. Galileo also dabbled in the probabilities of dice games around 1630. However, these early efforts were not pursued by others and the subject did not come to life until the Pascal-Fermat correspondence began. This correspondence examined the subject in a systematic manner and stimulated other mathematicians to look into the subject, and for these reasons Pascal and Fermat are frequently called the founders of probability theory.

In the century and a half following Fermat and Pascal, such names as Huygens, Leibnitz, Bernoulli, and de Moivre appeared in connection with probability, although the emphasis still remained on problems arising in gambling houses. Interest in the subject reached a high point in 1812 with the publication of *Théorie Analytique des Probabilitiés* by Pierre de Laplace (1749–1827). In this work, Laplace removed probability theory from the gaming rooms and showed it to be a serious field of mathematical study with applications to many areas of science.

Laplace's work was based on what is now called the "classical definition" of probability. In this setting, one considers an experiment or trial

Pierre-Simon Laplace
David Smith Collection

having a finite number of possible outcomes. Examples of such experiments that are traditionally considered are the flipping of coins, rolling of dice, drawing of cards from a deck, and so on. The basic assumption made is that each of the outcomes is "equally likely" to occur. If certain outcomes of an experiment are regarded as "favorable," then the probability p that the outcome on any particular execution of the experiment will be favorable is defined to be the *relative frequency* of favorable outcomes among all possible outcomes, that is, $p = f/n$ where n is the total number of possible outcomes and f is the number of favorable outcomes. For example, if a pair of dice is rolled, there are 36 possible outcomes. An outcome that is regarded as favorable to those who perform this experiment frequently is that of rolling a 7, that is, the sum of the spots on the upper faces of the dice is 7. Since there are 6 ways in which this favorable outcome can occur, as shown in the table, we conclude that the probability of rolling 7 is $\frac{6}{36} = \frac{1}{6}$.

Die I	1	2	3	4	5	6
Die II	6	5	4	3	2	1

In the historical development of probability theory, there were many disputes concerning the philosophical problem of what probability really means. Writers of the past tried to give a scientific bent to their intuitive feeling of what "probable" meant to them. What does it mean to say that the probability of getting heads on the flip of a coin is $\frac{1}{2}$? It certainly does not mean that whenever the coin is flipped twice, there will be exactly 1 head and 1 tail. (Along these lines, there is the classic story of the doctor who assured a patient suffering from a dread disease known to be fatal in 99 out of 100 cases of a complete cure because this was the hundredth such case the doctor had treated and the previous 99 had all died.) The assertion that the probability of getting heads on the flip of a coin is $\frac{1}{2}$ can be viewed as a "long haul" statement. That is, if the coin were actually flipped a very large number of times, then the ratio of the number of heads obtained to the total number of flips would very likely be close to $\frac{1}{2}$. The probability of an event thus measures the relative frequency of the event's occurrence in a great number of trials. The point is that probability can not be interpreted as predicting behavior on a specific trial. If a coin happens to come up heads on 42 consecutive flips, no assertion can be made about the 43rd flip except that it is just as likely to come up heads as it is tails. However, there are people who will argue that tails is more likely to occur on the 43rd flip since the "law of averages" must assert itself. This is just as naive as saying that heads is more likely than tails since the coin has a preference for heads. If the coin is fair (that is, not weighted to favor either side), then each of these arguments is equally spurious.

Although this classical definition of probability in terms of relative frequency is quite adequate for computing probabilities in simple situations such as games of chance, it is quite restrictive and does not apply in many important and interesting situations. This is due to the fact that this definition requires that the outcomes of the experiments considered must be equally likely and finite in number. In 1933, A. N. Kolmogorov (1903–) published a basic tract entitled *Foundations of Probability* in

A. N. Kolmogorov
Sovfoto

which he gave the modern, broader definition of probability in terms of an axiomatic structure called a probability space. This definition not only included the classical definition as a special case, but also permitted the probabilistic consideration of many new problems that did not fit the classical setting. Since then, probability theory has blossomed into a deep and powerful area of mathematics with applications in many diverse areas such as atomic physics, economics, engineering, and social science, as well as other fields of mathematics such as number theory, mathematical statistics, and potential theory.

Kolmogorov's axioms for a probability space are stated in Section 5.2 for the restricted case of finite sets. It is necessary for us to make this latter restriction here since the development of probability theory for infinite sets requires the use of measure and integration theory, a subject that is beyond the scope of this book. Nevertheless, this setting will include the classical one and will also allow the consideration of experiments in which the outcomes are not equally likely. Moreover, it should provide the reader with some flavor of modern probability theory. There is a brief discussion of probability theory for infinite sets at the end of Section 5.2 that indicates some of the difficulties that arise in this situation.

5.1. SAMPLE SPACES

In the application of probability theory to the study of a particular experiment or trial, the first step is usually to construct a set, called the *sample space* of the experiment, in such a way that the outcomes of interest correspond in a one-to-one manner with the elements of the set. That is, each outcome of the experiment is represented by some element of the sample space and each element of the sample space corresponds uniquely to an outcome of the experiment. In this section, we shall discuss several examples of sample spaces and introduce some terminology that will be used in the subsequent sections of this chapter.

Consider the simple experiment of flipping a coin and recording how it lands. If we rule out such freakish behavior as the coin landing on edge or being lost down a hole, then there are just two possible outcomes, namely, "heads" or "tails." Thus, we may take as the sample space for this experiment the set $S_1 = \{H,T\}$, where H represents the outcome heads and T stands for tails. Actually, any set of two elements would be adequate, but this notation seems quite appropriate. If the experiment consists of flipping a coin twice and recording how it lands each time, then a different sample space is needed since the outcomes are not represented by the elements of S_1. A natural way to represent the outcome of getting heads on the first flip and tails on the second flip is to use the ordered pair (H,T). Thus, the set $S_2 = \{(H,H),(H,T),(T,H),(T,T)\}$ is a good choice for the sample space of this experiment. Note that S_2 is just the

Cartesian product $S_1 \times S_1$. A sample space that represents all of the outcomes obtained by flipping a coin n times is

$$S_n = \underbrace{S_1 \times \cdots \times S_1}_{n \text{ factors}},$$

the set of ordered n-tuples where each term in an n-tuple is either an H or a T. S_n has 2^n elements, one element for each outcome of the experiment.

If an experiment consists of recording 1 roll of a die, then there are 6 possible outcomes, each outcome corresponding to the number of spots on the top face of the die when it comes to a stop. Thus, the set $S_1 = \{1,2,3,4,5,6\}$ will serve as a sample space. If the experiment consists of rolling 2 dice at once, or equivalently, rolling 1 die twice, then there are 36 possible outcomes and the sample space S_2 of this experiment is the Cartesian product of S_1 with itself:

$$S_2 = S_1 \times S_1 = \{(i,j) : i = 1,2,3,4,5,6,\ j = 1,2,3,4,5,6\}.$$

Each ordered pair of S_2 corresponds to one outcome of the experiment and conversely. This set can be pictured as a set of points in the plane as in Figure 5.1. For example, the point (2,5) corresponds to obtaining a 2 on the first die and a 5 on the second die.

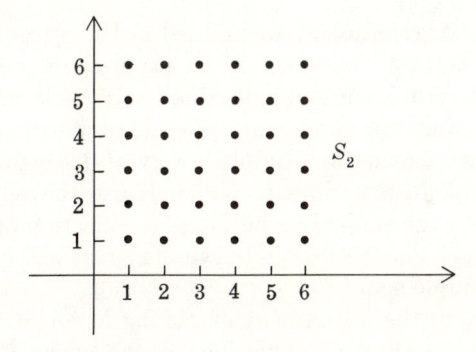

Figure 5.1

It is important to realize that the choice of a sample space related to a given experiment is not unique, but depends on the type of outcome that is of interest. For example, consider the experiment of drawing a single card from an ordinary deck of 52 playing cards and noting the suit of the card that is drawn. A natural choice of sample space would be a 52 element set such that each card is represented by some element of the set. However, we actually do not need such a large set. The set $\{s,c,h,d\}$, where s represents spades, c clubs, and so on, would be an adequate sample space for this experiment. If we consider an experiment in which we are only concerned with the rank of the drawn card, then the set {K,Q,J,10,9,8,7,6, 5,4,3,2,1} is a satisfactory sample space. Or if a problem depended only on whether the drawn card was a face card or not, then a two element sample space would suffice. There is no single correct choice of sample space to represent a given experiment. However, it must contain sufficiently many elements to represent all of the outcomes needed to solve the problems at hand.

Although we will be concerned almost exclusively with finite sample spaces (that is, a sample space with a finite number of elements) in the remainder of the chapter, it is worth noting that infinite sample spaces can arise through simply described experiments. As an example of an experiment having an infinite sample space, consider the act of throwing a dart at a circular dart board. To formulate a mathematical setting for this experiment, we idealize the situation by removing the experiment from the physical world and regarding it as a conceptual experiment. Thus, we assume the dart thrower never misses the dart board and that the point of the dart is small enough to cover exactly one point on the board. With these assumptions, we can then take a circular disc in the plane of the same radius as the target for our sample space, where each point of the disc corresponds to the outcome that the dart will land at the corresponding point of the target.

In the classical terminology mentioned in the introductory remarks to this chapter, certain outcomes of an experiment were regarded as "favorable." This term is undoubtedly due to the early association probability had with gambling. Since any type of outcome can be considered favorable by someone, this was simply a way of distinguishing a certain set of outcomes of an experiment. We shall use the nonpartisan term "event" in place of the subjective notion of a collection of favorable outcomes. That is, each subset of a finite sample space will be referred to as an *event* in the sample space.

For example, in the experiment of casting a pair of dice, one event of interest is that of rolling a 7. This event is the subset $E = \{(i,j) \in S_2 : i + j = 7\}$ of the sample space S_2 as shown in Figure 5.2. Note that E is the set of those points in S_2 that lie on the line $x + y = 7$.

In any sample space S there are two events distinguished by special appellations. One of these is the empty set $\emptyset$, which is called the *impossible event*. This event could be described facetiously in the experiment of rolling a pair of dice by saying it is the event that corresponds to rolling a sum of 37 with the two dice. An event in the opposite extreme is the *certain event*, which is the whole set S. This could be described in the dice experiment as the event of rolling a sum greater than one with the two dice.

The usual operations of set algebra allow us to combine events in a

5.1. SAMPLE SPACES

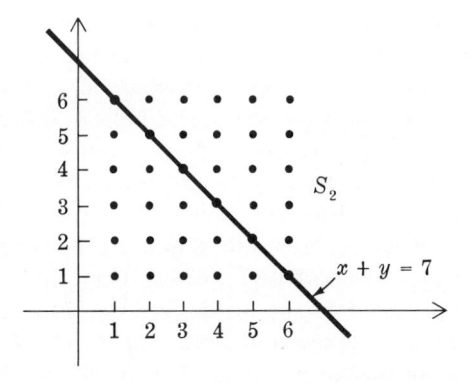

Figure 5.2

sample space to get new events. Thus, if E and F are events in the sample space S, then the union $E \cup F$, the intersection $E \cap F$, and the complement E' of E are also events. E' is called the *complementary event* of E. Two events E and F are called *mutually exclusive* if $E \cap F = \emptyset$, that is, if they are disjoint subsets of S. Notice that E and E' are always mutually exclusive events.

For example, consider the experiment of flipping a coin 3 times in succession. The associated sample space S of this experiment consists of all ordered triples of elements of the set $S_1 = \{H, T\}$. Let E_i be the event of getting exactly i heads out of the 3 flips for $i = 0, 1, 2, 3$. For instance, $E_2 = \{(H,H,T), (H,T,H), (T,H,H)\}$. Clearly, E_i and E_j are mutually disjoint whenever $i \neq j$. $E = E_0 \cup E_1$ can be described as the event of getting *at most* one head, and the complementary event E' is the event of getting *at least* two heads (note that $E' = E_2 \cup E_3$).

In the experiment of casting a pair of dice, if E is the event of rolling a 7 and F is the event that the second die comes up 4, then the event $E \cup F$ can be described by saying that either the sum of the dice is 7 *or* the second die is 4. Also, the event $E \cap F$ of rolling a 7 *and* having the second die come up 4 consists of the single element (3,4).

Sampling

One type of experiment or procedure that occurs frequently in probability is that of drawing a sample of a specified size from a given population or set of objects. A traditional setting for sampling problems is the urn full of balls. Admittedly, the drawing of a number of balls from an urn does not sound very exciting and may appear somewhat artificial, but the fact is that this setting is a prototype for sampling of all kinds. Selecting cards from a deck, or money from a box, or members from a TV audience, or items from an assembly line, and so on, does not change the basic mathematical nature of the sampling and, as far as probability is concerned, will be no more exciting than drawing balls from urns. The experiment of the flipping of a coin, for example, can also be viewed as drawing a ball from an urn that contains 2 balls of different colors. Similarly, rolling a die is simply the selection of a number from 1 to 6 and is conceptually the same as drawing one of 6 different balls from an urn.

There are two basic ways in which a sample can be drawn, *with replacement* and *without replacement*. Sampling balls from an urn *with* replacement means that as each ball is drawn from the urn, its description is noted and it is returned to the urn before the next ball is selected. In essence, sampling with replacement is the repetition of the same experiment, namely that of drawing a single ball from the given collection of balls, several times in succession. For after a ball is selected, it is replaced, and thus the same circumstances prevail for the next draw that existed on the previous draw. Similarly, if a poll is taken by interviewing people on the street, it is conceivable that the same person could be questioned more than once and the person interviewed is thus returned to the population being polled before the next person is approached. This is also sampling with replacement.

On the other hand, in drawing a sample of balls from an urn *without* replacement, once a ball is removed from the urn it remains out of the urn and cannot be drawn again during the sampling process. The dealing of a bridge hand is an example of sampling without replacement from the set of 52 playing cards in the deck. Or, if items are removed from a production line for quality inspection, this will be sampling without replacement as long as the items are not returned after testing.

Example 5.1. An urn contains 3 red balls and 5 white balls. Describe a sample space for the sampling of 2 balls (a) with replacement, and (b) without replacement.

One possible way to describe the outcomes of drawing 2 balls from the urn is to simply record the colors of the balls as they are selected. For example, (R,W) could denote the outcome that a red ball was drawn first and a white ball second. This would yield the 4-element sample space {(R,R), (R,W), (W,R), (W,W)}. However, while it is adequate for some experiments, this sample space is frequently not the most convenient to use. For example, it does not take into account the number of balls involved, nor does it indicate which type of sampling is being performed.

More informative sample spaces can be constructed if we regard the balls as 8 distinct objects. In drawing 2 balls with replacement, we have 8 possibilities for the first ball and 8 choices for the second ball. Each outcome can thus be represented by an ordered pair (b_1, b_2) where b_1 and b_2 denote the first and second balls respectively drawn from the set B of

8 balls. Notice that since we are sampling with replacement, it is possible to have $b_1 = b_2$. The sample space for (a) is thus seen to be the set

$$S_1 = \{(b_1,b_2): b_1 \in B, b_2 \in B\} = B \times B.$$

Note that S_1 has $8^2 = 64$ elements since B has eight elements.

When the drawing is done without replacement, we may still represent outcomes by ordered pairs (b_1,b_2), but now we have the restriction that $b_1 \neq b_2$ since the first ball is not replaced in the urn. Thus the sample space for (b) is

$$S_2 = \{(b_1,b_2): b_1 \in B, b_2 \in B, b_1 \neq b_2\}.$$

We see that S_2 has $8 \cdot 7 = 56$ elements since b_1 can be chosen in 8 ways while only 7 choices are available for b_2.

Let us consider the event described by saying "the second ball drawn is red" for both types of sampling, with replacement and without replacement. If R denotes the set of red balls and W the set of white balls in the urn, then R has 3 elements, W has 5 elements, and $R \cup W = B$. When the drawing is done with replacement, then the described event corresponds to the subset of S_1,

$$E_1 = \{(b_1,b_2) \in S_1: b_1 \in B, b_2 \in R\}.$$

Since there are 8 choices for b_1 and 3 choices for b_2, we see that E_1 is a 24-element subset of S_1. However, when the drawing is done without replacement, then the described event corresponds to the subset of S_2,

$$E_2 = \{(b_1,b_2): b_2 \in R, b_1 \in B, b_1 \neq b_2\}.$$

The fact that $b_1 \neq b_2$ means that while b_2 can be any one of the 3 red balls, b_1 can be only one of the 7 remaining balls in B. Thus, E_2 is a 21-element subset of S_2. That is, there are 21 ordered pairs (b_1,b_2) such that $b_1 \in B$, $b_2 \in R$, and $b_1 \neq b_2$.

In the above example, we viewed the drawing as being performed by removing the balls one at a time in order. It is possible to think of sampling without replacement as being done in one operation, say by reaching into the urn and removing the balls simultaneously. This viewpoint leads quite naturally to a different choice of sample space than the one considered in the above example. For instead of ordered pairs (b_1,b_2) of balls, we would simply have subsets $\{b_1,b_2\}$ of B. Since there are $\binom{8}{2} = 28$ ways of selecting 2 objects from a set of 8 objects, this sample space would have 28 elements. Note this is just half the number of elements contained in S_2 above, which is to be expected since each set $\{b_1,b_2\}$ can be ordered in two ways, namely (b_1,b_2) and (b_2,b_1).

EXERCISES

5.1. Describe a sample space S for the experiment of flipping a coin three times in succession. List all the elements contained in the following events:
   (a) The event E that at least 2 heads occur.
   (b) The event F that at most 2 heads occur.
   (c) The event G that at least 1 head and at least 1 tail occur.
   (d) The event $E \cap F$.
   (e) The event $F \cap G$.

5.2. Describe a sample space S for the experiment of rolling a die 3 times in succession. How many elements do the following events contain?
   (a) The event E of rolling a sum of at least 17.
   (b) The event F of rolling a sum of 7.
   (c) The event G that the first roll is 1 and the third roll is at most 2.
   (d) The event $F \cap G$.

5.3. Consider the experiment of spelling a 5-letter word, where a 5-letter word is defined to be any ordered 5-tuple of letters drawn from the alphabet with replacement. Describe a sample space S for this experiment and determine the number of elements of S. How many elements are in the event described by saying that the word begins and ends with the same letter?

5.4. An urn contains 3 red balls and 5 white balls, and 3 balls are drawn in succession without replacement. Let E_i be the event that the ith ball drawn is red for $i = 1, 2, 3$. By using set operations, express the following events in terms of E_1, E_2, and E_3.
   (a) All 3 balls are red.
   (b) At most 1 ball is red.
   (c) At least 1 ball is red.
   (d) All 3 balls are the same color.

5.5. Describe a sample space for the experiment of flipping a coin and rolling a die.

5.6. An urn contains 10 red balls and 5 white balls. Assuming the balls are distinguishable, describe sample spaces for the drawing of 3 balls with replacement and without replacement. Determine the number of elements in the event "the first ball is red and the third ball is red" in these two cases.

5.7. Suppose 3 balls are drawn without replacement from an urn containing 5 red balls, 7 white balls, and 8 blue balls. Describe a sample space for this experiment. How many elements belong to the event described by the phrase "there is 1 ball of each color?"

5.2. PROBABILITY SPACES

Intuitively, the probability of an event is a measure of how likely it is to occur. This measurement is done by assigning to each event E in a sample space S a numerical value between 0 and 1. The closer this value is to 1, the more likely the event is to occur, and we interpret a value near 0 to mean that the event is not very likely to occur. Thus, we have a function whose domain of definition is a collection of subsets of S and whose values lie in the set of real numbers. Not every function that assigns a real number to each event is a probability measure however. There are conditions that must be satisfied and these are given in the following definition. It will be assumed throughout this chapter that all sample spaces under consideration are not empty.

Definition. A *probability measure* on a finite sample space S is a real-valued function P defined on the collection of all subsets of S that satisfies the following conditions:

(P-1) $0 \leq P(E) \leq 1$ for each event E in S.
(P-2) $P(S) = 1$.
(P-3) If E and F are mutually exclusive events in S (that is, $E \cap F = \emptyset$), then

$$P(E \cup F) = P(E) + P(F).$$

A *probability space* (S,P) is a sample space S together with a specific probability measure P on S.

The following result establishes some simple consequences of this definition that will be useful.

Proposition 5.1. If (S,P) is a probability space, then:

(P-4) $P(E') = 1 - P(E)$ for any event E in S.
(P-5) $P(\emptyset) = 0$.
(P-6) $P(E) \leq P(F)$ whenever E and F are events in S such that $E \subset F$.
(P-7) $P\left(\bigcup_{i=1}^{n} E_i\right) = \sum_{i=1}^{n} P(E_i)$ whenever $E_1, \cdots, E_n$ are events in S such that $E_i \cap E_j = \emptyset$ for all $i \neq j$.

PROOF.

(P-4) Since E and E' are mutually exclusive events, it follows from (P-3) and (P-2) that

$$P(E) + P(E') = P(E \cup E') = P(S) = 1.$$

Consequently, $P(E') = 1 - P(E)$ for each event E in S.

(P-5) This is an immediate consequence of (P-4) and (P-2) since $S' = \emptyset$.

(P-6) If E and F are events such that $E \subset F$, then $F = E \cup (F \cap E')$, and the events E, $(F \cap E')$ are mutually exclusive. Therefore, since $P(F \cap E') \geq 0$ by (P-1), it follows that

$$P(E) \leq P(E) + P(F \cap E') = P(F).$$

(P-7) This result can be derived from (P-3) by a straightforward application of mathematical induction. We leave the details to the reader.

Property (P-7) of a probability space (S,P), which is often referred to as the *finite additivity* of the probability measure P, has the following useful consequence. If $E = \{x_1, \cdots, x_n\}$ is any event in S, then $P(E) = P(\{x_1\}) + \cdots + P(\{x_n\})$ since the singleton subsets $\{x_1\}, \cdots, \{x_n\}$ of E are mutually exclusive events in S. In other words, the *probability $P(E)$ of an event E in S is the sum of the probabilities of the outcomes that comprise the event E*. Symbolically, we write this statement as

$$P(E) = \sum_{x \in E} p(x) \qquad (*)$$

where p is the real-valued function defined on S by $p(x) = P(\{x\})$ for each $x \in S$. The function p, which completely determines the probability measure P through (*), is called the *point probability* corresponding to P. The following result characterizes functions that are point probabilities corresponding to probability measures on finite sets.

Proposition 5.2. If S is a finite set, then a real-valued function p on S is the point probability corresponding to a probability measure P on S if and only if:

(a) $0 \leq p(x) \leq 1$ for each $x \in S$;

(b) $\sum_{x \in S} p(x) = 1$.

PROOF. It is clear that if p is the point probability corresponding to a probability measure P on S, then p has properties (a) and (b). Conversely, if p is a real-valued function defined on S and if P is defined on each subset E of S by (*) above, then P is a probability measure on S. Conditions

(P-1) and (P-2) for a probability measure follow immediately from properties (a) and (b) for p. To verify (P-3), suppose E and F are disjoint subsets of S. Then, clearly, the sum of values $p(x)$ as x ranges over $E \cup F$ is the same as the sum of the numbers $p(x)$ as x ranges over E plus the sum obtained by adding the numbers $p(x)$ as x ranges over F. That is, if $E \cap F = \emptyset$, then

$$P(E \cup F) = \sum_{x \in E \cup F} p(x) = \sum_{x \in E} p(x) + \sum_{x \in F} p(x)$$
$$= P(E) + P(F).$$

Hence, condition (P-3) is satisfied and P is therefore a probability measure on S.

Example 5.2. Consider the experiment of rolling a die and the associated sample space $S = \{1,2,3,4,5,6\}$. The definition of a probability measure on S depends on the nature of the die involved. We shall consider two cases, a fair die and a biased or "loaded" die.

Suppose first that the die is fair so that each value of the die is "equally likely" to occur on a given roll of the die. This assumption means that the point probability p_1 determining the probability measure P_1 on S should satisfy

$$p_1(1) = p_1(2) = p_1(3) = p_1(4) = p_1(5) = p_1(6).$$

Since p_1 must also satisfy

$$p_1(1) + p_1(2) + p_1(3) + p_1(4) + p_1(5) + p_1(6) = 1$$

we see that $p_1(i) = \frac{1}{6}$ for each $i \in S$. If $E = \{1,6\}$ is the event of rolling either 1 or 6, then, in this case, $P_1(E) = \frac{1}{6} + \frac{1}{6} = \frac{1}{3}$.

Now suppose the die is biased so that certain sides are favored when it is rolled. For example, suppose that rolling either 1 or 6 is twice as likely to occur as rolling either 2, 3, 4, or 5. (One way to achieve such a bias is to shave the opposing 1 and 6 faces slightly. A die of this type is referred to as a "1 − 6 flat" since the die has been flattened in the 1 − 6 direction.) Let us determine the appropriate probability measure P_2 on S for this situation. The assumptions on the die imply that the point probability p_2 corresponding to P_2 must satisfy

$$p_2(1) = p_2(6) = 2p_2(2) = 2p_2(3) = 2p_2(4) = 2p_2(5).$$

This, together with the fact that the sum of the values of p_2 must be 1, forces us to conclude that

$$p_2(1) = p_2(6) = \tfrac{1}{4},$$
$$p_2(2) = p_2(3) = p_2(4) = p_2(5) = \tfrac{1}{8}.$$

In this case, the probability of the event $E = \{1,6\}$ is

$$P_2(E) = \tfrac{1}{4} + \tfrac{1}{4} = \tfrac{1}{2}.$$

In practical applications of probability theory, the values of a probability measure or equivalently a point probability for a particular experiment are assigned on the basis of physical evidence, or in the absence of such evidence, simply on intuition. If a coin is suspected to be biased, for example, then a weight analysis or a count of the number of heads obtained during a large number of flips could be performed to determine what the appropriate values of $p(\text{H})$ and $p(\text{T})$ should be. However, it is not really in the realm of probability theory to ascertain what values are correct in a specific instance. It is only after such information is given that the mathematical theory of probability comes into play. Indeed, in the application of any mathematical theory, the initial step is to construct a mathematical model to represent the physical situation. Frequently, it is impossible to take all features of the situation into account and only the most cogent points are considered. It is not in the purview of mathematics to question the model but only to draw conclusions concerning the model within the framework of the theory. The success of the application of mathematics depends greatly on how well the physical situation is idealized, but we shall not go into these considerations here.

In many situations, such as simple games of chance, a reasonable assumption frequently made is that each outcome is equally likely to occur. Indeed, the ubiquitous phrase "at random" simply means that each outcome of some activity is equally likely to occur. Of course, an unsuspecting player might discover this assumption to have disastrous consequences, but unless some information to the contrary is available, this assumption is generally satisfactory.

If S is a sample space, then the point probability p on S corresponding to this hypothesis of equal likelihood must satisfy $p(x) = p(y)$ for all x and y in S. Since the sum of the numbers $p(x)$ as x ranges over S must be 1, we clearly have that $p(x) = 1/n$ for all x in S where n is the number of elements in S. If P is the probability measure on S corresponding to p and if E is an event in S, then

$$P(E) = \sum_{x \in E} p(x) = \frac{k}{n},$$

where k is the number of elements of E. That is, the *equally likely probability measure* on S satisfies

$$P(E) = \frac{\text{number of elements in } E}{\text{number of elements in } S}.$$

This corresponds to the so-called classical definition of probability mentioned in the introductory remarks to this chapter. Since, in this case, the probability of an event depends only on the number of elements involved, problems involving equally likely probability measures are essentially counting problems. This by no means makes them all easy to solve, for sophisticated counting techniques can be difficult to master. But it should be realized that such problems are related to a particular aspect of probability theory, and students who become frustrated by intricate counting problems should not take it as an indication that they have no understanding of probability theory at all.

In the following, it will be assumed that the reader is familiar with the counting techniques developed in Chapter 4.

Example 5.3. A fair coin is flipped 7 times. What is the probability of getting exactly 4 heads?

As sample space, we take the set S of 7-tuples where each component of a 7-tuple is either an H or a T. That is, S is the Cartesian product of $\{H,T\}$ with itself 7 times. S has $2^7 = 128$ elements and we take P to be the equally likely probability measure on S. Let E denote the event of getting 4 heads. We must determine the number of elements in E. In order for a 7-tuple to be in E, 4 of its terms must be H and the other 3 terms T. Thus, there are as many ways of getting 4 heads on 7 flips as there are ways of choosing 4 elements out of 7 elements, and that is $\binom{7}{4} = 35$. Thus $p(E) = \frac{35}{128}$.

Example 5.4. A poker hand consists of 5 cards dealt from a standard deck of 52 playing cards. An appropriate sample space for the experiment of dealing a poker hand is the set S of all 5-element subsets of the set of 52 cards. This sample space S has $\binom{52}{5} = 2{,}598{,}960$ elements. By our earlier calculations in Chapter 4 (see Example 4.2), we have that the event F in S of dealing a full house has 1872 elements. If we assume that each poker hand is equally likely to be dealt, then the probability of dealing a full house is

$$P(F) = \frac{1872}{2{,}598{,}960} \approx 0.0007,$$

where P denotes the equally likely probability measure on S.

The additivity properties (P-3) and (P-7) of a probability measure fail if the sets are not mutually exclusive. The correct relation for the case of 2 events is as follows.

Proposition 5.3. If (S,P) is a probability space and E, F are events in S, then
$$P(E \cup F) = P(E) + P(F) - P(E \cap F).$$

PROOF. $E \cup F$ can be expressed as $E \cup F = (E \cap F') \cup (E' \cap F) \cup (E \cap F)$ and the 3 events $E \cap F'$, $E' \cap F$, $E \cap F$ are mutually exclusive

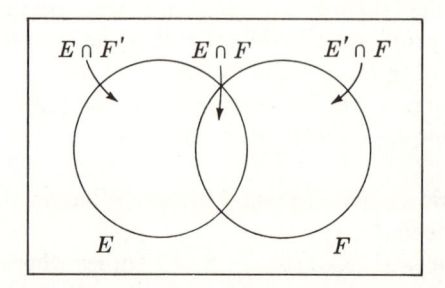

Figure 5.3

in pairs. (See Figure 5.3.) Thus, from (P-7), we have
$$P(E \cup F) = P(E \cap F') + P(E' \cap F) + P(E \cap F).$$
Adding $P(E \cap F)$ to both sides and observing that $P(E \cap F') + P(E \cap F) = P(E)$ and $P(E' \cap F) + P(E \cap F) = P(F)$, we get
$$P(E \cup F) + P(E \cap F) = P(E) + P(F)$$
which implies the desired result.

Example 5.5. One card is drawn from a deck of 52 playing cards. Find the probability that it is either a spade or a face card (that is, a king, queen, or jack).

We take as sample space S a 52-element set representing the cards of the deck, and we assume that each card is equally likely to be drawn. Thus, we are dealing with the equally likely probability measure P on S. Let E be the event that a spade is drawn and let F be the event that a face card is drawn. Then since there are 13 spades, 12 face cards, and 3 cards that are both spades and face cards, we have
$$\begin{aligned} P(E \cup F) &= P(E) + P(F) - P(E \cap F) \\ &= \tfrac{13}{52} + \tfrac{12}{52} - \tfrac{3}{52} \\ &= \tfrac{11}{26}. \end{aligned}$$

The analogue to Proposition 5.3 for n events is also valid. More specifically, suppose that $E_1, \cdots, E_n$ are events in a probability space (S,P). Define the numbers $S_1, \cdots, S_n$ as follows:

$$S_1 = \sum_{i=1}^{n} P(E_i) = P(E_1) + \cdots + P(E_n),$$

that is, S_1 is the sum of the probabilities of the events $E_1, \cdots, E_n$.

$$S_2 = P(E_1 \cap E_2) + P(E_1 \cap E_3) + \cdots + P(E_{n-1} \cap E_n),$$

that is, S_2 is the sum of the probabilities of the events $E_i \cap E_j$ as i, j range from 1 to n with $i < j$. In general, we define S_k to be the sum of the values of P on all possible intersections of k events from among the given events $E_1, \cdots, E_n$. Then the following formula is valid

$$P(E_1 \cup \cdots \cup E_n) = S_1 - S_2 + \cdots + (-1)^{k+1} S_k \\ + \cdots + (-1)^{n+1} S_n. \quad (**)$$

Note that this formula reduces to the formula in Proposition 5.3 in case $n = 2$. For $n = 3$ it takes the form

$$P(E_1 \cup E_2 \cup E_3) = P(E_1) + P(E_2) + P(E_3) \\ - P(E_1 \cap E_2) - P(E_1 \cap E_3) - P(E_2 \cap E_3) \\ + P(E_1 \cap E_2 \cap E_3).$$

The general formula (**) can be established by reasoning very similar to that used in the proof of the Inclusion-Exclusion Principle (see Section 4.4). Consequently, we shall omit the details here.

Example 5.6. (Matching) Suppose two identical stacks of n different cards are shuffled and the results of the shuffling compared by turning the cards face up one at a time. If two cards that are the same are turned up at the same time, we say that a *match* (or *rencontre*) occurs. We wish to calculate the probability that there will be at least one match.

If the cards in one stack are numbered consecutively from 1 to n, then the corresponding cards in the other stack can be viewed as defining a permutation of the set $\{1,2, \cdots ,n\}$. The set S of all such permutations can thus serve as sample space. The assumption that each arrangement of cards is equally likely to occur corresponds to choosing the equally likely probability measure P on S. If S_i denotes the event that there is a match at the ith card, then we wish to compute the probability of the event $E = S_1 \cup \cdots \cup S_n$. The number of elements in E was previously calculated to be

$$n! \left(1 - \frac{1}{2!} + \frac{1}{3!} + \cdots + (-1)^{n+1} \frac{1}{n!} \right)$$

by using the Principle of Inclusion-Exclusion (see Example 4.8). Therefore, we have

$$P(E) = 1 - \frac{1}{2!} + \frac{1}{3!} - \cdots + (-1)^{n+1}\frac{1}{n!}.$$

Those readers with calculus background may recall that the Maclaurin series of the function e^x is

$$e^x = 1 + x + \frac{x^2}{2!} + \frac{x^3}{3!} + \cdots + \frac{x^n}{n!} + \cdots.$$

In particular, for $x = -1$ we have

$$e^{-1} = \frac{1}{2!} - \frac{1}{3!} + \frac{1}{4!} - \cdots + (-1)^n \frac{1}{n!} + \cdots.$$

Thus, we have the approximation

$$P(E) \approx 1 - e^{-1} = 0.63212 \cdots.$$

This approximation is quite good even for relatively small values of n. For example, if $n = 5$, then $P(E) = 0.633 \cdots$ and the above approximation is accurate to two decimal places. In general, for a given n, it can be seen that

$$|P(E) - (1 - e^{-1})| < \frac{1}{(n+1)!}.$$

For $n = 10$, for example, $1 - e^{-1}$ will approximate $P(E)$ to seven decimal places.

Interpreted in terms of probability, this yields a rather surprising conclusion. The probability of getting at least one match is virtually the same regardless of the number n of cards involved. That is, if we have 10 cards, 52 cards, or 100 cards, the probability of getting at least one match is approximately $1 - e^{-1} = 0.63212 \cdots$.

EXERCISES

5.8. Let $S = \{a,b,c\}$ and set $p(a) = \frac{1}{4}$, $p(b) = \frac{1}{3}$. Assign a value to $p(c)$ such that p is a point probability on S. Find the corresponding probability of the event $E = \{a,b\}$.

5.9. Let (S,P) be a probability space and let E, F be events in S with $P(E) = \frac{1}{3}$, $P(F) = \frac{1}{4}$, $P(E \cap F) = \frac{1}{6}$. Find $P(E \cup F)$, $P(E' \cap F)$, and $P(E \cap F')$. (See Exersise 5.16.)

5.10. If the letters in the word *math* are rearranged, what is the probability that the new word will start with t? Assume that each arrangement is equally likely. [Answer: $\frac{1}{4}$.]

5.11. If the letters in the word ILLINOIS are rearranged, what is the probability that the three I's will appear consecutively?

[Answer: $\frac{1}{560}$.]

5.12. A die is rolled twice. Assuming that each of the 36 outcomes is equally likely to occur, find the probability that:
   (a) The value of the first roll is greater than that of the second.

[Answer: $\frac{5}{12}$.]

   (b) Either the sum is 7 or the first roll comes up 6. [Answer: $\frac{11}{36}$.]
   (c) The first roll comes up even and the second comes up odd.

[Answer: $\frac{1}{4}$.]

5.13. Suppose a coin is flipped 4 times and that each of the 16 outcomes is equally likely. Find the probability that:
   (a) The first flip is heads and the second tails. [Answer: $\frac{1}{4}$.]
   (b) There are more heads than tails. [Answer: $\frac{5}{16}$.]
   (c) There is at least 1 head and at most 1 tail. [Answer: $\frac{5}{16}$.]
   (d) The first flip is heads or exactly 2 heads occur. [Answer: $\frac{11}{16}$.]

5.14. Two balls are drawn without replacement from an urn containing 3 red balls and 7 white balls. Suppose each pair is equally likely to be drawn. What is the probability that both balls are white?

[Answer: $\frac{7}{15}$.]

5.15. Suppose 5 men check their hats at a nightclub. At the end of the evening, their hats are returned in random order. What is the probability that:
   (a) At least 1 man gets his own hat.
   (b) Exactly 1 man gets his own hat.

5.16. Show that if E and F are events in S and P is a probability measure on S, then

$$P(E \cap F') = P(E) - P(E \cap F).$$

5.17. Show that if (S,P) is a probability space and $E_1, E_2, \cdots, E_n$ are events in S, then

$$P\left(\bigcup_{i=1}^{n} E_i\right) \leq \sum_{i=1}^{n} P(E_i).$$

Infinite Probability Spaces

We conclude this section with a few brief remarks concerning probability measures on infinite sample spaces. The definition given at the beginning of this section is altered in two ways if the sample space S is not assumed to be finite. One change concerns the additivity property (P-3). In order to obtain a fruitful theory, it is necessary to have an analogue of (P-3) for infinitely many events. This is known as the countable additivity property of P and is stated as follows:

(P-3)' If $E_1, E_2, \cdots, E_n, \cdots$ are events in S such that $E_i \cap E_j = \emptyset$ whenever $i \neq j$, then

$$P\left(\bigcup_{n=1}^{\infty} E_n\right) = \sum_{n=1}^{\infty} P(E_n).$$

Conditions (P-1) and (P-2) of the definition remain as stated. The second change in the definition involves the domain of P.

In the case of a finite sample space, a probability measure is defined on the collection of *all* subsets of the space. For infinite sample spaces, this requirement is too restrictive. To see why this is so, let us consider the dart throwing experiment mentioned in the last section. We shall assume that the target is a disc D of area 1 and that the dart is just as likely to land at one point of the target as any other. More precisely, these assumptions mean that a probability measure P on D should have the following properties:

(a) If A and B are congruent subsets of the disc D, then

$$P(A) = P(B).$$

(b) If R is a rectangular region contained in D, then

$$P(R) = (\text{Area of } R).$$

(The use of rectangular subregions in the statement of (b) is not essential; any type of subregion for which the area is readily computable such as circular or triangular regions would do just as well.)

It can be shown that there does not exist a function P defined on the set of *all* subsets of D such that P satisfies (P-1), (P-2), (P-3)' as well as (a) and (b). The proof that no such function exists is not easy and will not be presented here. Basically, the point is that too many demands have been placed on the function P and any attempt to satisfy all of them is doomed to failure.

If condition (P-3)' is replaced by the weaker condition (P-3), the existence of a function P satisfying the resulting set of requirements can be established. However, the use of condition (P-3)' is so essential to the development of a significant theory for infinite probability spaces that its replacement by (P-3) is not acceptable. A much better alternative is to relax the requirement that the function P be defined for *all* subsets of D. If this is done, it can be shown that there exists a function P_D with properties (P-1), (P-2), (P-3)', (a), (b) such that the domain of P_D is a collection $\mathfrak{F}_D$ of subsets of D that includes all rectangular subregions of D.

5.2. PROBABILITY SPACES

The domain of a probability measure P on a sample space S cannot be devoid of any structure. For instance, it is necessary to consider unions, intersections, and complements of events. Consequently, if $\mathfrak{F}$ is to be the domain of P, then $\mathfrak{F}$ should be a class of subsets of S that is closed with respect to the set operations of union, intersection and complementation. Also, in order for the countable additivity property (P-3)′ to make sense, we require that $\bigcup_{n=1}^{\infty} E_n \in \mathfrak{F}$ whenever $E_n \in \mathfrak{F}$ for each positive integer n.
A nonempty set $\mathfrak{F}$ of subsets of S with the above properties is called a *σ-field* of subsets of S. Of course, the set of all subsets of the set S is a σ-field, but as we have already remarked, this σ-field is not always a suitable domain for a probability measure on S. In the case of the dart experiment mentioned above, the domain of the appropriate probability measure P_D is called the σ-field of *Lebesgue measurable* subsets of the disc D and P_D is called the *Lebesgue measure* on D. These names are in honor of one of the founders of measure and integration theory, Henri Lebesgue (1875–1941).

The general definition of a probability space can now be stated. A *probability space* is a triple $(S,\mathfrak{F},P)$ consisting of a (possibly infinite) set S, a σ-field $\mathfrak{F}$ of subsets of S, and a real-valued function P defined on $\mathfrak{F}$ with properties (P-1), (P-2), (P-3)′. The sets in $\mathfrak{F}$ are called *events*. Note that this definition of probability space reduces to that given at the beginning of this section if S is finite and $\mathfrak{F}$ is the set of all subsets of S since (P-3)′ is clearly equivalent to (P-3) in this case.

As we have already noted in our discussion of the dart throwing experiment, the appropriate probability measure P_D is not defined on all subsets of the disc D, but rather only on the Lebesgue measurable sets. Any region R in D with a geometrically computable area such as a polygonal or circular region is a Lebesgue measurable set in D and the probability measure satisfies

$$P_D(R) = \text{Area of } R$$

for each such region R.

The consideration of the Lebesgue probability measure is by no means restricted to a circular disc D of area 1. For example, if S is any polygonal region in the plane with area 1, there is a probability measure P_S on S which behaves like the measure P_D. That is, P_S is not defined for all subsets of S, but for subregions R of S with geometrically computable areas, we have

$$P_S(R) = \text{Area of } R.$$

For this reason, this probability measure on S is called the Lebesgue measure on S. If the area of such a region S is not 1, we can divide the area measure by the area of S to obtain a probability measure P on S; in this case, if R is a subregion of S with a geometrically computable area, then

$$P(R) = \frac{\text{Area of } R}{\text{Area of } S}.$$

We shall now illustrate the use of such measures in certain problems involving infinite sample spaces.

Example 5.7. *(The Waiting Problem)* Two persons agree to meet at a prescribed place sometime between noon and 1 P.M. one day. However, it turns out that due to other commitments neither person is able to wait for the other for more than 15 minutes. What is the probability that the 2 persons will meet during that 1-hour period?

To compute this probability, let us suppose that x and y denote the arrival times for the first and second person, respectively. Since both persons can arrive at any time during the 1-hour period prescribed, x and y can assume all values from 0 to 1 with equal likelihood. Thus, the "unit square"

$$S = \{(x,y) \in \mathbf{R}^2 \colon 0 \leq x \leq 1, 0 \leq y \leq 1\}$$

provides a sample space for this problem. (See Figure 5.4.) Since each person will wait for the other for a maximum of only $\frac{1}{4}$ of an hour, the

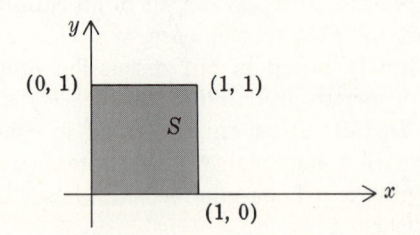

Figure 5.4

2 persons will meet if and only if $|x - y| \leq \frac{1}{4}$. Thus, the event E in S described by "the 2 persons will meet" is $E = \{(x,y) \in S \colon -\frac{1}{4} \leq x - y \leq \frac{1}{4}\}$. (See Figure 5.5.) For the complementary event E', we clearly have

$$P_S(E') = \tfrac{1}{2}(\tfrac{3}{4})(\tfrac{3}{4}) + \tfrac{1}{2}(\tfrac{3}{4})(\tfrac{3}{4}) = \tfrac{9}{16}.$$

Hence, $P_S(E) = \tfrac{7}{16}$.

Example 5.8. If two numbers x and y are selected at random in the interval from 0 to 2, what is the probability that the sum of the squares of these two numbers will not exceed 1 and that their sum will not be less than 1?

5.2. PROBABILITY SPACES

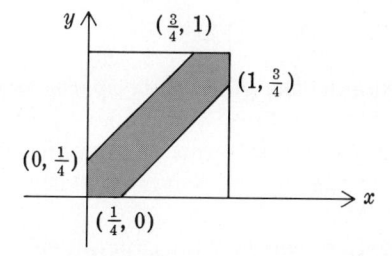

Figure 5.5

Since x and y may assume all values from 0 to 2 with equal likelihood, an appropriate sample space for this problem is the set $S = \{(x,y) \in \mathbf{R}^2: 0 \leq x \leq 2, 0 \leq y \leq 2\}$. We wish to compute the probability of the event $E = \{(x,y) \in S: x^2 + y^2 \leq 1, x + y \geq 1\}$. (See Figure 5.6.) The area

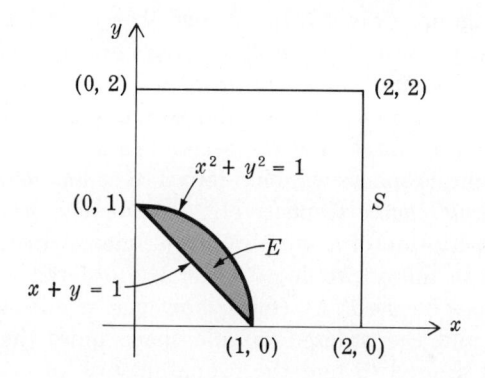

Figure 5.6

of E is $\pi/4 - \frac{1}{2}$; consequently, since the area of S is 4, we have

$$P(E) = \frac{\frac{\pi}{4} - \frac{1}{2}}{4} = \frac{\pi - 2}{16}.$$

EXERCISES

5.18. Two points C and D are selected at random from a line segment AB of length 1. Find the probability that the point C is closer to D than A. [Answer: $\frac{3}{4}$.]

5.19. A line segment of length 1 is broken into 3 pieces. Find the probability that the resulting line segments are the sides of a triangle. [Answer: $\frac{7}{8}$.]

5.3. SOME SAMPLING PROBLEMS

In this section, we examine the probability spaces associated with different types of sampling procedures. Since many problems in elementary probability can be interpreted in terms of sampling, this also supplies us with techniques for setting up and solving a fairly broad range of such problems.

As our basic model, we consider the experiment of drawing balls from an urn. We assume that each ball is just as likely to be selected as any other on the draw of a single ball. As mentioned previously, there are two ways in which a sample of more than one ball can be drawn from the urn, without replacement and with replacement. These two types of sampling are vitally different and an essential part of the given information in any sampling problem is the specification of which of the two types is being considered.

When k balls are drawn from an urn, the resulting sample can be viewed in two ways, either as simply a set, where the order in which the balls are drawn is ignored, or as an ordered k-tuple, the result of noting which ball was drawn first, second, and so on. We call the first kind of sample an *unordered sample* and the second kind an *ordered sample*. The corresponding sample spaces will be referred to as *unordered sample spaces* and *ordered sample spaces* respectively. Frequently, it is not explicitly stated in the description of a sampling experiment which kind of sample is intended and in many problems either the ordered or the unordered sample space may be used. At times, however, events are described in terms of order and the ordered sample space must then be used. For example, if it is desired to find the probability of an event described in such terms as "the second ball is red," then clearly the unordered sample space is not appropriate. On the other hand, for an event of the type "one of the balls is red," either sample space could be used. We shall give some detailed examples and some recommendations along these lines shortly.

We now wish to define probability measures on these various sample spaces. Let us first consider sampling without replacement. Suppose k balls are drawn without replacement from an urn containing n balls. Then the ordered sample space, which we will denote by S_0, consists of all ordered k-tuples $(b_1, b_2, \cdots, b_k)$ where $b_i \neq b_j$ if $i \neq j$, and b_i represents the ball selected on the ith draw. Note that S_0 contains $n(n-1) \cdots (n-k+1)$ elements. The equally likely probability measure on S_0, which we denote by P_0, is appropriate for this experiment since on the first draw each ball is equally likely to be chosen, on the second draw each of the remaining $n-1$ balls is just as likely to be selected, and so on.

Example 5.9. Suppose that an urn contains a total of n balls, r of which are red while the remaining $n - r$ are white. If 3 balls are drawn without replacement, find the probability that the second ball is red.

Since order is involved in the description of the event in question, we must use the ordered probability space (S_0, P_0) just discussed. We wish to find $P_0(E)$ where E is the event consisting of all ordered triples (b_1, b_2, b_3), where b_2 denotes a red ball. There is no restriction on b_1 or b_3. How many such triples are there? b_2 can represent any of r red balls, and once b_2 is fixed, there remain $n - 1$ possibilities for b_1 and $n - 2$ possibilities for b_3. Thus, E has $r(n - 1)(n - 2)$ elements, and since P_0 is the equally likely probability measure we obtain

$$P_0(E) = \frac{r(n-1)(n-2)}{n(n-1)(n-2)} = \frac{r}{n}.$$

Observe that this is the same as the probability of getting a red ball on a single draw, which is not surprising since no conditions were imposed on the first and third balls.

The following way of computing $P_0(E)$ is a bit more cumbersome, but it may seem more natural. We first note that there are n choices for the ball b_1. After b_1 has been selected, there are either r or $r - 1$ choices available for b_2 depending on whether b_1 was white or red. After b_2 is selected, there remain $n - 2$ choices for b_3. Thus, the number of elements in E is obtained by adding the number of possible triples (b_1, b_2, b_3) with b_1 white and b_2 red to the number of such triples with both b_1 and b_2 red. That is, the number of elements of E is

$$(n - r)r(n - 2) + r(r - 1)(n - 2) = r(n - 1)(n - 2).$$

Thus, we again arrive at the conclusion that $P_0(E) = r/n$.

We now return to the experiment of drawing k balls from an urn containing n balls and examine the unordered probability space associated with sampling without replacement. In this case, the unordered sample space, which we denote by S_u, consists of all sets $\{b_1, b_2, \cdots, b_k\}$ of k of the n balls. S_u has $\binom{n}{k}$ elements since this is the number of ways k objects can be selected from a set of n objects. Since each set of k elements can be arranged in $k!$ different ways, we see that there is a direct relationship between the unordered sample space S_u and the ordered sample space S_0. For example, in the case $k = 3$, each element $\{b_1, b_2, b_3\}$ in S_u corresponds to the following $3! = 6$ elements in S_0:

$$(b_1, b_2, b_3), \; (b_1, b_3, b_2), \; (b_2, b_1, b_3), \; (b_2, b_3, b_1), \; (b_3, b_2, b_1), \; (b_3, b_1, b_2).$$

In general, each element $\{b_1, b_2, \cdots, b_k\}$ of S_u corresponds to the $k!$ possible k-tuples in S_0 obtained by ordering the given set in $k!$ possible ways. If each element in S_0 is assumed equally likely to occur, then it follows that each element of S_u is also equally likely to occur and we may use the equally likely probability measure P_u on S_u.

We shall now consider an example in which both the ordered and unordered sample spaces can be used.

Example 5.10. An urn contains 9 red balls and 6 white balls, and 5 balls are drawn without replacement. Find the probability that exactly 2 of the balls are red.

We shall first make use of the ordered sample space S_0 to solve the problem. The described event corresponds to the subset E_0 of S_0 consisting of all ordered 5-tuples $(b_1, b_2, b_3, b_4, b_5)$ in which 2 terms represent red balls and 3 represent white balls. Now there are $\binom{5}{2} = 10$ arrangements of 5 balls in which 2 balls are red and the other 3 are white. In any particular one of these arrangements (for example, (red, white, red, white, white)), there are 9 ways to choose the first red ball and 8 ways to choose the second red ball. Also, there are 6 ways of choosing the first white ball, 5 ways of choosing the second white ball, and 4 ways of choosing the third white ball. Hence, the probability that exactly 2 of the 5 balls are red is

$$P_0(E_0) = \frac{\binom{5}{2} \cdot 9 \cdot 8 \cdot 6 \cdot 5 \cdot 4}{15 \cdot 14 \cdot 13 \cdot 12 \cdot 11} = \frac{240}{1001} \approx 0.239.$$

Let us now solve the problem by using the unordered sample space S_u. The described event then corresponds to the subset E_u of S_u consisting of all sets of 5 balls such that 2 of them are red and 3 are white. Since there are $\binom{9}{2} = 36$ ways of choosing 2 red balls from the 9 available, and $\binom{6}{3} = 20$ ways of choosing 3 of the 6 white balls, we see that

$$P_u(E_u) = \frac{\binom{9}{2}\binom{6}{3}}{\binom{15}{5}} = \frac{240}{1001}.$$

In the preceding example, each method of solving the problem led to the same answer. The final value for the probability of an event does not depend on the particular choice of probability space employed to arrive at the answer. However, it might be noted that the use of the unordered sample space led to a simpler computation. If sampling is done without replacement, and if the event in question allows a choice of using either the ordered or unordered sample space, it is usually the case that computations will be simpler if the unordered sample space is chosen.

In Example 5.10, we actually derived a special case of the following general result. If k balls are drawn without replacement from an urn contatining r red balls and w white balls, then the probability that exactly m of the balls are red is

$$\frac{\binom{r}{m}\binom{w}{k-m}}{\binom{r+w}{k}}.$$

This is established by an argument similar to that used in Example 5.10 and the details are left to the reader.

Example 5.11. Thirteen cards (a bridge hand) are dealt from a standard deck of 52 cards. What is the probability that exactly 4 of the cards are aces or kings?

There are 8 cards that are either aces or kings and 44 other cards. If we view the aces and kings as red balls and the remaining cards as white balls, then an application of the above formula gives us the probability

$$\frac{\binom{8}{4}\binom{44}{9}}{\binom{52}{13}} = \frac{100{,}529}{1{,}286{,}390} \approx 0.078.$$

We now consider sampling *with* replacement. Although we again have two possible viewpoints, ordered and unordered samples, it is preferable to think in terms of ordered samples. One reason for this is that sampling with replacement is conveniently viewed as the repeated performance of a simple experiment, namely, the drawing of a single object. In effect, the act of drawing k balls from an urn can be pictured as drawing a single ball from each of k urns, each containing the same distribution of balls. (The subject of repeated trials will be discussed later in Section 5.5.) However, the most important observation to make concerning sampling with replacement is that the equally likely probability measure is frequently appropriate on the space of ordered samples, but not on the space of unordered samples. This is illustrated in the following example.

Example 5.12. Suppose 2 balls are drawn with replacement from an urn containing 2 balls, 1 white and 1 red. If each ball is equally likely to be selected on a single draw, then it is natural to take the four possible outcomes represented by the ordered pairs (R,R), (R,W), (W,R), (W,W) as equally likely to occur. Thus, the probability that the 2 balls will be of different colors is $\frac{1}{2}$. However, if the colors of the drawn balls are recorded without regard to order, then there are only three outcomes: {R}, {R,W}, and {W}. [*Note:* {R} = {R,R} and {R,W} = {W,R}.] But these outcomes are not equally likely since we have seen that the probability of drawing different colored balls is $\frac{1}{2}$, not $\frac{1}{3}$. In the unordered sample space, the fact that balls of different colors can be obtained in two different ways is not brought out. Thus, the equally likely probability measure is not appropriate on the unordered sample space in this case.

The possible confusion between the two types of sample spaces that can arise in sampling with replacement can be avoided by viewing experiments in such a way that one is led naturally to work with the ordered sample space. For example, in rolling several dice, we can assume the dice to be of different colors, or we can think of a single die being rolled several times in succession. With such precautions, the problem of the unordered sample space in sampling with replacement need not arise.

Example 5.13. An urn contains 9 red balls and 6 white balls. If 5 balls are drawn with replacement, what is the probability that exactly 2 of the balls are red? (See Example 5.10.)

The ordered sample space S for this problem consists of all 5-tuples $(b_1, b_2, b_3, b_4, b_5)$ where each b_i may represent any of the 15 balls. Thus, S has 15^5 elements and we may use the equally likely probability measure P on S. The event E in question consists of all 5-tuples in which 2 terms represent red balls and 3 terms represent white balls. There are $\binom{5}{2} = 10$ arrangements of 5 balls in which 2 are red and 3 are white. In any particular one of these arrangements, there are 9 ways to choose the first red ball and 9 ways to choose the second red ball, since the sampling is done with replacement. Also, there are 6 ways to obtain each white ball in a given arrangement. Thus, we have

$$P(E) = \binom{5}{2} \frac{9^2 6^3}{15^5} = \frac{144}{625} \approx 0.231.$$

Notice that this is slightly smaller than the probability obtained in Example 5.10 when sampling was done without replacement.

In general, if k balls are drawn with replacement from an urn containing r red balls and w white balls, then the probability that exactly m of the drawn balls are red is

$$\binom{k}{m} \frac{r^m w^{k-m}}{(r+w)^k}.$$

The verification of this formula is left to the reader as an exercise.

We shall now consider problems known as *occupancy problems* that are closely related to sampling. If k balls are dropped at random into n hitherto empty boxes, then questions arise concerning the number of boxes occupied by certain numbers of balls. A sample space for this experiment can be constructed as follows. Let us label the boxes $b_1, b_2, \cdots, b_n$, and regard the toss of a ball into a box as the selection of a box. Note that this amounts to sampling with replacement since several balls may occupy the same box. Each distribution of the balls in the boxes corresponds to a k-tuple where the ith term of the k-tuple indicates which box was selected on the ith toss. (For example, if three balls are tossed into three boxes, then (b_2,b_3,b_1) indicates that the first ball went into the second box, the second ball went into the third box, and the third ball went into the first box.) Since each of the k balls can occupy any of n boxes, the sample space of k-tuples has n^k elements.

We shall now consider two examples of occupancy problems.

Example 5.14. Three balls are dropped at random into 3 boxes. Find the probability that all 3 balls end up in the same box.

We assume that each possible distribution of balls is equally likely to occur. The sample space described above has $3^3 = 27$ elements and the event E under consideration consists of the triples (b_1,b_1,b_1), (b_2,b_2,b_2), and (b_3,b_3,b_3). Hence $P(E) = \frac{3}{27} = \frac{1}{9}$.

Example 5.15. (The Birthday Problem) What is the probability that in a group of n people at least two of them have the same birthday? Assume that no one was born on February 29 of a leap year and that each of the 365 days is equally likely to be a birthday.

This can be regarded as an occupancy problem where the names of the n people are dropped into 365 boxes, one box for each day of the year. We take as a sample space the set S of all n-tuples where each term in the n-tuple is an integer from 1 to 365. S has $(365)^n$ elements. The event E that at least 2 people have the same birthday consists of all n-tuples in which 2 or more terms are the same. The complementary event E' is the set of n-tuples whose terms are all different and we see that

$$P(E) = 1 - P(E') = 1 - \frac{365 \cdot 364 \cdots (365-n+1)}{(365)^n}.$$

The values of $P(E)$ for various values of n are given in the accompanying table.

n	5	10	15	20	22	23	30	40	50
$P(E)$	0.027	0.117	0.253	0.411	0.476	0.507	0.706	0.891	0.970

It might be somewhat surprising to note that if a mere 23 people are involved, the probability that at least 2 of them have the same birthday is greater than $\frac{1}{2}$, and if 50 people are present, the event is almost a certainty. A little reflection, however, will lessen the surprise. Note that no particular date is fixed and any day may serve as a common birthday. It is this lack of restriction that permits the high probabilities. Upon first hearing the problem, our intuition perhaps thinks in terms of another problem, that is, if one of the n people declares his birthday, what is the probability that at least 1 other person will have that date as his birthday? A computation will show that the answer to this problem is much smaller and may come closer to what we feel should happen.

EXERCISES

5.20. A game is played by A and B as follows: A flips a coin and wins if either he gets heads on the first flip or he gets heads on 2 out of 3 flips. Otherwise B wins. What is the probability that A wins?
[Answer: $\frac{5}{8}$.]

5.21. Five dice are thrown. Find the probability that:
 (a) Exactly 3 sixes are obtained. [Answer: $250/6^5$.]
 (b) At least 3 sixes are obtained. [Answer: $276/6^5$.]

5.22. A card is drawn from a deck of playing cards. Then it is replaced in the deck and a second card is drawn. Find the probability of the following events:
 (a) At least 1 card is a heart. [Answer: $\frac{7}{16}$.]
 (b) The first card is red and the second card is a spade.
[Answer: $\frac{1}{8}$.]
 (c) The first card is an ace and exactly 1 card is a diamond.
[Answer: $\frac{3}{104}$.]

5.23. Solve Exercise 5.22 for the case in which the cards are drawn without replacement. [Answer: (a) $\frac{15}{34}$, (b) $\frac{13}{102}$, (c) $\frac{1}{34}$.]

5.24. Urn I contains 2 white balls, 3 red balls, and 2 blue balls and Urn II contains 3 white, 1 red, and 4 blue balls. Find the probability that if 1 ball is taken from each urn:
 (a) Both balls are the same color. [Answer: $\frac{17}{56}$.]
 (b) At least 1 ball is blue. [Answer: $\frac{9}{14}$.]
 (c) One ball is blue and the other is red. [Answer: $\frac{1}{4}$.]

5.25. A shelf contains 100 light bulbs of which 20 are defective.
   (a) A shopper selects 2 light bulbs at random. What is the probability that both are free from defects? [Answer: $\frac{316}{495}$.]
   (b) Another shopper comes after 50 light bulbs have been purchased and selects 2 bulbs from the remaining supply. What is the probability that both are free from defects?
   [Answer: $\frac{316}{495}$.]
5.26. Six balls are drawn without replacement from an urn containing 5 red balls, 10 white balls, and 15 blue balls. Find the probability that:
   (a) There are 2 balls of each color. [Answer: $\frac{30}{377}$.]
   (b) The first ball is red, the third ball is white, and the fifth ball is blue. [Answer: $\frac{25}{812}$.]
5.27. Solve Exercise 5.26 for the case in which drawing is done with replacement. [Answer: (a) $\frac{5}{72}$, (b) $\frac{1}{36}$.]
5.28. In a bridge game, North is dummy and South notes that between them North and South have all the spades except the K, 9, 4. What is the probability that the K is unguarded? [Answer: $\frac{1}{4}$.]
5.29. Three balls are dropped into 3 boxes. What is the probability that box 1 is empty? What is the probability that exactly 1 box is empty? [Answer: $\frac{8}{27}$, $\frac{2}{3}$.]
5.30. Four players are dealt poker hands. Find the probability that each player is dealt an ace. (See Example 4.4.)
5.31. Suppose that a pair of dice is rolled and that the 2 dice are considered indistinguishable. List the 21 elements of the corresponding *unordered* sample space. If the dice are fair, assign the correct probability to each of these outcomes.

5.4. CONDITIONAL PROBABILITY; BAYES' THEOREM

Suppose that a pair of dice is rolled in the following manner. One die is thrown on a table in full view while the second die is placed in a closed box and then rattled inside the box. The lid of the box is then raised and the sum of the dice is observed. If we wish to guess the sum of the dice before the lid of the box is raised, the knowledge of the value of the first die will clearly affect our guess. We will certainly not guess a sum of 10 if the first die comes up 3. Thus, the information provided by the first die will influence our judgment as to the total outcome, and evidently the probability that the sum will be a certain number is altered by our knowledge of the first die. For example, the probability of rolling a sum of 8 with a pair of fair dice is $\frac{5}{36}$. However, if the value of the first die is known to be 3, then the sum depends only on the die in the box. In effect, our interest in the sample space of outcomes for the experiment of rolling a pair of dice is narrowed to the subset consisting of those outcomes for which the first die is a 3. (See Figure 5.7.) We see that the probability of rolling 8 given the condition that the first die is 3 is $\frac{1}{6}$ since (3,5) is the only outcome yielding a sum of 8 among 6 possible outcomes having 3 as the value of the first die.

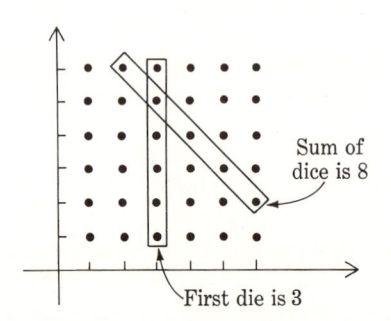

Figure 5.7

These are simple instances of the notion of conditional probability. We have found the conditional probability of an event E (rolling an 8) given the information that an event F (first die is 3) has occurred. The precise definition of this notion now follows.

Definition. Let (S,P) be a probability space and let E and F be events in S where $P(F) \neq 0$. Then the *conditional probability of E given F* is denoted by $P(E|F)$ and is defined to be

$$P(E|F) = \frac{P(E \cap F)}{P(F)}.$$

The knowledge that F has occurred means that in considering an event E we are interested in those elements of E that belong to F (that is, $E \cap F$) and the conditional probability of E given F is a measurement of the portion of E that lies in F. In fact, if we define a function P_F on the set of all subsets of S by

$$P_F(E) = P(E \cap F), \quad E \subset S$$

then it is easy to verify that P_F has all of the defining properties of a probability measure on S except that $P_F(S) = P(F)$ need not be 1. Consequently, if we divide the expression defining $P_F(E)$ by $P(F)$, we obtain an expression defining a probability measure on S. This probability measure on S is precisely the function that assigns to each event E in S its conditional probability $P(E|F)$.

5.4. CONDITIONAL PROBABILITY; BAYES' THEOREM

In the case of the equally likely probability measure, the conditional probability of E given F is precisely the ratio of the number of elements of E that belong to F to the number of elements of F. Thus, in the above dice example, if E is the event "rolling an 8" and F is the event "the first die is 3," then

$$P(E|F) = \tfrac{1}{6}$$

since F has six elements and $E \cap F$ has one element.

We shall now consider some other examples involving conditional probabilities.

Example 5.16. A man has 2 drawers of socks. The first drawer has 8 black socks and 4 brown socks and the second drawer has 4 black socks and 6 brown socks. He takes 1 sock from each drawer. Assuming that each sock is equally likely to be selected, what is the probability that they are both black given that they match in color?

There are $12 \cdot 10 = 120$ possible pairs of socks obtainable in this way and we have assumed that each is equally likely to be selected. Let E be the event that both socks are black and let F be the event that the socks match in color. Note that $E \cap F = E$. Now

$$P(E) = \frac{8 \cdot 4}{120} = \frac{32}{120}$$

since there are 8 ways to choose a black sock from the first drawer and 4 ways to get a black sock from the second drawer. Also,

$$P(F) = \frac{8 \cdot 4 + 4 \cdot 6}{120} = \frac{56}{120}.$$

Thus,

$$P(E|F) = \tfrac{32}{56} = \tfrac{4}{7}.$$

There are times when it is easy to compute the conditional probability $P(E|F)$ of certain events in a direct manner. If the definition of conditional probability is written in the form

$$P(E|F)P(F) = P(E \cap F),$$

this leads to a method of calculating probabilities of events which we illustrate by an example.

Example 5.17. Suppose 2 urns are given and that Urn I contains 3 red balls and 5 white balls while Urn II contains 4 red balls and 3 white balls. An experiment is performed in two stages. First, one of the urns is chosen at random, and then a ball is drawn from the selected urn. Let us find the probability that this drawing results in a red ball.

This compound experiment can be represented by a "tree diagram." (See Figure 5.8.) At the first stage of the experiment, an urn is selected.

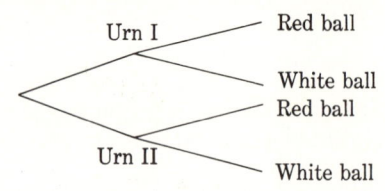

```
                Urn I  —— Red ball
                       \
                        —— White ball
                        —— Red ball
                       /
                Urn II —— White ball
```

Figure 5.8

Let U_1 be the event that the ball is taken from Urn I and let U_2 be the event that the ball is taken from Urn II. If E is the event that the drawn ball is red, we have $E = (E \cap U_1) \cup (E \cap U_2)$ and the two events $E \cap U_1$ and $E \cap U_2$ are mutually exclusive. Hence

$$P(E) = P(E \cap U_1) + P(E \cap U_2)$$
$$= P(E|U_1)P(U_1) + P(E|U_2)P(U_2).$$

Since the urn is selected at random, we know that $P(U_1) = P(U_2) = \frac{1}{2}$. The conditional probabilities involved here are easy to compute, for once an urn is specified we are dealing with the drawing of 1 ball from a single urn. Since 3 of the 8 balls in Urn I are red, we have $P(E|U_1) = \frac{3}{8}$. Similarly, $P(E|U_2) = \frac{4}{7}$. Hence

$$P(E) = \frac{3}{8} \cdot \frac{1}{2} + \frac{4}{7} \cdot \frac{1}{2} = \frac{53}{112}.$$

This procedure can be viewed in the following manner. Let us assign each branch of the tree diagram a probability as shown in Figure 5.9. Then there are 2 paths leading from left to right along which a red ball can be obtained. If we multiply the probabilities on the branches of each

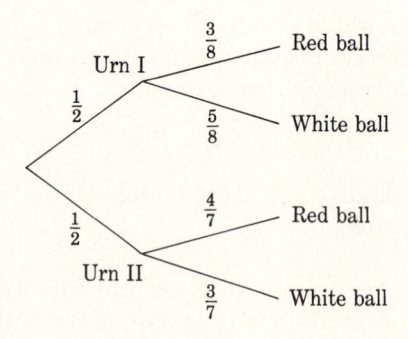

```
                      3/8
              Urn I  ———— Red ball
                   \
              1/2   5/8
                    ———— White ball
                    4/7
              1/2   ———— Red ball
                   /
              Urn II
                    3/7
                    ———— White ball
```

Figure 5.9

path and then add, we get the probability of drawing a red ball. Similarly, the probability of drawing a white ball is

$$\tfrac{5}{8} \cdot \tfrac{1}{2} + \tfrac{3}{7} \cdot \tfrac{1}{2} = \tfrac{59}{112}.$$

The method illustrated above can be stated in the following form.

Proposition 5.4. If (S,P) is a probability space and if S is the union of disjoint subsets $F_1, F_2, \cdots, F_n$ such that $P(F_i) \neq 0$ for each i, then

$$P(E) = \sum_{i=1}^{n} P(E|F_i)P(F_i)$$

for any event E in S.

PROOF. The collection $\{F_1, \cdots, F_n\}$ is a partition of S and the situation described in the hypothesis is pictured in Figure 5.10. For each $E \subset S$ it

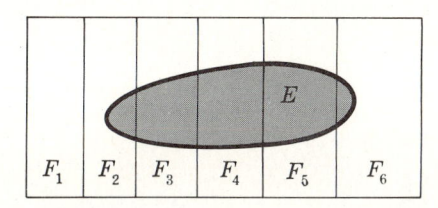

Figure 5.10

is true that $E = \bigcup_{i=1}^{n} (E \cap F_i)$ and that the events $E \cap F_i$, $i = 1, 2, \cdots, n$ are mutually exclusive in pairs. Thus, by the finite additivity property (P-7) of P, we have

$$P(E) = \sum_{i=1}^{n} P(E \cap F_i).$$

By applying the relation $P(E \cap F_i) = P(E|F_i)P(F_i)$ for each i, we obtain the desired result.

The formula $P(E \cap F) = P(E|F)P(F)$ can be extended to more than two events. For example,

$$P(E_1 \cap E_2 \cap E_3) = P(E_1)P(E_2|E_1)P(E_3|E_1 \cap E_2).$$

The associated tree diagrams can be constructed accordingly as illustrated in the following example.

Example 5.18. A supposed quality control expert states that a proposed quality control system consists of 3 inspection phases. The probability of a defective item being discovered at the first phase is 0.2, at the second phase is 0.3, and at the third phase is 0.5. Therefore, he reasons, it will be impossible for a defective item to remain undetected by the system since the probability of it being discovered is $0.2 + 0.3 + 0.5 = 1$.

The conclusion is, of course, ridiculous. For if the first inspection phase were later improved to 0.25 effectiveness instead of 0.20, then according to his reasoning the probability of a defective item being discovered would be 1.05. A high probability indeed!

The correct probability can be calculated as follows. Consider a defective item and let E_i denote the event that it is discovered at the ith inspection phase and let F_i be the event that it successfully eludes the ith inspection phase, where $i = 1, 2, 3$. The various possible fates awaiting the item can be represented by the tree diagram in Figure 5.11 where

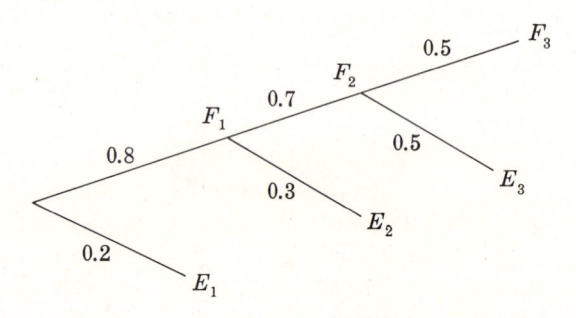

Figure 5.11

the given probabilities are placed at the appropriate locations. The probability that the quality control system will fail to discover the item is thus seen to be

$$P(F_1 \cap F_2 \cap F_3) = P(F_1)P(F_2|F_1)P(F_3|F_1 \cap F_2)$$
$$= (0.8)(0.7)(0.5)$$
$$= 0.28.$$

Therefore, the probability that the defective item will be discovered is $1 - 0.28 = 0.72$, a long way from certainty.

There is a simple consequence of Proposition 5.4 known as Bayes' Theorem that has interesting applications to computing probabilities.

Proposition 5.5. *(Bayes' Theorem)* If (S,P) is a probability space and if S is the union of disjoint subsets $F_1, \cdots, F_n$ of S such that $P(F_i) \neq 0$ for

$i = 1, 2, \cdots, n$, then for any event E in S with $P(E) \neq 0$,

$$P(F_j|E) = \frac{P(F_j)P(E|F_j)}{\sum_{i=1}^{n} P(F_i)P(E|F_i)}$$

for $j = 1, 2, \cdots, n$.

PROOF. By definition of conditional probability, we have

$$P(F_j|E) = \frac{P(F_j \cap E)}{P(E)} = \frac{P(F_j)P(E|F_j)}{P(E)}$$

and by Proposition 5.4, the denominator $P(E)$ of the preceding expression can be written as

$$P(E) = \sum_{i=1}^{n} P(F_i)P(E|F_i).$$

Consequently, Bayes' formula follows.

We shall now describe one general setting in which Bayes' Theorem is often applied. Suppose that every outcome in a sample space S corresponding to a certain experiment is in one of n mutually exclusive events $F_1, \cdots, F_n$ in S. The probabilities $P(F_1), \cdots, P(F_n)$ are assumed to be known prior to performing the experiment and are therefore called *a priori* probabilities. The experiment is then performed and it is observed that the event E occurs. Bayes' Theorem then enables us to compute the so-called *a posteriori* probabilities $P(F_1|E), \cdots, P(F_n|E)$, that is, the probabilities $P(F_i|E)$ that the event F_i has occurred given that the event E is known to have occurred. Of course, the application of Bayes' Theorem in such a situation requires a knowledge of the conditional probabilities $P(E|F_1), \cdots, P(E|F_n)$. However, it often happens that these latter probabilities are readily computable, and in such cases, Bayes' Theorem is a useful device for computing the *a posteriori* probabilities.

We shall now consider a simple application of Bayes' Theorem of the sort described above.

Example 5.19. Two boxes of 50 items each are received from 2 different manufacturers, but the shipping labels have been lost. In the past, 10 percent of the items from Manufacturer I have been defective while 20 percent of the items from Manufacturer II have been defective. One of the boxes is opened and an item is removed. Upon inspection, it is found to be defective. What is the probability that the opened box came from Manufacturer II?

Notice that this is just a ball and urn problem of the same basic structure as in Example 5.17. However, the viewpoint has been reversed. Instead of asking for the probability that the selected item is defective (or that the ball is red), we are told that it is defective and want the probability that it came from a certain manufacturer (or urn).

Let F_1 be the event that the item came from Manufacturer I and F_2 the event that it came from Manufacturer II. Since the boxes were equally likely to be chosen, we have *a priori* probabilities $P(F_1) = P(F_2) = \frac{1}{2}$. Let E be the event that the selected item was defective. The given percentages tell us that

$$P(E|F_1) = \tfrac{1}{10}, \qquad P(E|F_2) = \tfrac{1}{5}.$$

Now an application of Bayes' Theorem yields

$$\begin{aligned} P(F_2|E) &= \frac{P(F_2)P(E|F_2)}{P(F_1)P(E|F_1) + P(F_2)P(E|F_2)} \\ &= \frac{\tfrac{1}{2} \cdot \tfrac{1}{5}}{\tfrac{1}{2} \cdot \tfrac{1}{5} + \tfrac{1}{2} \cdot \tfrac{1}{10}} \\ &= \tfrac{2}{3}. \end{aligned}$$

Thus, the knowledge of the sampled item increased the probability that the opened box came from Manufacturer II from $\frac{1}{2}$ to $\frac{2}{3}$.

EXERCISES

5.32. A pair of dice is rolled. Find the conditional probability of rolling a 7 given that:
  (a) The sum is odd. [Answer: $\tfrac{1}{3}$.]
  (b) The sum is even. [Answer: 0.]
  (c) At least one of the dice is a 4. [Answer: $\tfrac{2}{11}$.]

5.33. A fair coin is flipped 5 times. Find the conditional probability that heads will occur on the first 3 flips given that at least 3 heads occur. [Answer: $\tfrac{1}{4}$.]

5.34. An unscrupulous poker player deals and manages to guarantee that his hand will contain the ace of spades. What is the probability that he will have 3 aces?

5.35. Two drawers contain black, brown, and blue socks as indicated.

	Black	Brown	Blue
Drawer I	6	3	2
Drawer II	3	4	4

A man selects one of the drawers at random and then selects a pair of socks at random from that drawer. What is the probability that both socks will be black? What is the probability that the socks will match in color? [Answer: $\frac{9}{55}, \frac{17}{55}$.]

5.36. Four balls are drawn with replacement from an urn containing 8 white balls and 4 red balls. Find the conditional probability that the third ball drawn is white given that 3 of the 4 balls are white. [Answer: $\frac{3}{4}$.]

5.37. Three boxes contain numbered discs as follows: Box I contains discs numbered from 1 to 5, Box II has discs numbered from 1 to 10, and Box III has discs numbered from 6 to 13. A person rolls a fair die and draws a disc from Box I if the die comes up 1, from Box II if the die comes up 2, and from Box III otherwise. What is the probability that an even numbered disc is drawn? [Answer: $\frac{29}{60}$.]

5.38. A student takes a true-false exam. If he knows the answer to a question, which is 60 percent of the time, he marks the correct answer. Otherwise, he flips a fair coin and answers true if the coin lands heads and false if the coin lands tails. What is the probability that he answers a given question correctly? Given that he answered a particular question correctly, what is the probability that he knew the answer? [Answer: $\frac{4}{5}, \frac{3}{4}$.]

5.39. Urn I contains 3 red balls and 2 white balls. Urn II contains 2 red balls and 5 white balls. An urn is selected at random and then a ball is drawn from that urn. Given that the drawn ball is red, what is the probability that it came from Urn I? [Answer: $\frac{21}{31}$.]

5.40. Consider the urns described in Exercise 5.39. Suppose that 1 ball is selected at random from Urn I and placed in Urn II, and then a ball is drawn from Urn II. If the drawn ball is white, what is the probability that the transferred ball was red? [Answer: $\frac{5}{9}$.]

5.41. In a group of 10 marksmen, 5 can hit a target with probability 0.6, 3 can hit a target with probability 0.7, and 2 can hit a target with probability 0.9.
   (a) One of the marksmen is selected at random to fire at the target. What is the probability he hits? [Answer: $\frac{69}{100}$.]
   (b) One of the marksmen fires and misses the target. What is the probability that he belongs to the first group? [Answer: $\frac{20}{31}$.]

5.5. INDEPENDENT EVENTS; REPEATED TRIALS

Suppose that E and F are events in the probability space (S, P) and suppose that $P(E) \neq 0$ and $P(F) \neq 0$. One way to say that the event E does not depend on, or is not influenced by, the event F is to say that the (unconditional) probability $P(E)$ of E is the same as the conditional probability $P(E|F)$ of E given F has occurred. If this is the case, then

$$P(E) = P(E|F) = \frac{P(E \cap F)}{P(F)},$$

that is,

$$P(E)P(F) = P(E \cap F).$$

We see that this in turn implies that F does not depend on the occurrence of E since

$$P(F|E) = \frac{P(E \cap F)}{P(E)} = \frac{P(E)P(F)}{P(E)} = P(F).$$

These considerations suggest the following definition for the independence of two events.

Definition. Two events E and F in the probability space (S,P) are *independent* if $P(E \cap F) = P(E)P(F)$. If E and F are not independent, they are called *dependent*.

Notice that this definition makes sense even if E or F has probability 0, in which case conditional probability is not defined. Also, observe that an event F for which $P(F) = 0$ is independent of any other event E in S since $P(E \cap F) = 0$ in this case.

If E and F are mutually exclusive events, then $P(E \cap F) = P(\emptyset) = 0$. If, in addition, these events are independent, then we must have either $P(E) = 0$ or $P(F) = 0$. Hence, since it is easy to find mutually exclusive events with nonzero probabilities, we see that independence is not a consequence of two events being mutually exclusive.

It is also possible to have independent events that are not mutually exclusive. For example, if a coin is flipped twice, then the event E of getting a head on the first flip and the event F of getting 1 head and 1 tail are independent since $P(E) = \frac{1}{2}$, $P(F) = \frac{1}{2}$, and $P(E \cap F) = \frac{1}{4}$; however, $E \cap F \neq \emptyset$ so that E and F are not mutually exclusive.

It is reasonable to expect that if the probability of an event E is not affected by an event F, then it should not be affected by the complementary event F'. The anticipated relation between independence and complementation is quite correct as the following result shows.

Proposition 5.6. If E and F are events in a probability space (S,P), then the following statements are equivalent:

(a) E and F are independent.
(b) E and F' are independent.
(c) E' and F are independent.
(d) E' and F' are independent.

PROOF. If E and F are independent, then

$$\begin{aligned} P(E \cap F') &= P(E) - P(E \cap F) \\ &= P(E) - P(E)P(F) \\ &= P(E)(1 - P(F)) \\ &= P(E)P(F'). \end{aligned}$$

Therefore, E and F' are independent events. The remaining implications can be established in a similar manner.

In many cases, it is obvious when two given events are independent. For example, suppose a playing card is drawn from a standard deck of playing cards, suppose E is the event that the drawn card is a club, and suppose F is the event that the drawn card is a queen. Since a club may or may not be a queen and since a queen may be of any suit, it seems intuitively clear that these events are independent. This is easily verified from the definition, for since $P(E) = \frac{1}{4}$, $P(F) = \frac{1}{13}$, and $P(E \cap F) = \frac{1}{52}$, we see that $P(E \cap F) = P(E)P(F)$.

However, there are many occasions when intuition is inadequate and computation is essential.

Example 5.20. Suppose a fair coin is flipped n times where $n \geq 3$. Let E be the event that at least 1 head and 1 tail are obtained. Let F be the event that there is at most 1 tail. The complementary event E' is the event that there are either no heads or no tails, and this can occur in only 2 ways. Thus,

$$P(E) = 1 - P(E') = 1 - \frac{2}{2^n}.$$

Also, we have

$$P(F) = P(0 \text{ tails}) + P(1 \text{ tail}) = \frac{1}{2^n} + \frac{n}{2^n}.$$

Finally, since $E \cap F$ is the event that there is exactly 1 tail and this tail can come up on any one of the n flips, it follows that

$$P(E \cap F) = \frac{n}{2^n}.$$

If these events are to be independent, it is necessary that

$$P(E \cap F) = \frac{n}{2^n} = \left(1 - \frac{1}{2^{n-1}}\right)\left(\frac{1+n}{2^n}\right) = P(E)P(F),$$

that is,

$$2^{n-1} = n + 1.$$

We see that if $n = 3$, this equation is satisfied, so that for 3 flips of a coin the events are indeed independent. However, if $n \geq 4$, then $2^{n-1} > n + 1$ (as may be verified by induction) and hence the events E and F are dependent if the coin is flipped 4 or more times.

Example 5.21. Suppose a pair of fair dice is rolled and consider the event E, the first die comes up 2, and the event F, both dice come up 2. Are E and F independent?

Instead of resorting to the definition, we can show E and F are dependent by a somewhat simpler computation. Observe that the conditional probability $P(E|F)$ is obviously equal to 1 since if both dice come up 2, then evidently the first die comes up 2. Also, we find that $P(E) = \frac{1}{6}$. It then follows that $P(E|F) \neq P(E)$; hence, the events E and F must be dependent.

EXERCISES

5.42. Two balls are drawn from an urn containing r red balls and w white balls. E is the event that the first ball drawn is red and F is the event that the second ball drawn is red. Determine whether or not E and F are independent when the drawing is done (a) with replacement, and (b) without replacement.

5.43. Three fair dice are rolled. Let E be the event that the first die is 2. Determine if the events E and F are independent when F is the event:
(a) All 3 dice are 2.
(b) All 3 dice have the same value.
(c) The sum of the dice is 10.

5.44. A poker hand is dealt. Consider the event E that the hand contains the ace of spades, the event F that the hand contains the king of clubs, and the event G that all 5 cards have the same suit.
(a) Are E and F independent?
(b) Are E and G independent?

5.45. Based on past performances, pole-vaulter A can clear 16 feet with probability $\frac{2}{3}$ and pole-vaulter B can clear 16 feet with probability $\frac{3}{4}$. If both attempt this feat (independently), what is the probability that at least one of them will be successful? [Answer: $\frac{11}{12}$.]

Repeated Trials

The notion of independence provides a way of motivating the proper construction of a probability space when several experiments are performed in succession. Suppose two experiments are performed and let

5.5. INDEPENDENT EVENTS; REPEATED TRIALS

(S_1, P_1), (S_2, P_2) be the probability spaces corresponding to these experiments. We assume that the outcomes of either of the experiments do not affect in any way the outcomes of the other so that they are, in an intuitive sense, independent of one another. If the experiments are now performed in succession, one after the other, the sample space S for this procedure will be the set of ordered pairs (x,y) where $x \in S_1$ and $y \in S_2$. That is, S is the Cartesian product $S_1 \times S_2$ of S_1 and S_2.

We wish to define a probability measure P on S that is related to P_1 on S_1 and P_2 on S_2. If p, p_1, p_2 denote the corresponding point probabilities, the appropriate relation between them is given by

$$p(x,y) = p_1(x) p_2(y)$$

for each $(x,y) \in S$. This can be motivated by arguing as follows. Each element $(x,y) \in S$ can be viewed as the intersection of two events in S, namely, $\{(x,y)\} = \{\{x\} \times S_2\} \cap \{S_1 \times \{y\}\}$. This is illustrated schematically in Figure 5.12. The assumption that the outcomes of the separate experiments do not affect each other is tantamount to saying that

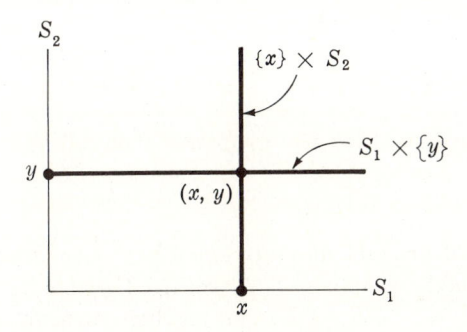

Figure 5.12

these events are independent, that is,

$$P((\{x\} \times S_2) \cap (S_1 \times \{y\})) = P(\{x\} \times S_2) P(S_1 \times \{y\}).$$

Since the probability of the event $\{x\} \times S_2$ is obviously determined only by $\{x\}$, it is quite natural to set $P(\{x\} \times S_2) = p_1(x)$. Similarly, it seems reasonable to have $P(S_1 \times \{y\}) = p_2(y)$. We thus conclude that

$$p(x,y) = P(\{x\} \times S_2) P(S_1 \times \{y\}) = p_1(x) p_2(y).$$

Although this appears to be the appropriate definition of p on S, it must still be verified that p is actually a point probability on S. This is easily done since p is clearly nonnegative and

$$P(S) = \sum_{y \in S_2} \sum_{x \in S_1} p(x,y) = \sum_{y \in S_2} p_2(y) \sum_{x \in S_1} p_1(x) = 1.$$

The probability measure P defined above on $S = S_1 \times S_2$ by p is called the *product measure* of P_1 and P_2. It is readily seen that if $E \subset S_1$ and $F \subset S_2$, then $P(E \times F) = P_1(E)P_2(F)$. Note that if P_1 and P_2 are the equally likely probability measures on S_1 and S_2, respectively, then the product measure P on $S = S_1 \times S_2$ is nothing more than the equally likely probability measure on S.

Example 5.22. As we mentioned in Example 5.2, one way of biasing a die is to shave the 1 and 6 faces slightly so that the die has a higher probability of coming up 1 or 6 than a fair die. Let us suppose that the probability of rolling a 1 or a 6 is $\frac{1}{4}$ each and the probability of rolling a 2, 3, 4, or 5 is $\frac{1}{8}$ each. If a pair of such dice are rolled, what is the probability of rolling a 7?

For the roll of a single die, we take the sample space $S_1 = \{1,2,3,4,5,6\}$. The given values define the point probability p_1 on S_1 as $p_1(1) = p_1(6) = \frac{1}{4}$, $p_2(2) = p_1(3) = p_1(4) = p_1(5) = \frac{1}{8}$. If $S = S_1 \times S_1$ and P is the corresponding product measure on S, and if $E = \{(1,6),(2,5),(3,4),(4,3),(5,2),(6,1)\}$ is the event in S of rolling a sum of 7, then we obtain

$$P(E) = \tfrac{1}{4} \cdot \tfrac{1}{4} + \tfrac{1}{8} \cdot \tfrac{1}{8} + \tfrac{1}{8} \cdot \tfrac{1}{8} + \tfrac{1}{8} \cdot \tfrac{1}{8} + \tfrac{1}{8} \cdot \tfrac{1}{8} + \tfrac{1}{4} \cdot \tfrac{1}{4} = \tfrac{3}{16}.$$

Observe this is greater than the probability of rolling a 7 with a pair of fair dice.

The notion of product measure extends to the product of any finite number of probability spaces $(S_1,P_1), (S_2,P_2), \cdots, (S_n,P_n)$. For if $S = S_1 \times S_2 \times \cdots \times S_n$, then the product measure P on S is given by the point probability p defined on S by

$$p(x_1,x_2,\cdots,x_n) = p_1(x_1)p_2(x_2) \cdots p_n(x_n),$$

where p_i is the point probability for the probability measure P_i ($i = 1, \cdots, n$). If E_i is an event in S_i, $i = 1, 2, \cdots, n$, then

$$P(E_1 \times E_2 \times \cdots \times E_n) = P_1(E_1)P_2(E_2) \cdots P_n(E_n).$$

The following examples indicate how probabilities can be computed by using products.

Example 5.23. Suppose 3 dart players have probabilities of hitting the bullseye of a target of $\frac{1}{4}, \frac{2}{5}, \frac{2}{3}$, respectively. If all 3 players throw simultaneously, what is the probability the first player misses and the other 2 players hit the bullseye?

5.5. INDEPENDENT EVENTS; REPEATED TRIALS

If E_i is the event that the ith player will hit the bullseye, then the desired probability is

$$P(E_1' \times E_2 \times E_3) = \tfrac{3}{4} \cdot \tfrac{2}{5} \cdot \tfrac{2}{3} = \tfrac{1}{5}.$$

Example 5.24. Suppose n urns each contain n balls such that in the kth urn there are k red balls and $n - k$ white balls. If 1 ball is drawn from each urn, then the probability that every ball is red is

$$\frac{1}{n} \cdot \frac{2}{n} \cdots \frac{n-1}{n} \cdot \frac{n}{n} = \frac{(n-1)!}{n^{n-1}}.$$

Bernoulli Trials

The term *Bernoulli trial*, used in honor of Jakob Bernoulli (1654–1705), refers to an experiment that has only two outcomes. Because of the early association between probability and games of chance, the two outcomes are traditionally called "success" and "failure." In general, these outcomes are not equally likely and we will let p denote the probability of success and $q = 1 - p$ denote the probability of failure.

Jakob Bernoulli
David Smith Collection

Suppose a particular Bernoulli trial is repeated n times. This gives rise to a sample space of 2^n distinct n-tuples, where each term of an n-tuple represents either success or failure. Let $B(k; n,p)$ denote the probability of getting exactly k successes ($k \leq n$) out of the n trials where p is the probability of success on a single trial. A formula for $B(k; n,p)$ can be obtained as follows. Consider a specific n-tuple consisting of k successes and $n - k$ failures. The probability assigned to this n-tuple is the product of the probabilities of the individual terms of the n-tuple, and since there are k

successes and $n - k$ failures, this is $p^k q^{(n-k)}$ where $q = 1 - p$. Since there are $\binom{n}{k}$ ways of obtaining such an n-tuple (the number of ways of selecting k terms out of n), we conclude that

$$B(k; n,p) = \binom{n}{k} p^k q^{n-k}.$$

Example 5.25. An urn containing red balls and white balls has twice as many red balls as white balls. What is the probability of getting at least 3 red balls when 5 balls are drawn with replacement?

The ratio of red balls to white balls gives us that the probability of drawing a red ball (success) on a single draw is $p = \frac{2}{3}$. The probability of the described event is thus

$$\begin{aligned} B(3; 5,\tfrac{2}{3}) &+ B(4; 5,\tfrac{2}{3}) + B(5; 5,\tfrac{2}{3}) \\ &= \binom{5}{3} (\tfrac{2}{3})^3 (\tfrac{1}{3})^2 + \binom{5}{4} (\tfrac{2}{3})^4 (\tfrac{1}{3}) + \binom{5}{5} (\tfrac{2}{3})^5 \\ &= \frac{192}{3^5} \\ &= \tfrac{64}{81}. \end{aligned}$$

Example 5.26. One of the problems that the gambler de Méré discussed with Pascal in the seventeenth century is the following. Which is more likely to occur: (a) at least 1 ace in a single roll of 4 dice, or (b) at least 1 double ace in 24 rolls of a pair of dice? (The term ace refers to the die coming up 1.) It is assumed the dice are all unbiased. de Méré felt intuitively that these events should have equal probabilities, yet his gambling experience led him to believe that the former event was actually more likely than the latter.

If we regard the roll of single die to be a Bernoulli trial where success refers to rolling an ace and failure refers to not rolling an ace, then the probability of success is $\frac{1}{6}$ and the probability of failure is $\frac{5}{6}$. The roll of 4 dice can be viewed as this Bernoulli trial performed 4 times. Therefore, the probability of getting no aces in the roll of 4 dice is

$$B(0; 4, \tfrac{1}{6}) = (\tfrac{5}{6})^4.$$

Thus, the probability of the event (a) of getting at least 1 ace in the roll of 4 dice is

$$1 - (\tfrac{5}{6})^4 = 0.517747.$$

The experiment in (b) can be similarly viewed as a repeated Bernoulli trial. In this case, our Bernoulli trial is the roll of a pair of dice and success refers to getting a double ace. The probability of success is $\frac{1}{36}$ in this case.

Thus, the probability of getting no double aces on 24 rolls of a pair of dice is

$$B(0; 24, \tfrac{1}{36}) = (\tfrac{35}{36})^{24}.$$

The probability of getting at least 1 double ace in 24 rolls is therefore

$$1 - (\tfrac{35}{36})^{24} = 0.491404.$$

Thus, de Méré's gambling instinct was correct, even though the difference in the probabilities is only 0.026343!

EXERCISES

5.46. One urn contains 8 red balls and 6 white balls, a second urn contains 5 red balls and 6 white balls, and a third urn contains 3 red balls and 7 white balls. One ball is drawn from each of the urns. What is the probability they are the same color?   [Answer: $\tfrac{93}{385}$.]

5.47. A pair of dice is rolled. One of the dice is fair and the other is a 1 − 6 flat as in Example 5.22.

(a) What is the probability of rolling 7? rolling 11?

[Answer: $\tfrac{1}{6}, \tfrac{1}{16}$.]

(b) What is the probability that the fair die will come up a greater value than the biased die?   [Answer: $\tfrac{5}{12}$.]

5.48. If a pair of fair dice is rolled 10 times, what is the probability of getting:

(a) At least one 7.   [Answer: $1 - (\tfrac{5}{6})^{10}$.]

(b) At least two 7's.

5.49. A multiple choice test consists of 10 questions with 4 possible answers given for each question. If exactly one of the choices is correct in each case and if a student guesses randomly on each question, what is the probability that he will answer at least 8 questions correctly?   [Answer: $109/4^9$.]

5.50. If 5 balls are tossed randomly into 3 boxes, find the probability that Box I receives:

(a) Exactly 3 balls.   [Answer: $\tfrac{40}{243}$.]

(b) At least 3 balls.   [Answer: $\tfrac{17}{81}$.]

5.51. Two teams play a 7-game series, the series ending as soon as one team wins 4 games. Assuming that the teams are evenly matched, find the probability that the series goes a full 7 games.   [Answer: $\tfrac{5}{16}$.]

5.52. (Banach's Matchbox Problem.) A man uses 2 matchboxes for smoking and reaches for one or the other at random. Suppose each matchbox held n matches originally. What is the probability that when the last match is taken from one of the boxes, the other box will contain exactly k matches? $\left[\text{Answer: } \binom{2n-k}{n} 2^{k-2n}.\right]$

5.53. A basketball player has a probability of $\frac{3}{4}$ of making a free throw. If he attempts 10 free throws, what is the most probable number of successes? [Answer: 8.]

5.54. An urn contains 10 red and 2 white balls. n balls are drawn with replacement from the urn. What is the smallest number of draws needed so that the probability of getting at least 1 white ball is greater than $\frac{1}{2}$? [Answer: $n = 4$.]

5.55. Three marksman can hit a target with probabilities $\frac{1}{2}$, $\frac{2}{3}$, $\frac{3}{4}$, respectively. All 3 fire at a target simultaneously and there are exactly 2 hits. What is the probability that the first marksman missed? [Answer: $\frac{6}{11}$.]

REMARKS AND REFERENCES

It is now rather common to find a unit on probability theory in most textbook series intended for the secondary schools. It is usually taken up in the fourth year of the college preparatory program. However, as is suggested by Sobel [6], a unit on probability theory can be used to great advantage in mathematics classes for slow learners as well. In this latter setting, the study of this subject provides an excellent opportunity to do remedial work with arithmetic skills and at the same time cover material that the students find interesting and prestigous.

The classroom discussion of probability theory should include experimentation in determining probabilities empirically. For example, it is a very instructive homework exercise to flip a coin or roll a die 100 times to determine the frequency of the various outcomes. The comparison of the combined results for the entire class should support the use of the equally likely probability measure in experiments of this sort, and the likely occurrence of long runs of a single outcome for some students should help to dispel some of the misconceptions concerning the meaning of statements such as "the probability that the coin comes up heads is $\frac{1}{2}$." Some students may also enjoy filing down two opposite faces of a die and then determining empirically a probability measure that would be appropriate for the experiment of rolling that die. Certainly, the probabilistic conclusion of the Birthday Problem (see Example 5.15) should also be tested in the classroom since students generally find this result to be surprising and interesting.

Many secondary school textbooks make use of the classical definition of probability in terms of relative frequency of "favorable" outcomes. There seems to be little advantage to this definition from a computational point of view and it has the distinct disadvantage of neglecting the modern theory and methods. Since the language and techniques of set algebra are familiar to students at this level, there is no good reason to avoid the modern definition of a probability measure as a function defined on the set of events in a sample space. This point of view is essential in more advanced treatments of the subject that the student may encounter in his college training and there is no point in handicapping the student with the restrictive and outmoded classical definition of probability.

Any discussion of probability measures on finite sample spaces must be preceded or accompanied by a careful and detailed study of systematic counting procedures such as those considered in Chapter 4. These techniques are often difficult for beginning students. However, the consideration of a wide variety of problems and exercises requiring systematic counting procedures can do much to help the students to gain facility with the techniques involved.

If time permits, some discussion of probability measures on infinite sample spaces is worthwhile. In this connection, examples in which the probability measure is related to the area function as in Examples 5.7 and 5.8 are particularly instructive and intuitively appealing.

We now single out a few of the many books on probability theory for further reading. The books by Gangolli and Ylvisaker [2] and Goldberg [3] are fairly elementary and augment the discussion in this chapter quite well. For a good supply of supplementary examples and problems, see the book by Lipschutz [4]. The books by Feller [1] and Parzen [5] contain advanced material for the reader who wishes to pursue the subject more seriously.

1. Feller, William, *An Introduction to Probability Theory and Its Applications*, Vol. I, Second Edition, New York: John Wiley & Sons, Inc., 1957.
2. Gangolli, R. and D. Ylvisaker, *Discrete Probability*, New York: Harcourt, Brace & World, Inc., 1967.
3. Goldberg, S., *Probability, An Introduction*, Englewood Cliffs, N.J.: Prentice-Hall, Inc., 1960.
4. Lipschutz, S., *Probability*, Schaum's Outline Series, New York: McGraw-Hill Book Company, Inc., 1968.
5. Parzen, E., *Modern Probability Theory and Its Applications*, New York: John Wiley & Sons, Inc., 1960.
6. Sobel, Max A., *Teaching General Mathematics*, Englewood Cliffs, N.J.: Prentice-Hall, Inc., 1967.

Chapter 6

BOOLEAN ALGEBRAS AND THEIR APPLICATIONS

In Propositions 1.1 and 1.2 in Chapter 1, we established a number of basic properties of union, intersection, and complementation on the set $P(X)$ of all subsets of a given set X. In the remarks following the proof of Proposition 1.2, we gave alternate derivations of two parts of that proposition based solely on the set equations established in Proposition 1.1 without reference to the defined meanings of the operations $\cup$, $\cap$, ′, or of the sets X, $\emptyset$. Thus, these alternate proofs referred only to the fact that the set of all subsets of X is an algebraic system with two binary operations $\cup$, $\cap$, a unary operation ′ and two special elements $\emptyset$, X subject to the basic properties (1) through (5) in Proposition 1.1. We also remarked that the remaining conclusions of Proposition 1.2 as well as all other such equations for $P(X)$ could be derived in a similar fashion. In other words, the entire theory of the operations of union, intersection, and complementation for $P(X)$ can be developed from the fact that $(P(X), \cup, \cap, ′)$ is an algebraic system subject to the equations in Proposition 1.1.

The structure and properties of the algebraic system $(P(X), \cup, \cap, ′)$ are similar in many basic respects to other useful and interesting algebraic systems that arise quite naturally in mathematics and its applications. For this reason, we shall now study an abstract algebraic system, called a Boolean algebra, which will include as special cases the concrete algebraic system $(P(X), \cup, \cap, ′)$ as well as other significant algebraic systems. Later in this chapter, we shall use a very simple, concrete Boolean algebra to study switching networks and the logical design of computers.

The basic ideas of what is now known as the theory of Boolean algebras were initially developed by George Boole (1815–1864). In 1854, he published his fundamental treatise on this subject entitled *Laws of Thought* in which algebraic techniques were used to analyze the rules of logic and the methods of deductive reasoning. The subsequent development of the theory of Boolean algebras provided considerable stimulus for

George Boole
New York Public Library Picture Collection

research in abstract algebra during the first half of the present century. Also, this theory has been shown to have interesting ramifications in other branches of mathematics such as analysis and topology. However, probably the single most important factor contributing to the current strong interest in the theory of Boolean algebras is the relevance of this subject to the design of electrical networks. This latter connection, which was first observed by Claude Shannon in 1938, has played a significant role in the development of electronic computers and modern network theory.

6.1. BASIC PROPERTIES OF BOOLEAN ALGEBRAS

We begin this section with the definition of the abstract structure that will be our main object of study in this chapter.

Definition. A *Boolean algebra* $(B, \vee, \wedge, ')$ is a set B together with binary operations $\vee$ and $\wedge$ on B and a unary operation $'$ on B such that:

(1) For all a, b, c in B,

$$a \vee (b \vee c) = (a \vee b) \vee c$$
$$a \wedge (b \wedge c) = (a \wedge b) \wedge c.$$ (The Associative Properties)

(2) For all a and b in B,

$$a \vee b = b \vee a$$
$$a \wedge b = b \wedge a.$$ (The Commutative Properties)

(3) There exist unique elements 0 and 1 in B such that $0 \neq 1$ and

$$a \vee 0 = a$$
$$a \wedge 1 = a$$

for all a in B.

(4) For all a in B,

$$a \vee a' = 1$$
$$a \wedge a' = 0.$$

(5) For all a, b, c in B,

$$a \vee (b \wedge c) = (a \vee b) \wedge (a \vee c) \quad \text{(The Distributive}$$
$$a \wedge (b \vee c) = (a \wedge b) \vee (a \wedge c). \quad \text{Properties)}$$

The elements 0 and 1 of B are called the *zero element* of B and the *unit element* of B, respectively. The operation $\vee$ will sometimes be referred to as *cup* (or *join*), while the term *cap* (or *meet*) will be used to refer to $\wedge$. The operation $'$ is called *complementation* and a' is called the *complement* of a.

We shall often denote a Boolean algebra $(B, \vee, \wedge, ')$ more simply by B when there is no need to emphasize the particular nature of the operations on B or when the meaning of these operations is clearly understood from the context.

The following examples serve to clarify the concepts introduced above.

Example 6.1. If $P(X)$ is the power set of a nonempty set X and if $\vee, \wedge, '$ are interpreted to be the set operations of union, intersection, and complementation, respectively, then $P(X)$ is a Boolean algebra by virtue of Proposition 1.1. Note that the empty set $\emptyset$ is the zero element and the set X is the unit element of this Boolean algebra.

Example 6.2. If X is a nonempty set and if B is the set consisting of the two elements $\emptyset$ and X, then B is closed with respect to the operations of union, intersection, and complementation on $P(X)$. It follows that $(B, \vee, \wedge, ')$ is a Boolean algebra with zero element $\emptyset$ and unit element X.

Example 6.3. Suppose that B_2 is the set consisting of the integers 0 and 1 and suppose that operations $\vee, \wedge, '$ are defined on B_2 by the following tables:

$\vee$	0	1
0	0	1
1	1	1

$\wedge$	0	1
0	0	0
1	0	1

	$'$
0	1
1	0

6.1. BASIC PROPERTIES OF BOOLEAN ALGEBRAS

For example, $1 \vee 0 = 1$, $1 \wedge 1 = 1$, and $0' = 1$. It is an easy matter to verify that $(B_2, \vee, \wedge, ')$ is a Boolean algebra by a case check of the defining properties (1) through (5) for the two elements of B_2. Also, it is easy to see that if X is any nonempty set, then the mapping $\varphi \colon B_2 \to \{\emptyset, X\}$ defined by

$$\varphi(0) = \emptyset, \qquad \varphi(1) = X$$

is an isomorphism between the Boolean algebra B_2 and the Boolean algebra defined in Example 6.2.

Example 6.4. Consider a set B_4 consisting of four elements denoted by $0, 1, a, b$. Define the operations $\vee, \wedge, '$ on B by the following tables:

$\vee$	0	1	a	b
0	0	1	a	b
1	1	1	1	1
a	a	1	a	1
b	b	1	1	b

$\wedge$	0	1	a	b
0	0	0	0	0
1	0	1	a	b
a	0	a	a	0
b	0	b	0	b

	$'$
0	1
1	0
a	b
b	a

Then $(B_4, \vee, \wedge, ')$ is a Boolean algebra with zero element 0 and unit element 1.

Example 6.5. Suppose that X is a nonempty set and that $B_F(X)$ is the set of all subsets A of X such that either A or its complement A' in X is a finite set. Then the set $B_F(X)$ is closed with respect to the operations of union, intersection, and complementation on the power set $P(X)$ of X. (The reader should verify this statement.) Therefore, since $\emptyset \in B_F(X)$, $X \in B_F(X)$ and since $(P(X), \cup, \cap, ')$ is a Boolean algebra, it follows that $(B_F(X), \cup, \cap, ')$ is a Boolean algebra with zero element $\emptyset$ and unit element X. Note that if X is a finite set, then $B_F(X) = P(X)$.

Example 6.6. Suppose that m is a positive integer greater than 1 and that $D(m)$ is the set of all positive integers that divide m. For example,

$$D(42) = \{1,2,3,6,7,14,21,42\},$$
$$D(12) = \{1,2,3,4,6,12\}.$$

Define operations $\vee, \wedge, '$ on $D(m)$ as follows. If a and b are elements of $D(m)$, then:

$$a \vee b = \text{least common multiple of } a \text{ and } b,$$
$$a \wedge b = \text{greatest common divisor of } a \text{ and } b,$$
$$a' = \frac{m}{a}.$$

For example, if $m = 42$, then $2 \vee 3 = 6$, $6 \wedge 14 = 2$, and $3' = 14$.

The operations $\vee$ and $\wedge$ on $D(m)$ defined above evidently satisfy properties (1) and (2) for a Boolean algebra, while the integers 1 and m play the roles of the zero and unit elements respectively in condition (3). The Distributive Properties (5) for a Boolean algebra may be verified for the operations $\vee$ and $\wedge$ on $D(m)$ by making use of the relationships that the notions of least common multiple and greatest common divisor have with the prime factorization of natural numbers. (See Proposition 3.13.)

However, $D(m)$ is not a Boolean algebra in general since property (4) concerning complements is not always satisfied. For example, if $m = 12$, then $6' = 2$ and $2 \vee 6 = 6$. But 6 is not the unit element of $D(12)$ and therefore $D(12)$ does not have the property that $a \vee a'$ equals the unit element for all a. Nor does $a \wedge a'$ equal the zero element in $D(12)$ when $a = 6$ since $6 \wedge 6' = 6 \wedge 2 = 2$. Thus, $(D(12), \vee, \wedge, ')$ is *not* a Boolean algebra.

The reason property (4) fails in $D(12)$ is that $12 = 2 \cdot 2 \cdot 3$ has the prime number 2 repeated in its prime factorization, allowing 6 and its complement to share a factor of 2. We now prove that the absence of repeated prime factors in m is a necessary and sufficient condition for $D(m)$ to have property (4), and hence for $D(m)$ to be a Boolean algebra. Positive integers having the property that they are the product of distinct prime numbers are frequently called *square-free*.

Proposition 6.1. $(D(m), \vee, \wedge, ')$ is a Boolean algebra if and only if the positive integer m is the product of distinct prime numbers.

PROOF. If m is the product of distinct prime numbers and if a is a factor of m, then a and its complement $a' = m/a$ have no prime factors in common. Also, every prime factor of m is either a prime factor of a or a prime factor of a'. Therefore, the least common multiple of a and a' is m, while the greatest common divisor of a and a' is 1. Thus, property (4) is satisfied and $D(m)$ is a Boolean algebra.

On the other hand, if m is not the product of distinct prime numbers, then there is a prime number p such that p^2 is a divisor of m. But then, the element p of $D(m)$ is such that p is a divisor of both p and its complement $p' = m/p$. It follows that the greatest common divisor of p and p' is p, and hence $p \wedge p'$ does not equal the zero element of $D(m)$. Therefore, property (4) for a Boolean algebra is not valid for $D(m)$ so that $D(m)$ is not a Boolean algebra.

For example, since $30 = 2 \cdot 3 \cdot 5$ and $462 = 2 \cdot 3 \cdot 7 \cdot 11$, we see from Proposition 6.1 that $D(30)$ and $D(462)$ are Boolean algebras. However, $D(28)$ is not a Boolean algebra since $28 = 2 \cdot 2 \cdot 7$.

EXERCISES

6.1. Construct tables for the operations $\vee$, $\wedge$, and $'$ on the Boolean algebras $D(15)$ and $D(30)$. Is $D(1470)$ a Boolean algebra?

6.2. If X is an infinite set, show that the sets $P(X)$ and $B_F(X)$ are distinct sets and that $B_F(X)$ is an infinite set.

The five axioms for a Boolean algebra involve equations such as $a \vee (b \vee c) = (a \vee b) \vee c$ and $a \wedge 1 = a$ in which a, b, c are arbitrary elements of the Boolean algebra. Equivalently, the symbols a, b, c appearing in these equations can be regarded as values for variables x, y, z whose domains are the Boolean algebra B. When this latter viewpoint is adopted, the equations mentioned in each of the axioms are viewed as identities, that is, as equations involving variables that are true when the variables involved are replaced by arbitrary elements of B. The expressions on each side of these identities combine these variables and the elements 0 and 1 by means of the operations $\vee$, $\wedge$, $'$. Such expressions are called Boolean formulas. More precisely, a *Boolean formula* is an expression obtained by combining variables $x, y, z, \cdots$ whose domains are a Boolean algebra B, and possibly the elements 0 and 1 of B, by applying the operations $\vee$, $\wedge$, and $'$ a finite number of times. For example, $(x \vee y') \wedge 1$ and $(x' \wedge y) \vee z$ are Boolean formulas.

Each Boolean formula on B defines a function whose range is B since each replacement of the variables in the formula by elements of B results in an element of B. Such functions defined by Boolean formulas will be called *Boolean functions*. Of course, different Boolean formulas may define the same Boolean function. For example, since a Boolean algebra B has the Distributive Properties (5), each of the Boolean formulas $x \wedge (y \vee z)$ and $(x \wedge y) \vee (x \wedge z)$ define the same Boolean function. That is, both of these formulas yield the same element of B when the variables x, y, z are replaced by any elements a, b, c of B.

Two Boolean formulas that define the same Boolean function are said to be *equivalent* and this relationship is designated by equating the two Boolean formulas. For example, since $x \wedge (y \vee z)$ and $(x \wedge y) \vee (x \wedge z)$ are equivalent Boolean formulas, we write

$$x \wedge (y \vee z) = (x \wedge y) \vee (x \wedge z).$$

Example 6.7. If x and y are variables whose domains are the Boolean algebra $P(X)$, where X is a given nonempty set, then $(x \wedge y)'$ is a Boolean formula on $P(X)$. If A and B are specific subsets of X, then for the values A, B of x, y, respectively, the Boolean function on $P(X)$ defined by this Boolean formula has value $(A \cap B)'$. Since $(A \cap B)' = A' \cup B'$ by the

De Morgan Property for sets (see Proposition 1.2), we see that the Boolean formula $x' \vee y'$ defines the same Boolean function as $(x \wedge y)'$. Thus, these two Boolean formulas on $P(X)$ are equivalent. (We shall soon see that $(x \wedge y)' = x' \vee y'$ when x and y are variables whose domains are *any* Boolean algebra B.)

Each of the defining properties (1) through (5) for a Boolean algebra can be regarded as a pair of equations in which Boolean formulas occur. Note that each equation in such a pair can be obtained from the other equation by simply interchanging $\vee$ and $\wedge$ and also interchanging 0 and 1 whenever they appear. For example, the equation $a \wedge 1 = a$ in (3) is obtained from $a \vee 0 = a$ by replacing $\vee$ by $\wedge$ and 0 by 1.

Two Boolean formulas are *dual* if each can be obtained from the other by replacing $\vee$ by $\wedge$, $\wedge$ by $\vee$, 0 by 1, and 1 by 0 whenever they occur. In this case, we refer to each of the Boolean formulas as the dual of the other. For example, the dual of the Boolean formula $x \wedge (y \vee 1)$ is the Boolean formula $x \vee (y \wedge 0)$. With this terminology, we see that the defining properties (1) through (5) of a Boolean algebra each consists of two equations, the second equation involving the duals of the Boolean formulas occurring in the first equation. This leads to the following important observation.

Duality Principle. If two Boolean formulas on a Boolean algebra B can be shown to be equivalent on the basis of defining properties (1) through (5) of a Boolean algebra, then the dual Boolean formulas are also equivalent.

For example, once it can be shown that $a \vee 1 = 1$ for all elements a of a Boolean algebra B, it will then follow from the Duality Principle that $a \wedge 0 = 0$ for all a in B. Similarly, the identity $a \wedge (a \vee b) = a$ will follow from the identity $a \vee (a \wedge b) = a$ for the same reason.

The following result is the abstract analogue of Proposition 1.2 which dealt with the specific Boolean algebra $P(X)$ of all subsets of a nonempty set X.

Proposition 6.2. If $(B, \vee, \wedge, ')$ is a Boolean algebra with zero element 0 and unit element 1 and if a, b are members of B, then:

(6) $a \vee a = a$ (The Idempotent Properties)
    $a \wedge a = a$.

(7) $0' = 1$
    $1' = 0$.

6.1. BASIC PROPERTIES OF BOOLEAN ALGEBRAS

(8) $a \vee 1 = 1$
      $a \wedge 0 = 0.$

(9) $a \vee (a \wedge b) = a$ (The Absorption Properties)
      $a \wedge (a \vee b) = a.$

(10)   If for some $c \in B$ it is true that
      $a \wedge c = b \wedge c$ and $a \vee c = b \vee c$,
      then $a = b$.

(11)   If $a \wedge b = 0$ and $a \vee b = 1$,
      then $b = a'$.   (Uniqueness of Complements)

(12) $(a \vee b)' = a' \wedge b'$
      $(a \wedge b)' = a' \vee b'.$ (The De Morgan Identities)

PROOF. In view of the Duality Principle, it is sufficient to prove only the first identity in each of (6), (7), (8), (9), and (12).

(6) $a = a \vee 0$ by (3)
            $= a \vee (a \wedge a')$ by (4)
            $= (a \vee a) \wedge (a \vee a')$ by (5)
            $= (a \vee a) \wedge 1$ by (4)
            $= a \vee a.$ by (3)

(7) $0' = 0' \vee 0$ by (3)
            $= 1.$ by (2), (4)

(8) $a \vee 1 = a \vee (a \vee a')$ by (4)
            $= (a \vee a) \vee a'$ by (1)
            $= a \vee a'$ by (6)
            $= 1.$ by (4)

(9) $a \vee (a \wedge b) = (a \wedge 1) \vee (a \wedge b)$ by (3)
                  $= a \wedge (1 \vee b)$ by (5)
                  $= a \wedge 1$ by (2), (8)
                  $= a.$ by (3)

(10) $a = a \vee (a \wedge c)$ by (9)
            $= a \vee (b \wedge c)$ since $a \wedge c = b \wedge c$
                                          by hypothesis
            $= (a \vee b) \wedge (a \vee c)$ by (5)
            $= (a \vee b) \wedge (b \vee c)$ since $a \vee c = b \vee c$
                                          by hypothesis
            $= b \vee (a \wedge c)$ by (2), (5)
            $= b \vee (b \wedge c)$ since $a \wedge c = b \wedge c$
                                          by hypothesis
            $= b.$ by (9)

(11) Since $a \wedge b = 0 = a \wedge a'$ and $a \vee b = 1 = a \vee a'$, the fact that $b = a'$ follows from (10).

(12) It follows from (11) that in order to prove $(a \vee b)' = a' \wedge b'$, it suffices to prove both $(a \vee b) \wedge (a' \wedge b') = 0$ and $(a \vee b) \vee (a' \wedge b') = 1$.

These equations can be established as follows:

$$\begin{aligned}
(a \vee b) \wedge (a' \wedge b') &= [(a' \wedge b') \wedge a] \vee [(a' \wedge b') \wedge b] && \text{by (2), (5)} \\
&= [b' \wedge (a \wedge a')] \vee [a' \wedge (b \wedge b')] && \text{by (2), (1)} \\
&= (b' \wedge 0) \vee (a' \wedge 0) && \text{by (4)} \\
&= 0 \vee 0 && \text{by (8)} \\
&= 0. && \text{by (3) or (6)} \\
(a \vee b) \vee (a' \wedge b') &= [(a \vee b) \vee a'] \wedge [(a \vee b) \vee b'] && \text{by (5)} \\
&= [(a \vee a') \vee b] \wedge [(b \vee b') \vee a] && \text{by (1), (2)} \\
&= (1 \vee b) \wedge (1 \vee a) && \text{by (4)} \\
&= 1 \wedge 1 && \text{by (2), (8)} \\
&= 1. && \text{by (3) or (6)}
\end{aligned}$$

The defining properties of a Boolean algebra together with the results of Proposition 6.2 provide a working basis for the algebraic simplification of Boolean formulas. As we shall see, such simplifications are quite useful in the applications of the theory. The following examples indicate how simplifications can be performed.

Example 6.8. The Boolean formula

$$(x \wedge y) \vee (x \wedge y') \vee (x' \vee y)$$

can be simplified as follows. We first note that

$$\begin{aligned}
(x \wedge y) \vee (x \wedge y') &= x \wedge (y \vee y') && \text{by (5)} \\
&= x \wedge 1 && \text{by (4)} \\
&= x. && \text{by (3)}
\end{aligned}$$

Thus,

$$\begin{aligned}
(x \wedge y) \vee (x \wedge y') \vee (x' \wedge y) &= x \vee (x' \vee y) \\
&= (x \vee x') \vee y && \text{by (1)} \\
&= 1 \vee y && \text{by (4)} \\
&= y \vee 1 && \text{by (2)} \\
&= 1. && \text{by (8)}
\end{aligned}$$

6.1. BASIC PROPERTIES OF BOOLEAN ALGEBRAS

Example 6.9.

$$\begin{aligned}
(x \wedge y') \vee [x \wedge (y \wedge z)'] \vee z \\
= (x \wedge y') \vee [x \wedge (y' \vee z')] \vee z & \quad \text{by (12)} \\
= (x \wedge y') \vee [(x \wedge y') \vee (x \wedge z')] \vee z & \quad \text{by (5)} \\
= (x \wedge y') \vee (x \wedge z') \vee z & \quad \text{by (1), (6)} \\
= (x \wedge y') \vee [(x \vee z) \wedge (z' \vee z)] & \quad \text{by (5)} \\
= (x \wedge y') \vee [(x \vee z) \wedge 1] & \quad \text{by (2), (4)} \\
= (x \wedge y') \vee (x \vee z) & \quad \text{by (3)} \\
= [(x \wedge y') \vee x] \vee z & \quad \text{by (1)} \\
= x \vee z. & \quad \text{by (9)}
\end{aligned}$$

Example 6.10. To simplify

$$(x \wedge y \wedge z) \vee (x \wedge y \wedge z') \vee (x' \wedge y \wedge z)$$

we can use the following device. First note that by using the Idempotent Property and the Associativity Property, the above Boolean formula can be written as

$$[(x \wedge y \wedge z) \vee (x \wedge y \wedge z')] \vee [(x \wedge y \wedge z) \vee (x' \wedge y \wedge z)].$$

Now the first bracketed expression is equivalent to

$$(x \wedge y) \wedge (z \vee z') = (x \wedge y) \wedge 1 = x \wedge y$$

and the second bracketed expression is equivalent to

$$(x \vee x') \wedge (y \wedge z) = 1 \wedge (y \wedge z) = y \wedge z.$$

Hence, the original Boolean formula is equivalent to

$$(x \wedge y) \vee (y \wedge z) = y \wedge (x \vee z).$$

We shall conclude this section with the discussion of a binary operation on a Boolean algebra that can be used to relate the theory of Boolean algebras to that of commutative rings with units. The latter type of algebraic system was considered in Section 2.2. Those readers who have not studied the material in Chapter 2 on commutative rings with units may choose to proceed directly to the exercises at the end of this section since the remaining sections of this chapter will make no use of the material to be discussed here.

If a and b are elements of a Boolean algebra B, we define the *Boolean sum* $a + b$ of a and b as follows:

$$a + b = (a \wedge b') \vee (a' \wedge b).$$

248 BOOLEAN ALGEBRAS AND THEIR APPLICATIONS

It is an easy matter to verify that this defines a binary operation $+$ on the Boolean algebra B, which we shall call *Boolean addition*, and that this operation has the following properties:

(i) For all a and b in B, $\quad a + b = b + a$.
(ii) For all a in B, $\quad a + 0 = a$.
(iii) For all a in B, $\quad a + a = 0$.

The following result establishes two other basic properties of Boolean addition.

Proposition 6.3. If a, b, c are elements of a Boolean algebra B, then:

(iv) $\quad a + (b + c) = (a + b) + c$
(v) $\quad a \wedge (b + c) = (a \wedge b) + (a \wedge c)$.

PROOF. To verify (iv), we first note that

(*) $(a + b) + c = \{[(a \wedge b') \vee (a' \wedge b)] \wedge c'\}$
$\qquad \vee \{[(a \wedge b') \vee (a' \wedge b)]' \wedge c\}$.

The second bracketed expression can be rewritten as follows:

$$
\begin{aligned}
[(a \wedge b') \vee (a' \wedge b)]' &= (a \wedge b')' \wedge (a' \wedge b)' & &\text{by (12)} \\
&= (a' \vee b'') \wedge (a'' \vee b') & &\text{by (12)} \\
&= (a' \vee b) \wedge (a \vee b') & &\text{by Exercise 6.4} \\
&= \{a' \wedge (a \vee b')\} \vee \{b \wedge (a \vee b')\} & &\text{by (5)} \\
&= (a' \wedge a) \vee (a' \wedge b') \vee (b \wedge a) \vee (b \wedge b') & &\text{by (5)} \\
&= 0 \vee (a' \wedge b') \vee (b \wedge a) \vee 0 & &\text{by (4)} \\
&= (a \wedge b) \vee (a' \wedge b'). & &\text{by (2), (3)}
\end{aligned}
$$

Therefore, (*) together with (5) and (1) yield

$(a + b) + c = \{[(a \wedge b') \vee (a' \wedge b)] \wedge c'\} \vee \{[(a \wedge b) \vee (a' \wedge b')] \wedge c\}$
$\qquad = [a \wedge b' \wedge c'] \vee [a' \wedge b \wedge c'] \vee [a \wedge b \wedge c] \vee [a' \wedge b' \wedge c]$.

In the last expression, the roles of a and c may be interchanged without changing the element represented by the expression. Hence, $(a + b) + c = (c + b) + a$ for all a, b, c in B. Thus (iv) holds since Boolean addition is commutative by (i).

In order to prove (v) we observe that

$a \wedge (b + c) = a \wedge [(b \wedge c') \vee (b' \wedge c)] = [a \wedge b \wedge c'] \vee [a \wedge b' \wedge c]$

6.1. BASIC PROPERTIES OF BOOLEAN ALGEBRAS

(by (5) and (1)) and also that

$$\begin{aligned}(a \wedge b) + (a \wedge c) &= [(a \wedge b) \wedge (a \wedge c)'] \vee [(a \wedge b)' \wedge (a \wedge c)] \\ &= [(a \wedge b) \wedge (a' \vee c')] \vee [(a' \vee b') \wedge (a \wedge c)] \\ &= [a \wedge b \wedge a'] \vee [a \wedge b \wedge c'] \vee [a' \wedge a \wedge c] \\ &\qquad \vee [b' \wedge a \wedge c] \\ &= [0] \vee [a \wedge b \wedge c'] \vee [0] \vee [a \wedge b' \wedge c] \\ &= [a \wedge b \wedge c'] \vee [a \wedge b' \wedge c].\end{aligned}$$

(We leave it to the reader to justify each step of the preceding computation.) Comparing the extreme right-hand expressions in the above equations, we conclude that (v) holds for all a, b, c in B.

Corollary. If B is a Boolean algebra, if $+$ is the operation of Boolean addition, and if $\cdot$ is the cap operation on B, then $(B,+,\cdot)$ is a commutative ring with unit element 1.

The proof of the corollary is left to the reader.

EXERCISES

6.3. Suppose a and b are elements of a Boolean algebra. Prove that:
    (a) If $a \vee b = a \wedge b$, then $a = b$.
    (b) If $a \wedge b = 1$, then $a = 1$ and $b = 1$.
    (c) If $a \vee b = 0$, then $a = 0$ and $b = 0$.

6.4. Prove that $(a')' = a$ for any element a of a Boolean algebra.

6.5. Show that:
    (a) $x \vee (x' \wedge y) = x \vee y$.
    (b) $(x \wedge y) \vee [x \wedge (x \wedge y)'] = x$.
    (c) $[x \vee (y \wedge x')] \wedge [y \vee (z \wedge y')] \wedge z = (x \vee y) \wedge z$.
    (d) $(a \wedge b \wedge c) \vee (a \wedge b \wedge c') \vee (a \wedge b' \wedge c) \vee (a' \wedge b \wedge c)$
                                         $= (a \wedge b) \vee (a \wedge c) \vee (b \wedge c)$.

6.6. Determine which pairs of the following Boolean formulas are equivalent:
    (a) $(x \wedge z) \vee (x' \wedge y)$.
    (b) $(x \wedge y') \vee z$.
    (c) $(x \vee y) \wedge (x' \vee z) \wedge (y \vee z)$.

6.7. Simplify each of the following:
    (a) $(a \wedge b \wedge c) \vee (a \wedge b' \wedge c) \vee (b' \vee c')$.
    (b) $[(a \wedge b') \vee (b \wedge c')]' \wedge (a \wedge b)$.
    (c) $(x \wedge y \wedge z) \vee (x \wedge y) \vee (x \wedge y \wedge w) \vee (x \wedge w)$
        $\vee (x \wedge z \wedge w) \vee (x \wedge w)$.

6.8. Suppose the relation $\leq$ is defined on a Boolean algebra B as follows: $a \leq b$ if and only if $a = a \wedge b$.
   (a) Show that $\leq$ is an order relation on B; that is, show that $\leq$ is:
      (i) reflexive: $a \leq a$ for all $a \in B$.
      (ii) antisymmetric: $a \leq b$ and $b \leq a$ imply $a = b$.
      (iii) transitive: $a \leq b$ and $b \leq c$ imply $a \leq c$.
   (b) Interpret the meaning of $\leq$ in the Boolean algebras $P(X)$ and $D(m)$.
   (c) Show that $a \leq b$ if and only if $a \wedge b' = 0$.
   (d) Show that $a \leq b$ if and only if $b' \leq a'$.
   (e) Show that $0 \leq a$ and $a \leq 1$ for all $a \in B$.
6.9. Prove that any Boolean algebra B that contains precisely four elements must be isomorphic to the Boolean algebra B_4 of Example 6.4. [*Hint:* If c and d denote the elements of B that are distinct from the zero and unit elements of B, show that $c' = d$ and $d' = c$.]
6.10. Suppose $F(B)$ denotes the set of all Boolean functions on a Boolean algebra B. Show that it is possible to define operations $\vee$, $\wedge$, and $'$ on $F(B)$ so that $(F(B), \vee, \wedge, ')$ is a Boolean algebra.

6.2. FINITE BOOLEAN ALGEBRAS

The material in this section is not used in the remainder of this chapter. Consequently, the reader may proceed directly to Section 6.3 if he wishes.

Some of the examples of Boolean algebras introduced in the preceding section contain only a finite number of elements. For instance, the Boolean algebras discussed in Examples 6.2 and 6.3 each consist of two elements, and the Boolean algebra defined in Example 6.4 contains four elements. Moreover, we observed in Example 6.6 that if m is a square-free positive integer, then $D(m)$ is a finite Boolean algebra. In view of such examples, it is natural to pose the following questions:

(1) Given a positive integer $n > 1$, is there a Boolean algebra B such that B contains precisely n elements?

(2) Do there exist nonisomorphic Boolean algebras with the same finite number of elements?

The results of this section will provide complete answers to both of these questions.

The answer to the first question is negative in general. In fact, it is not difficult to show that there does not exist a Boolean algebra B that contains precisely three elements. For, if we assume that B consists of only three elements, then one of these elements must be the zero element

0 and another must be the unit element 1. This leaves one element, call it a, that is distinct from both 0 and 1. Now $a' \in B$, yet:

(1) $a' \neq 0$ since $a' = 0$ would imply $a = 1$.
(2) $a' \neq 1$ since $a' = 1$ would imply $a = 0$.
(3) $a' \neq a$ since $a' = a$ would imply $a = a \vee a = a \vee a' = 1$.

Therefore, no Boolean algebra exists that contains precisely three elements. More generally, we have the following result.

Proposition 6.4. If n is a positive integer greater than 1 and if there is a Boolean algebra B that contains precisely n elements, then n must be an even number.

PROOF. Since B contains more than one element and since $a \wedge a' = 0$, it is true that $a \neq a'$ for all $a \in B$. Define two subsets B_0 and B_1 of B as follows: Place the zero element in B_0 and the unit element in B_1. If $B \neq \{0,1\}$, choose $a \in B$ such that $a \neq 0$ and $a \neq 1$. Place a in B_0 and a' in B_1. If $B \neq \{0,1,a,a'\}$, choose $b \in B$ such that $b \neq 0$, $b \neq 1$, $b \neq a$, $b \neq a'$. Place b in B_0 and b' in B_1. Proceeding in this way, we eventually exhaust all of the elements of B since B is finite. Hence, each element of B is placed in one of the two disjoint sets B_0 or B_1. Since B_0 and B_1 obviously contain the same number of elements, we conclude that B must have an even number of elements.

Although the preceding result is interesting and simple to prove, its conclusion can be considerably sharpened with a more complete analysis of the structure of finite Boolean algebras. Our next objective is to carry out such an analysis. As we shall see, there is a strong analogy between the structure theory for finite Boolean algebras and the divisibility theory for positive integers. To emphasize this analogy, we shall borrow terminology and descriptive titles for results from the theory of numbers in the development that follows.

Definition. If a and b are elements of a Boolean algebra B, then a *divides* b, written $a|b$, if $a = a \wedge b$. If $p \in B$ and $p \neq 0$, then p is *prime* if $a|p$ implies that $a = 0$ or $a = p$.

Note that if the positive integer m is the product of distinct prime numbers and if a and b are elements of the Boolean algebra $D(m)$ of positive divisors of m, then the defined meaning of $a|b$ reduces to its usual meaning. That is, $a|b$ if and only if $b = k \cdot a$ for some positive integer k. Also, the prime elements of $D(m)$ are just the prime factors of m. On the other hand, if a and b are elements of the Boolean algebra $P(X)$ (or $B_F(X)$) for some nonempty set X, then $a|b$ if and only if a is a subset of b.

252   BOOLEAN ALGEBRAS AND THEIR APPLICATIONS

In this case, the prime elements are just the singletons in X, that is, the subsets of X that contain precisely one element.

The following result lays the foundation for factorization in a finite Boolean algebra.

Lemma 6.1. If a and b are elements of a Boolean algebra B, then $a|b$ if and only if $b = a \vee b$. If $a|b$, if $a \neq 0$, and if $a \neq b$, there is a unique $c \in B$ such that $b = a \vee c$, $a \wedge c = 0$, $c \neq 0$ and $c \neq b$.

PROOF. If $a|b$, then $a \vee b = (a \wedge b) \vee b = b$ by the Absorption Property (see (9) in Proposition 6.2). On the other hand, if $a \vee b = b$, then $a \wedge b = a \wedge (a \vee b) = a$ for the same reason, that is, $a|b$.

If $a|b$, define $c = b \wedge a'$. Then

$$a \vee c = a \vee (b \wedge a') = (a \vee b) \wedge (a \vee a') = (a \vee b) \wedge 1 = a \vee b = b$$

and also $a \wedge c = a \wedge (b \wedge a') = (a \wedge a') \wedge b = 0 \wedge b = 0$. If $c = b$, then $b = b \wedge a'$ so that $a = b \wedge a = (b \wedge a') \wedge a = b \wedge (a' \wedge a) = 0$, contrary to the assumption that $a \neq 0$; consequently, $c \neq b$. If $c = 0$, then

$$a = a \vee (b \wedge a') = (a \vee b) \wedge (a \vee a') = b \wedge 1 = b$$

contrary to the assumption that $a \neq b$; consequently, $c \neq 0$. The uniqueness of c follows from (10) of Proposition 6.2.

Lemma 6.2. *(Existence of Prime Factorization)* If B is a finite Boolean algebra and if a is a nonzero element of B, there exist primes $p_1, \cdots, p_n$ in B such that $a = p_1 \vee \cdots \vee p_n$.

PROOF. If a is a prime element, we are finished. If a is not prime, then Lemma 6.1 shows that there exist nonzero elements a_1 and a_2 such that $a = a_1 \vee a_2$ and $a_1 \wedge a_2 = 0$. If a_1 and a_2 are both prime elements, we have obtained the desired representations of a. If not, we can apply Lemma 6.1 again, and so on. All elements obtained in this way are distinct and nonzero. Consequently, this procedure must terminate since B is finite, and it is clear that the process terminates only when $a = p_1 \vee \cdots \vee p_n$ where each p_i is a prime element in B.

Lemma 6.3. *(Euclid's Lemma)* If p, a, b are elements of a Boolean algebra, if p is prime, and if $p|a \vee b$, then $p|a$ or $p|b$.

PROOF. Suppose p does not divide a. Since $p|a \vee b$, it follows that $p = p \wedge (a \vee b) = (p \wedge a) \vee (p \wedge b)$. Also $(p \wedge a)|p$ and $p \wedge a \neq p$,

6.2. FINITE BOOLEAN ALGEBRAS

so that $p \wedge a = 0$. But then $p = (p \wedge a) \vee (p \wedge b) = p \wedge b$, that is, $p|b$.

Corollary. If p is prime and $p|(a_1 \vee \cdots \vee a_n)$, then $p|a_i$ for some i.

Lemma 6.4. *(Uniqueness of Prime Factorization)* If $P = \{p_1, \cdots, p_n\}$ and $Q = \{q_1, \cdots, q_m\}$ are two sets of distinct primes in a Boolean algebra B and if $p_1 \vee \cdots \vee p_n = q_1 \vee \cdots \vee q_m$, then $P = Q$.

PROOF. Since $p_i | q_1 \vee \cdots \vee q_m$ for $i = 1, 2, \cdots, n$, it follows that each such p_i divides some q_j. But q_j is prime and $p_i \neq 0$; hence, each p_i coincides with some q_j. Thus, $P \subset Q$. A similar argument applied to the q_j's shows that $Q \subset P$. Therefore, $P = Q$.

The preceding results enable us to sharpen the conclusion of Proposition 6.4 and to provide the following complete answer to the first question posed at the beginning of this section.

Proposition 6.5. If B is a finite Boolean algebra, then B contains 2^n elements where n is the number of prime elements of B.

PROOF. By virtue of Lemmas 6.2 and 6.4, each nonzero element of B can be uniquely represented in terms of prime elements. Thus, each element in B either has a given prime element in this representation or it does not. Consequently, if B contains n prime elements, then B contains precisely $\underbrace{2 \cdots 2}_{n \text{ terms}} = 2^n$ elements.

We now establish a representation theorem for finite Boolean algebras that will provide the answer to the second question posed at the beginning of this section.

Proposition 6.6. If B is a finite Boolean algebra that contains precisely n prime elements, and if $q_1, \cdots, q_n$ are n distinct prime numbers, then B is isomorphic to the Boolean algebra $D(m)$ of positive divisors of $m = q_1 \cdots q_n$.

PROOF. Suppose that $\{p_1, \cdots, p_n\}$ is the set of prime elements in B. Define a mapping $\varphi: B \to D(m)$ as follows:

(1) Define $\varphi(0) = 1$.
(2) If $a \neq 0$ and if $\{p_{k_1}, \cdots, p_{k_r}\}$ is the unique set of prime elements of B such that $a = p_{k_1} \vee \cdots \vee p_{k_r}$, define $\varphi(a) = q_{k_1} \cdots q_{k_r}$.

Note that since $q_{k_1}, \cdots, q_{k_r}$ are distinct prime numbers, $\varphi(a)$ is the least common multiple of these numbers. Since each positive divisor of m is the product of a unique subset of the set of prime numbers $\{q_1, \cdots, q_n\}$ and since the prime factorization of elements of B is unique, we see that φ is a one-to-one mapping of B onto $D(m)$.

If a and b are elements of B and if $P_a = \{p_{k_1}, \cdots, p_{k_r}\}$, $P_b = \{p_{h_1}, \cdots, p_{h_s}\}$ are the sets of prime divisors of a, b respectively, then $P_a \cup P_b$ is the set of prime divisors of $a \vee b$, while $P_a \cap P_b$ is the set of prime divisors for $a \wedge b$. From this it follows that $\varphi(a \vee b) = \varphi(a) \vee \varphi(b)$ and that $\varphi(a \wedge b) = \varphi(a) \wedge \varphi(b)$ for all a, b in B. Consequently, φ is an isomorphism between the Boolean algebras B and $D(m)$ (see Exercise 6.11 below).

Corollary 1. If B and $\bar{B}$ are Boolean algebras with the same finite number of elements, then B and $\bar{B}$ are isomorphic.

PROOF. This result is an immediate consequence of the fact that there is a positive square-free integer m such that both B and $\bar{B}$ are ismorphic to $D(m)$.

Corollary 2. Every finite Boolean algebra B is isomorphic to the Boolean algebra $P(X)$ of all subsets of a finite set X.

PROOF. If B contains precisely n prime elements, if m is the product of n distinct prime numbers $p_1, \cdots, p_n$ and if $X = \{p_1, \cdots, p_n\}$, then a verification similar to that used in the proof of Proposition 6.6 shows that the mapping $\psi: P(X) \to D(m)$ defined by

(1) $\psi(\emptyset) = 1$,
(2) $\psi(\{p_{k_1}, \cdots, p_{k_r}\}) = p_{k_1} \cdots p_{k_r}$,

is an isomorphism between $P(X)$ and $D(m)$. Hence, since B is isomorphic to $D(m)$ for some square-free integer m, it follows that B is isomorphic to the Boolean algebra $P(X)$ of all subsets of some finite set X.

REMARK. In connection with Corollary 2, it should be pointed out that if B is an infinite Boolean algebra, then it is not necessarily true that B is isomorphic to the Boolean algebra $P(X)$ of *all* subsets of some infinite set X. For example, if **N** is the set of all natural numbers, then the Boolean algebra $B_F(\mathbf{N})$ cannot even be put into one-to-one correspondence with $P(X)$ for any set X since it is not difficult to verify that $B_F(\mathbf{N})$ is countable while $P(X)$ is either finite or uncountable for any nonempty set X (see Proposition 1.11). However, in 1936, M. H. Stone proved that for *any* Boolean algebra $(B, \vee, \wedge, ')$, there is a set X and a subset $\bar{B}$ of the

Boolean algebra $(P(X),\cup,\cap,')$ of all subsets of X such that:

(a) $\bar{B}$ is closed in $P(X)$ with respect to the operations of union, intersection, and complementation.

(b) $(B,\vee,\wedge,')$ and $(\bar{B},\cup,\cap,')$ are isomorphic Boolean algebras.

In other words, any Boolean algebra B is isomorphic to a Boolean algebra $\bar{B}$ of subsets of a set X and it may happen that $\bar{B}$ is a proper subset of $P(X)$.

EXERCISES

6.11. Suppose that φ is a one-to-one function from a Boolean algebra B onto a Boolean algebra $\bar{B}$. Prove that φ is an isomorphism if φ has any *two* of the following three properties:
   (a) $\varphi(a \vee b) = \varphi(a) \vee \varphi(b)$ for all a and b in B.
   (b) $\varphi(a \wedge b) = \varphi(a) \wedge \varphi(b)$ for all a and b in B.
   (c) $\varphi(a') = \varphi(a)'$ for all a in B.

6.12. If B is a Boolean algebra and if $a \in B$, then a is *prime factorable* in B if either $a = 0$ or there exist prime elements $p_1, \cdots, p_n$ in B such that $a = p_1 \vee p_2 \vee \cdots \vee p_n$. The set $\{p_1, \cdots, p_n\}$ is called a set of *prime factors of a*.
   (a) Prove that if a is prime factorable in B, the set of prime factors of a is unique.
   (b) If X is an infinite set, find all elements of the Boolean algebra $P(X)$ that are prime factorable.
   (c) If X is a nonempty set, prove that every element of the Boolean algebra $B_F(X)$ is prime factorable if and only if X is finite.
   (d) Suppose that B is a Boolean algebra and that B has the property that every element of B or its complement is prime factorable. Prove that B is isomorphic to the Boolean algebra $B_F(X)$ where X is the set of prime elements of B.

6.3. SWITCHING NETWORKS

In this section, we shall indicate how the algebraic properties of a Boolean algebra can be used to facilitate the logical design of certain electrical networks. As we shall see in the next section, these electrical networks can be constructed to perform simple computational functions. In fact, modern digital computers make use of many such networks as components, and an understanding of the logical design of these networks will provide the reader with some inkling of how computers work.

We shall only consider networks involving two-state switches, that is, switches for which there are precisely two mutually exclusive states, one allowing current to pass (closed state) and one not allowing current to pass (open state). The particular construction of a two-state switch is of no concern to the discussion that follows. Modern technology has developed several tiny and highly reliable devices that operate as two-state switches, but no understanding of these devices is required here. In fact, a common household light switch is a perfectly adequate model of a two-state switch to keep in mind for the analysis of networks that follows.

There are two very simple networks, one called a *basic parallel network* and the other called a *basic series network*, that will serve as building blocks for the type of networks we wish to consider. Each consists of a pair of two-state switches connected by wires to an input terminal and an output terminal as in Figure 6.1.

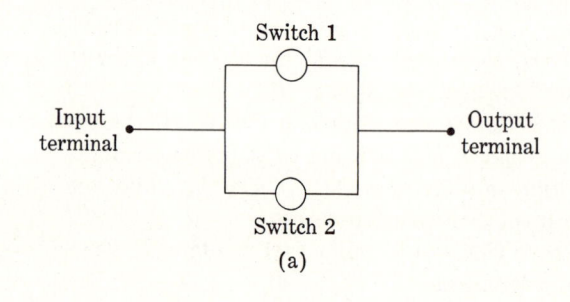

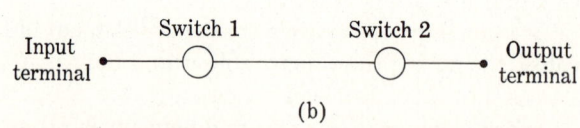

Figure 6.1 (a) Basic Parallel Network. (b) Basic Series Network.

In a basic parallel network, current can pass from the input terminal to the output terminal if and only if either Switch 1 *or* Switch 2 (or both) is closed. On the other hand, current reaching the input terminal of a basic series network can proceed to the output terminal if and only if both Switch 1 *and* Switch 2 are closed. Because of these operational features, basic parallel networks are sometimes called OR-*networks*, while basic series networks are often referred to as AND-*networks*.

Basic parallel and series networks may be combined to obtain more complex networks. This is done by cutting one of the wires or removing a switch in a given network, and then connecting the two resulting wire ends to the input and output terminals of a new OR-network or a new AND-network. This procedure is illustrated in Figure 6.2. Observe that

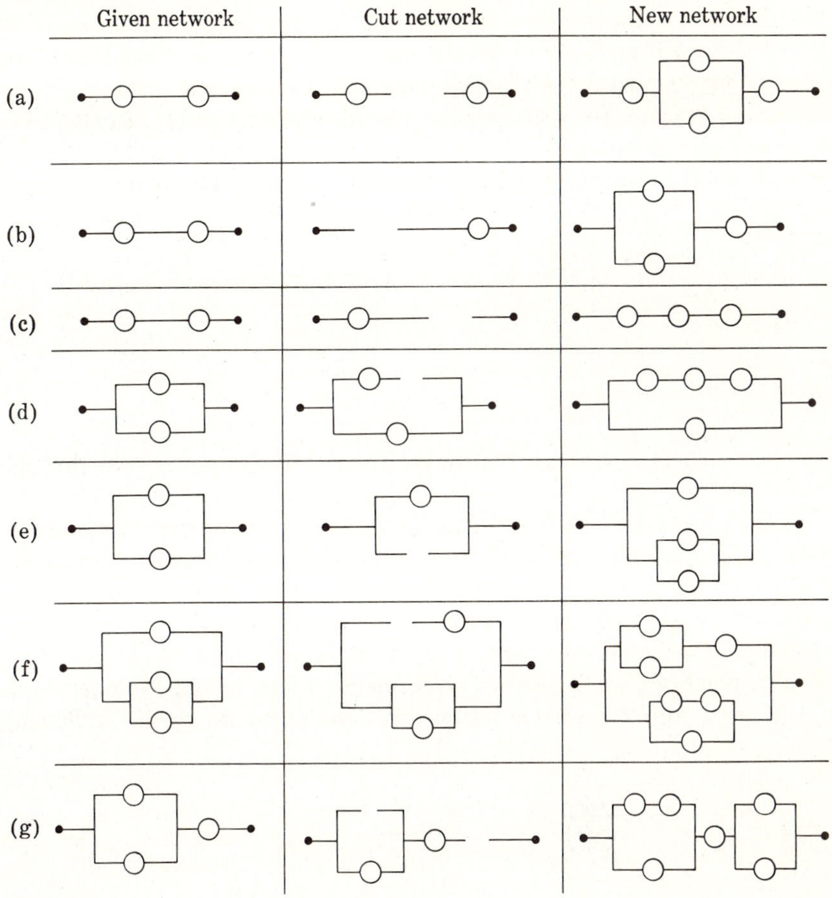

6.3. SWITCHING NETWORKS   257

Figure 6.2

the new network in (b) is the given network in (g) and that the new network in (e) is the given network in (f); consequently, we may regard the new networks in (f) and (g) as being built from the given basic networks (e) and (b) respectively by successively inserting new basic parallel or series networks. We shall refer to networks that can be obtained in this way as *series-parallel networks*.

We now introduce suitable notation for dealing with series-parallel networks. Switches in a network will be denoted by lower case letters x, y, and so on. If two switches x and y are connected so as to form a basic parallel network, this network will be denoted by $x \vee y$. A basic series network with switches x and y will be denoted by $x \wedge y$. It frequently happens that two or more switches in a network will be open or closed

258 BOOLEAN ALGEBRAS AND THEIR APPLICATIONS

simultaneously. Since such switches are always either all open or all closed, they will be denoted by the same letter. On the other hand, two switches might operate so that they are always in opposite states—when one switch is open, the other will be closed, and conversely. In this case, if one switch is denoted by a letter x, then the other switch will be denoted by x'. Note that the networks $x \vee x'$ and $x \wedge x'$ (see Figure 6.3) are such

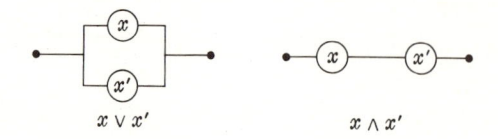

$x \vee x'$ $\quad\quad\quad x \wedge x'$

Figure 6.3

that current can always pass through $x \vee x'$ but can never pass through $x \wedge x'$.

The preceding notational conventions concerning switches and the basic networks were selected to facilitate the application of Boolean algebras to the study of series-parallel networks. To understand the connection between these two subjects, it is necessary to recall some facts concerning the Boolean algebra B_2 introduced in Example 6.3. The Boolean algebra B_2 consists of two elements 0 and 1, and the operations of cup, cap, and complementation are defined on B_2 by the following tables:

$\vee$	0	1		$\wedge$	0	1			$'$
0	0	1		0	0	0		0	1
1	1	1		1	0	1		1	0

A two-state switch x may be regarded as a variable with domain B_2 if we agree that:

(a) $x = 1$ means that the switch x is closed (that is, current may pass through x).

(b) $x = 0$ means that the switch x is open (that is, current cannot pass through x).

If x and y each denote two-state switches and if x and y are also regarded as variables with domain B_2, then we see that the notations x', $x \vee y$, $x \wedge y$ introduced above *for switches* are consistent with the meaning of these notations *for variables on B_2*. More explicitly, if the switches x and y are connected in a basic parallel network, then current will pass through the network if and only if x or y is closed. But this corresponds precisely to the statement that if x and y are variables with domain B_2 then $x \vee y$ assumes the value 1 if and only if $x = 1$ *or* $y = 1$. Similarly, if the switches x and y are connected in a basic series network, then current will pass through the network if and only if x *and* y are closed. The corresponding statement for variables with domain B_2 is that $x \wedge y$ assumes the value 1 if and only if $x = 1$ *and* $y = 1$. Finally, the convention that the switch x is closed if and only if the switch x' is open corresponds to the statement that $x = 1$ if and only if $x' = 0$ for the variable x on B_2.

The above conventions may be easily extended to associate a Boolean formula with any given series-parallel network of two-state switches. For example, the network in Figure 6.4 is described by the Boolean formula

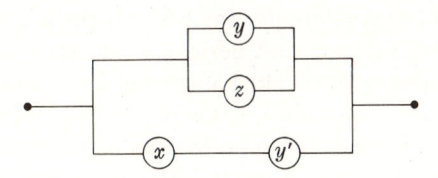

Figure 6.4

$(y \vee z) \vee (x \wedge y')$. On the other hand, a Boolean formula on the Boolean algebra B_2 determines a series-parallel network of two-state switches. For example, the Boolean formula $(x \wedge z) \vee (y \wedge z) \vee (y' \wedge x)$ describes the network in Figure 6.5. If the Boolean formula contains terms such as

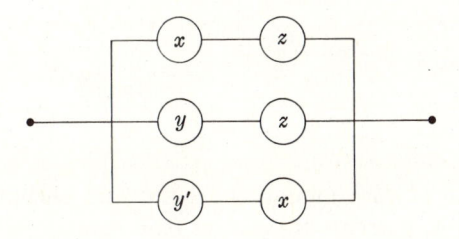

Figure 6.5

$(x \vee y)'$, this term should be replaced by $x' \wedge y'$ in order to obtain the associated series-parallel network. Similarly, terms of the form $(x \wedge y)'$ should be replaced by $x' \vee y'$. For example, given the Boolean formula $[x' \wedge (y \wedge z)'] \vee x$, we first replace $(y \wedge z)'$ by $(y' \vee z')$ to obtain the equivalent Boolean formula $[x' \wedge (y' \vee z')] \vee x$ for which the associated network is shown in Figure 6.6.

Suppose that N_1 and N_2 denote two series-parallel networks. It is natural to say that these two networks are "equivalent" or "perform the same function" if for *any* choice of open or closed positions for the two-

260 BOOLEAN ALGEBRAS AND THEIR APPLICATIONS

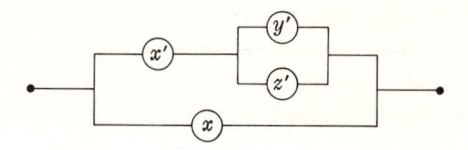

Figure 6.6

state switches involved in N_1 and N_2, it is true that either current will pass through both N_1 and N_2 or else current will pass through neither N_1 nor N_2. A problem which is of considerable interest in the design of electrical networks can be stated as follows: Given a series-parallel network N_1 which is known to perform a certain electrical function, find a series-parallel network N_2 that is simpler than N_1 (in the sense that N_2 involves fewer switches than N_1) and that performs the same function as N_1.

The correspondence established above between networks and Boolean formulas permits us to paraphrase this problem in the following equivalent form: Given a Boolean formula, find an equivalent Boolean formula that is simpler (in the sense that it involves fewer terms). For example, since the Absorption Property for Boolean algebras states that $x \lor (x \land y) = x$, we see that the network N_1 corresponding to $x \lor (x \land y)$ can be replaced by the simpler network N_2 containing only the single switch x. (See Figure 6.7.) Similarly, the network $x \lor x'$ can be replaced

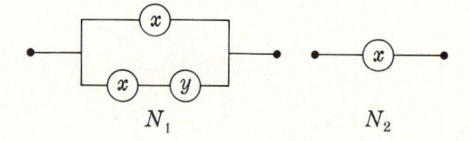

Figure 6.7

simply by a piece of wire joining the input and output terminals since $x \lor x' = 1$; that is, current will always pass through this network.

Thus, the problem of simplifying the design of a series-parallel network that performs a certain electrical function is equivalent to the problem of simplifying a Boolean formula that determines a certain Boolean function. As we have seen in Section 6.1, the theory of Boolean algebras provides a useful tool for simplifying Boolean formulas, and hence, for simplifying switching networks. The following examples illustrate the techniques involved.

Example 6.11. We wish to simplify the network shown in Figure 6.8. The top portion of the network has the associated Boolean formula $(x \lor z) \land y$ and the bottom portion is given by $[(y \land w) \lor w] \land y$.

6.3. SWITCHING NETWORKS

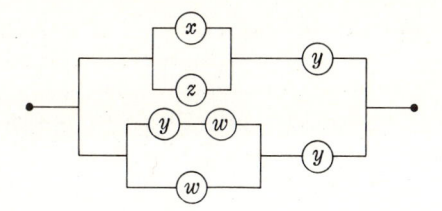

Figure 6.8

Thus, we see that the Boolean formula for the entire network is

$$[(x \vee z) \wedge y] \vee \{[(y \wedge w) \vee w] \wedge y\}.$$

To simplify this, we first note that $(y \wedge w) \vee w = w$ by the Absorption Property, so that the above expression becomes

$$[(x \vee z) \wedge y] \vee (w \wedge y).$$

An application of the Distributive Property now gives us the equivalent Boolean formula

$$(x \vee z \vee w) \wedge y.$$

Thus, the network in Figure 6.8 is equivalent to the network in Figure 6.9.

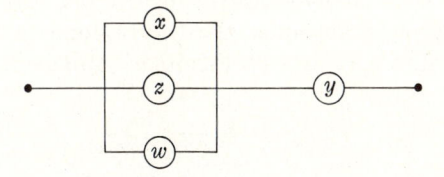

Figure 6.9

Example 6.12. The network in Figure 6.10 is determined by the Boolean formula

$$\{(z \wedge x) \vee [x \wedge (y \vee z')]\} \wedge (z \vee x) \wedge (z \vee y).$$

This formula can be simplified as follows:

$$\begin{aligned}
\{(z \wedge x) &\vee [x \wedge (y \vee z')]\} \wedge (z \vee x) \wedge (z \vee y) \\
&= \{(z \wedge x) \vee [(x \wedge y) \vee (x \wedge z')]\} \wedge (z \vee x) \wedge (z \vee y) \\
&= \{[(x \wedge z) \vee (x \wedge z')] \vee (x \wedge y)\} \wedge \{z \vee (x \wedge y)\} \\
&= \{[x \wedge (z \vee z')] \vee (x \wedge y)\} \wedge \{z \vee (x \wedge y)\} \\
&= \{x \vee (x \wedge y)\} \wedge \{z \vee (x \wedge y)\} \\
&= (x \wedge z) \vee (x \wedge y) \\
&= x \wedge (z \vee y).
\end{aligned}$$

262 BOOLEAN ALGEBRAS AND THEIR APPLICATIONS

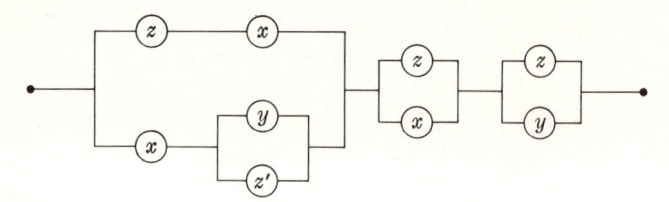

Figure 6.10

Therefore, the network in Figure 6.10 may be replaced by the simpler network in Figure 6.11 that performs the same function.

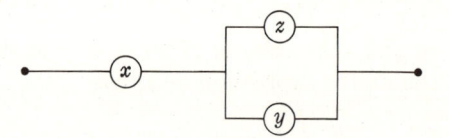

Figure 6.11

EXERCISES

6.13 Draw the networks corresponding to the following Boolean formulas:
   (a) $\{x \vee (y' \wedge z) \vee (x \wedge z')\} \vee (y \wedge z)$.
   (b) $x \vee [y' \vee (y \wedge z)' \vee z']$.

   Find simpler equivalent networks for (a) and (b).

6.14. Write the Boolean formulas that correspond to the Boolean networks in Figure 6.12 and find simpler equivalent networks.

(a)

(b)

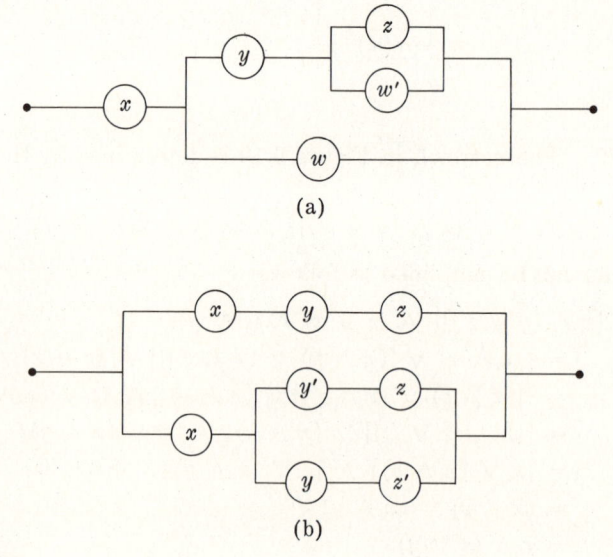

Figure 6.12

6.15. Design and simplify a series-parallel network that will allow current to pass if and only if three switches x, y, z are all closed or precisely one of these switches is closed.

6.16. The network in Figure 6.13 is *not* a series-parallel network. Find and simplify a series-parallel network that is equivalent to the given network.

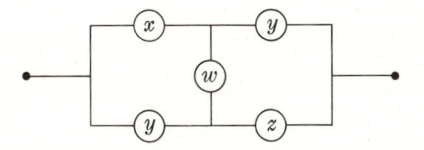

Figure 6.13

6.17. Each man on a four-man committee has control of a two-state switch with which he can record his vote on an issue. A closed switch records a "yes" vote while an open switch records a "no" vote. In case of a tie vote, the chairman's vote is counted twice to break the tie. Design and simplify a network which will allow current to pass if and only if the issue passes.

6.4. THE DESIGN OF SOME SIMPLE COMPUTER NETWORKS

Our next objective will be to indicate how two-state switch networks can be designed to perform addition of nonnegative integers. For the sake of simplicity, we shall restrict our attention to the addition of two integers from 0 to 7 so that the sum will always be an integer from 0 to 14. We shall see that once a network has been designed to add integers in this range, it is a simple matter to design a network that will handle larger integers by simply duplicating certain subnetworks in the design.

The ordinary decimal (that is, base 10) representation of nonnegative integers requires the use of the symbols $0, 1, 2, \cdots, 9$ as digits. This representation is not convenient for our purposes since two-state switches can only be in two positions, open or closed. For this reason, the binary (that is, base 2) representation of nonnegative integers is employed here since this representation requires the use of only two symbols 0 and 1 as digits.

Recall that a nonnegative integer n has binary representation $(a_k a_{k-1} \cdots a_1 a_0)_2$ or simply $a_k a_{k-1} \cdots a_1 a_0$ if the base is clear from the context, if

$$n = a_0 2^0 + a_1 2^1 + \cdots + a_{k-1} 2^{k-1} + a_k 2^k \qquad (*)$$

where $a_i = 0$ or $a_i = 1$ for $i = 0, 1, \cdots, k$ and $a_k = 1$ if n is not zero.

For example, since
$$10 = 0 \cdot 2^0 + 1 \cdot 2^1 + 0 \cdot 2^2 + 1 \cdot 2^3$$
$$13 = 1 \cdot 2^0 + 0 \cdot 2^1 + 1 \cdot 2^2 + 1 \cdot 2^3$$

it follows that the binary representation of 10 is 1010 and that of 13 is 1101. The number a_i in (*) is called the 2^i-*binary digit* of n for $i = 0$, $1, \cdots, k$. The binary representations of the integers between 0 and 15 are listed in Table 6.1.

Table 6.1

Integer	Binary Representation	Integer	Binary Representation
0	0	8	1000
1	1	9	1001
2	10	10	1010
3	11	11	1011
4	100	12	1100
5	101	13	1101
6	110	14	1110
7	111	15	1111

Any integer n between 0 and 15 may be represented by setting four switches x_0, x_1, x_2, x_3 in an appropriate position. The integer 0 is represented by opening all four switches, 1 is represented by closing x_0 and opening x_1, x_2, x_3, 2 is represented by closing x_1 and opening x_0, x_2, x_3, and so on. In general, an integer n such that $0 \leq n \leq 15$ is represented by closing the switch x_i if the 2^i-binary digit of n is 1; otherwise, x_i is left open ($i = 0, 1, 2, 3$). For example, the integer 9 is represented by closing switches x_0 and x_3 and opening x_1 and x_2 since the binary representation of 9 is 1001.

In order to develop a network which will perform the addition of two integers m, n between 0 and 7, it is necessary to carefully analyze the addition procedure for such integers represented in base 2. Let us begin the analysis by considering an example. Suppose that we add 5 and 7 by using their binary representations $5 = (101)_2$ and $7 = (111)_2$. The sequence of steps involved is displayed below:

```
     Step 1              Step 2                Step 3
       1                   1 1                   1 1
      101                  101                   101
     +111                 +111                  +111
     ----                 ----                  ----
        0                   00                  1100
 "1 + 1 = 0 and       "1 + 0 + 1 = 0       "1 + 1 + 1 = 11"
  1 to carry"         and 1 to carry"
```

6.4. THE DESIGN OF SOME SIMPLE COMPUTER NETWORKS 265

Since $(1100)_2 = 12$, we have obtained the desired sum. More generally, if m and n are integers such that $0 \leq m, n \leq 7$ and if $m = (a_2 a_1 a_0)_2$, $n = (b_2 b_1 b_0)_2$ then the sum $m + n$ is obtained as follows.

STEP 1. Add a_0 to b_0 to obtain the 2^0-binary digit s_0 of $m + n$ and the first carry c_1. $s_0 = 1$ if either $a_0 = 1$ or $b_0 = 1$ but not both; otherwise, $s_0 = 0$. $c_1 = 1$ if and only if $a_0 = b_0 = 1$.

STEP 2. Add c_1, a_1, b_1 to obtain the 2^1-binary digit s_1 of $m + n$ and the second carry c_2. $s_1 = 1$ if $a_1 = b_1 = c_1 = 1$ or if exactly one of the numbers a_1, b_1, c_1 is 1; otherwise, $s_1 = 0$. $c_2 = 1$ if at least two of the numbers a_1, b_1, c_1 are 1; otherwise, $c_2 = 0$.

STEP 3. Add c_2, a_2, b_2 to obtain the 2^2-binary digit s_2 and the 2^3-binary digit s_3. $s_2 = 1$ if $c_2 = a_2 = b_2 = 1$ or if exactly one of the numbers c_2, a_2, b_2 is 1; otherwise, $s_2 = 0$. $s_3 = 1$ if at least two of the numbers c_2, a_2, b_2 is 1; otherwise, $s_3 = 0$.

On the basis of this analysis of the addition process for the integers m and n given in binary representation, we can now design the desired network as follows. Suppose that the switches x_2, x_1, x_0 are used to represent the integer m ($0 \leq m \leq 7$) as indicated above and that y_2, y_1, y_0 are the corresponding switches used to represent the integer n ($0 \leq n \leq 7$). The two "carries" c_1 and c_2 as well as the binary digits s_3, s_2, s_1, s_0 of $m + n$ are to be regarded as "output terminals" to which current will or will not flow from an input source depending on the particular setting of the switches. Thus, these output terminals can be in one of two mutually exclusive states, *charged* (that is, current can reach the terminal) or *uncharged* (that is, current cannot reached the terminal). Consequently, we may regard $c_1, c_2, s_3, s_2, s_1, s_0$ as variables with domain B_2 and interpret the statement that a particular one of these outputs is charged as equivalent to the statement that the corresponding variable takes on the value 1. Finally, z_1 and z_2 will denote switches whose positions are determined by the carry outputs c_1 and c_2 respectively as follows: z_i is closed if and only if c_i is charged ($i = 1, 2$). Thus, the switches z_1 and z_2 reflect the output terminal states c_1 and c_2.

The three-step procedure for addition given above can now be rephrased in terms of switch positions and output states, and hence in terms of Boolean functions, as follows:

(1) s_0 is charged if either x_0 or y_0 but not both are closed, that is,

$$s_0 = (x_0 \wedge y_0') \vee (x_0' \wedge y_0).$$

266   BOOLEAN ALGEBRAS AND THEIR APPLICATIONS

c_1 is charged if and only if x_0 and y_0 are closed, that is,

$$c_1 = x_0 \wedge y_0.$$

(2) s_1 is charged if either x_1, y_1, z_1 are all closed or if exactly one of these switches is closed, that is,

$$s_1 = (x_1 \wedge y_1 \wedge z_1) \vee (x_1 \wedge y_1' \wedge z_1') \vee (x_1' \wedge y_1 \wedge z_1') \vee (x_1' \wedge y_1' \wedge z_1).$$

c_2 is charged if at least two of the switches x_1, y_1, z_1 are closed, that is,

$$c_2 = (x_1 \wedge y_1 \wedge z_1) \vee (x_1 \wedge y_1 \wedge z_1') \vee (x_1 \wedge y_1' \wedge z_1) \vee (x_1' \wedge y_1 \wedge z_1).$$

(3) s_2 is charged if either x_2, y_2, z_2 are all closed or if exactly one of these switches is closed, that is,

$$s_2 = (x_2 \wedge y_2 \wedge z_2) \vee (x_2 \wedge y_2' \wedge z_2') \vee (x_2' \wedge y_2 \wedge z_2') \vee (x_2' \wedge y_2' \wedge z_2).$$

s_3 is charged if at least two of the switches x_2, y_2, z_2 are closed, that is,

$$s_3 = (x_2 \wedge y_2 \wedge z_2) \vee (x_2 \wedge y_2 \wedge z_2') \vee (x_2 \wedge y_2' \wedge z_2) \vee (x_2' \wedge y_2 \wedge z_2).$$

The Boolean formulas for s_0 and c_1 in (1) correspond to the networks in Figure 6.14. The Boolean formulas for s_1 and c_2 in (2) can be rewritten

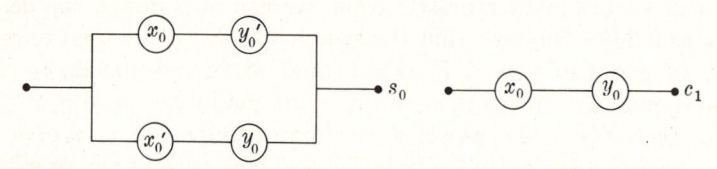

Figure 6.14

as follows:

$$s_1 = (x_1 \wedge [(y_1 \wedge z_1) \vee (y_1' \wedge z_1')]) \vee (x_1' \wedge [(y_1 \wedge z_1') \vee (y_1' \wedge z_1)]),$$
$$c_2 = (x_1 \wedge y_1) \vee ([(x_1 \wedge y_1') \vee (x_1' \wedge y_1)] \wedge z_1).$$

The networks corresponding to the latter Boolean formulas for s_1 and c_2 are shown in Figure 6.15. Finally, since the Boolean formulas for s_2 and s_3 in (3) can be obtained from those in (2) for s_1 and c_2 respectively by increasing each subscript by one, the same change in subscripts in Figure 6.15 will yield networks for s_2 and s_3.

The addition of two integers m, n between 0 and 7 can now be accomplished by the following sequence of steps:

(1) Set the switches x_2, x_1, x_0 to represent m and set the switches y_2, y_1, y_0 to represent n.

(2) Place a source of current at the input terminals of the networks for s_0 and c_1 and note the output states of s_0 and c_1.

6.4. THE DESIGN OF SOME SIMPLE COMPUTER NETWORKS

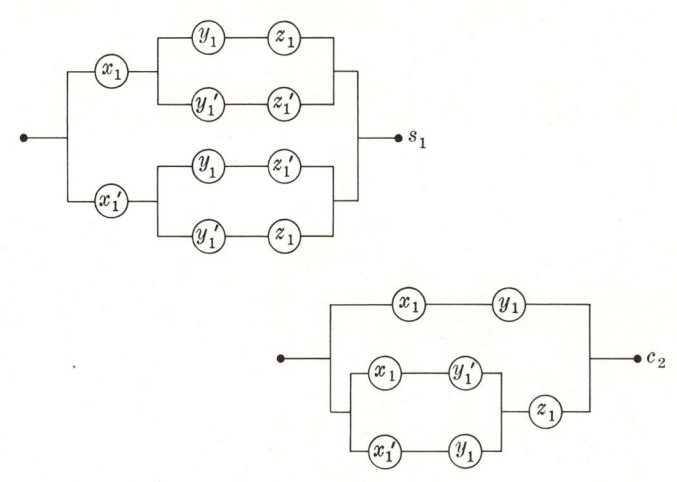

Figure 6.15

(3) Set the switch z_1 in accordance with the state of the output c_1 (that is, z_1 is closed if and only if c_1 is charged).

(4) Place a source of current at the input terminals of the networks for s_1 and c_2 and note the output states of s_1 and c_2.

(5) Set the switch z_2 in accordance with the state of the output c_2.

(6) Place a source of current at the input terminals of the networks for s_2 and s_3 and note the output states of s_2 and s_3.

(7) Use the outputs s_0, s_1, s_2, s_3 to determine the base 2 representation of $m + n$.

The various networks considered above can be combined into a single network with output terminals s_0, s_1, s_2, s_3. The types of two-state switches that are actually used in computers permit the carry outputs c_1, c_2 to "set" the switches z_1, z_2 directly. Moreover, since the networks for s_2, s_3 are identical in design to those for s_1, c_2 respectively, it is clear that the combined network may be augmented by further subnetworks identical to those for s_1, c_2 to permit the addition of integers with more than three binary digits.

It is a simple matter to build a crude working model of the "binary adder" described above from such inexpensive materials as a dry cell battery, six flashlight sockets and lights, a few electrical connector clips, some light, insulated copper wire and a piece of board or plywood of adequate size to accommodate the components. The schematic diagram given in Figure 6.16 provides a suitable "layout" for such a project. The two-state switches for this model each consist of a cluster of three electrical clips. A switch is closed by connecting the clip on the left of the cluster to

268 BOOLEAN ALGEBRAS AND THEIR APPLICATIONS

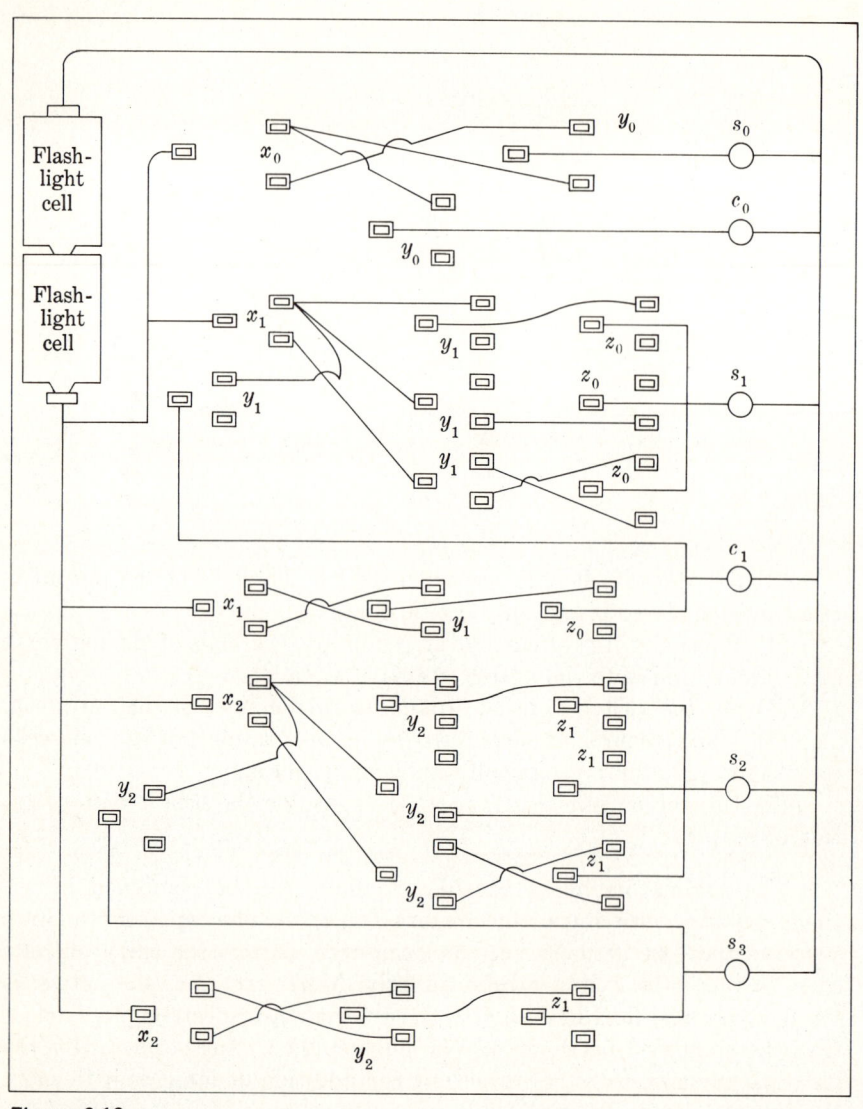

Figure 6.16

the top clip on the right by a short piece of wire, while a switch is opened by connecting the clip on the left to the bottom clip on the right in a similar way. The circles on the right-hand edge of the diagram represent flashlight sockets. This "computer" is operated in accordance with the seven step procedure given above.

The binary adder described above is operated by following a sequence of instructions based on our analysis of the binary addition process.

Modern electronic computers also require a sequence of instructions, usually referred to as a program or algorithm. However, unlike our binary adder for which the instructions are carried out by the operator, the modern computer stores and executes its programs internally. Needless to say, the typical program dictating the operation of a computer is a good deal more complicated than our program for the operation of our binary adder. Nevertheless, even the most complicated program is arrived at by the same careful analysis of the numerical operations to be performed.

EXERCISES

6.18. Analyze the schematic diagram in Figure 6.16 to verify that it does provide one design for a binary adder.

6.19. Trace the flow of current through the binary adder when it is used to compute:
 (a) $5 + 7$.
 (b) $4 + 6$.
 (c) $0 + 3$.

6.20. Design a "computer" which can be used to subtract an integer n from an integer m provided that $0 \leq n \leq m \leq 15$.

APPENDIX: A SIMPLE HUMAN COMPUTER—A CLASSROOM DEMONSTRATION

We have seen that devices with a very simple function such as two-state switches can be combined into a network that can perform the relatively sophisticated function of binary addition. In the development of the modern computer much of the genius involved centers around the ability to combine such very simple, highly reliable physical components into networks that can perform functions of much greater complexity than those of the individual components. This basic feature of computer design can be portrayed very effectively by an interesting classroom demonstration devised by D. C. Engelbart. (See "Games that teach the fundamentals of computer operation," *IRE Transactions on Electronic Computers* March, 1961, pp. 31–41.) His idea is to use the students themselves as the "components" of a computer. Each student is assigned the very simple task of either raising or lowering a card in accordance with a simple set of instructions, yet the "network" of students operates as a binary adder! Because we have found this demonstration to be instructive and stimulating to groups of secondary school students, we shall discuss its presentation in some detail.

Nineteen students and the teacher are required as components for this human computer. Ten of these students constitute the "arithmetic

270   BOOLEAN ALGEBRAS AND THEIR APPLICATIONS

unit," the remaining nine students comprise the "logic unit" and the teacher acts as the "signal source." The various components should be seated as the diagram below indicates with the students in the arithmetic unit facing the students in the logic unit and the remainder of the class or audience.

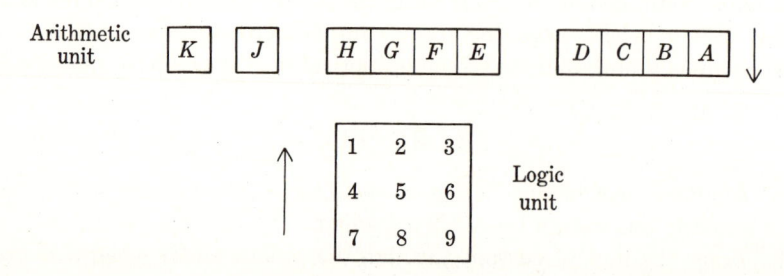

Arithmetic unit | K | J | | H | G | F | E | | D | C | B | A | ↓

```
    1  2  3
              Logic
    4  5  6   unit
    7  8  9
```

Each student is given a large card which he is to hold in either up or down position as the computer functions. The back side of each card (that is, the side faced by the card holder) should contain the letter or number given to the corresponding student in the seating chart above as well as the instruction code given in Table 6.2. This table also gives num-

Table 6.2

Card	Instruction Code	Front Side
1	Watch A, E	UP
2	Watch 1, K	UP
3	Watch A, E	DOWN
4	Watch 1, K	DOWN
5	Watch 3, 4	UP
6	Watch 3, K	DOWN
7	Watch 5	OPPOSITE
8	Watch 6, 7	UP
9	Watch 2, 8	DOWN
A	Watch B	1
B	Watch C	2
C	Watch D	4
D	Watch Dummy Down	8
E	Watch F	1
F	Watch G	2
G	Watch H	4
H	Watch J	8
J	Watch 9	16
K	Watch 5	

bers that should be written on the front side of some of the cards in characters large enough to be read by the audience.

We shall now explain the instruction code given to the numbered components. Each such component watches one or two other components and he is to adjust his card position according to the card positions of the components he watches. The reaction of a numbered component to the card positions of the components he watches is determined by the UP, DOWN, or OPPOSITE instruction code given on the back face of his card in accordance with the following rules:

(1) A numbered component with instruction code UP is to hold his card in up position when both of the components he is watching are holding their cards in up position; otherwise the numbered component is to hold his card in down position.

(2) A numbered component with instruction code DOWN is to hold his card in down position when both of the components he is watching are holding their cards in down position; otherwise, he is to hold his card up.

(3) The numbered component with instruction code OPPOSITE holds his card in the position opposite to that of the component he watches.

The teacher should take care to see that each numbered component understands the above instructions and knows the location of the component(s) he is to watch. It should be emphasized that the numbered components are to react *whenever necessary* to any changes in card positions of the components they watch.

We now proceed to explain the instruction codes for the lettered components. These components do not change their card positions in the same way as the numbered components. Rather, the lettered components hold a particular position until they get a signal from the "signal source," the teacher. Then they react in accordance with the special instructions given below.

The signal source repeats a two-part signal such as "one-two" or "bonk-bleep" at regular time intervals to operate the computer. When the first part of the signal is given by the signal source (say "one" or "bonk"), each lettered component should note the card position of the element he is assigned to watch. Then, when the second part of the signal is given (say "two" or "bleep"), each lettered component assumes the card position held at the time the first part of the signal was given by the component he is assigned to watch. Thus, if a lettered component has his card in the same position as that of the component he is watching when the first part of the signal is given, then he remains in that position when the second part of the signal is given (even if the component he watches changes his card position when the second part of the signal is given).

The lettered component D is given the special instruction to watch

"Dummy Down." This means that he is to pretend that he is watching a component that always has his card held down. This means that if the computer begins operation with the card held by D in up position, then D is to move his card to down position when the second part of the first signal is given. On the other hand, if the computer begins operation with the card held by D in down position, then D is to hold his card in that position for the remainder of the computer's operating cycle.

After the teacher has made certain that all of these instructions are clearly understood by the components, the human computer is ready to begin its operation.

The first step is to "load" the computer with the numbers (between 0 and 15) that are to be added by setting the lettered elements A through D in suitable positions to represent the first number and then using the lettered elements E through H in a similar way to represent the second number. For example, if the first number is 3, then components A and B should hold their cards up while C and D hold their cards down since the front side of the card A displays 1 and the front side of card B displays 2 to the audience. Similarly, if the second number is 6, then components F, G should hold their cards up while E, H should hold their cards down since the front of card F displays 2 and the front of card G displays 4 to the audience. It should now be apparent that these initial card settings simply constitute binary representations of the numbers to be added.

After the card positions of the components have been set as indicated above, the teacher should ask all the other components to assume the card positions dictated by their particular instructions and the card positions of the element or elements they have been instructed to watch.

The signal source, that is, the teacher, now gives a series of five two-part signals. Sufficient time should be given between the two-part signals and between each part of a given two-part signal to allow the components to react with care. After the last of the five signals is given, the card positions of the components E, F, G, H, J should represent the sum of the numbers loaded into the computer at the outset. For example, if the computer is loaded with 3 and 6 as indicated above, then after the last signal has been given, the components E and H should be holding their cards in up position (displaying 1 and 8 to the audience) while F, G, and J should be holding their cards in down position.

It often happens that this human computer fails to produce the correct result on the first trial due to a "malfunction" on the part of one or more of the components. In this case, the teacher should repeat the operation and check the card positions after each two-part signal. Table 6.3 gives the correct card position after each two-part signal for the case in which the numbers 3 and 6 are added. (U indicates UP while a blank indicates DOWN.)

Table 6.3

Components	Initial Setting	Position after Signal				
		1	2	3	4	5
A	U	U				
B	U					
C						
D						
E		U	U			U
F	U	U			U	
G	U			U		
H			U			U
J		U			U	
K			U	U		
1		U				
2						
3	U	U	U			U
4		U	U	U		
5		U	U			
6	U	U	U	U		U
7	U			U	U	U
8	U			U		U
9	U			U		U

In view of our discussion of the binary adder network in the last section, the logical function of some of the components in this computer should be fairly clear. For example, the instructions for the UP, DOWN, and OPPOSITE elements indicate that they act as AND-networks, OR-networks, and "opposite" switches respectively. However, a complete logical analysis of the human computer would require a discussion of certain devices that we have not considered here.

REMARKS AND REFERENCES

Many secondary schools are presently developing programs to provide computer oriented training for their students. The introduction of "time sharing" systems promises to make computing facilities available in the majority of secondary schools in the near future. Courses in computer programming are now taught in many secondary schools and it is evident that such courses are to become a standard part of the secondary school program. Responsibility for such courses should naturally rest with the mathematics staff and it is therefore imperative that secondary school mathematics teachers prepare themselves in computer related mathematics.

In an attempt to respond to these needs, many universities have added courses in computer programming to their lists of required courses for secondary mathematics education students. Although this is a step in the right direction, it is also desirable that teachers gain some appreciation for the manner in which computers function internally as well. In this chapter, we have attempted to provide some orientation in this direction. Of course, the treatment of this topic presented here is too brief and incomplete to be considered an introduction to the logical design of computer networks. However, after reading the discussion in this chapter, the interested student should find more extensive treatments of the subject such as those found in the books by Hohn [2] and Whitesitt [5] to be quite accessible. A brief, but illuminating discussion of computers written at a level suitable for secondary school students can be found in the book by Young [6]. Also, a number of interesting *Scientific American* articles on the capabilities, construction, and applications of computers have been reprinted in [3].

The study of the algebraic properties of Boolean algebras provides an excellent source of enrichment and independent study material for secondary school students because of the many comparisons and contrasts that can be made with the algebraic properties of the number systems. Our development of the isomorphism theorem for finite Boolean algebras in Section 6.2 demonstrates rather clearly how some results concerning Boolean algebras can be developed through analogies with more familiar number theoretic properties of the natural number system.

Boolean algebras also play a significant role in logic and very readable accounts of this aspect of the subject can be found in the books by Arnold [1] and Stoll [4].

1. Arnold, B. H., *Logic and Boolean Algebra*. Englewood Cliffs, N.J.: Prentice-Hall, Inc., 1962.
2. Hohn, Franz, *Applied Boolean Algebra*, Second Edition. New York: The Macmillan Company, 1966.
3. *Mathematics in the Modern World*, Readings from the *Scientific American* with introductions by Morris Kline. San Francisco: W. H. Freeman and Company, 1968.
4. Stoll, R. R., *Set Theory and Logic*. San Francisco: W. H. Freeman and Company, 1963.
5. Whitesitt, J. E., *Boolean Algebra and Its Application*. Reading, Mass.: Addison-Wesley Publishing Company, 1961.
6. Young, F. H., *The Nature of Mathematics*. New York: John Wiley & Sons, Inc., 1968.

Chapter 7
GRAPH THEORY

The mathematical theory of graphs began in 1735 with the now famous solution of the so-called Königsberg Bridge Problem by the great mathematician, Leonhard Euler (1707–1783). This problem can be simply

Leonhard Euler
Library of Congress

described as follows. The eastern European city of Königsberg (now in Russia and called Kaliningrad) is situated around a fork in what is now called the Pregolya River with part of the city located on an island in the river. In Euler's time, the various parts of the city were connected by seven bridges as shown in Figure 7.1. The problem is: Is it possible to plan a walk in the city that will cross each bridge once and only once? (The walk may begin at any point in the city and it need not end at its starting point.) Euler gave an elegant proof of the fact that no such walk is possi-

GRAPH THEORY

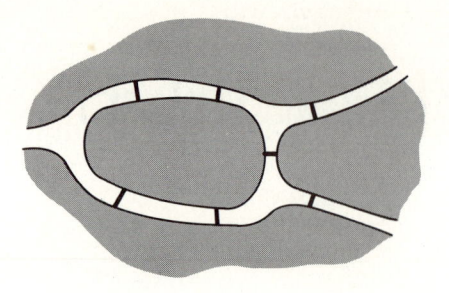

Figure 7.1

ble and we shall consider his proof below as a part of our motivation for the study of graphs.

In 1752, Euler also discovered and proved the formula $V - E + F = 2$, which still bears his name, relating the numbers V of vertices, E of edges, and F of faces in a simple polyhedron. (This formula is mentioned in a letter written by Descartes in 1640, and there is some evidence that it may even have been known to the ancient Greeks; however, Euler was the first to publish the result and to provide a proof.) In Section 7.3, we shall prove Euler's formula in the context of connected planar graphs. As we shall see, this formula will play an important role in the development of a number of results in this chapter.

Euler's results remained in splendid isolation for nearly a century before further interest in the subject was stimulated by popular puzzles and problems such as Hamilton's Game and the famous unsolved Four Color Problem, both of which are discussed below. Later, this interest was further stimulated by the growing realization that many more practical problems concerning molecular and crystalline structure, the design of electrical networks, linear programming, genetics, probability, and so on, led quite naturally to considerations in graph theory. In recent years, graph theory has evolved into a mature branch of mathematics with numerous applications to other disciplines. However, despite its growing utility as a scientific tool, the appealing nature of the puzzles and problems that lie at the historical roots of the subject continues to be a great source of stimulation to the beginner.

7.1. THE KÖNIGSBERG BRIDGE PROBLEM—A PRELUDE TO GRAPH THEORY

As a motivation for our development of some basic concepts in graph theory in Section 7.2, we shall now sketch Euler's original solution of the Königsberg Bridge Problem and then show how this problem can be

7.1. THE KÖNIGSBERG BRIDGE PROBLEM

formulated and solved more directly by placing it within the context of graph theory.

Suppose that we begin by labeling the various land masses and bridges in the Königsberg Bridge Problem as in Figure 7.2. With the land masses

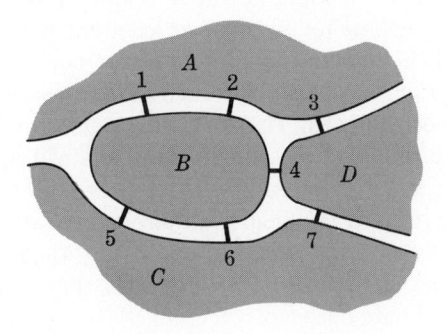

Figure 7.2

labeled in this way, a walk in the city can be described by a series of letters written in juxtaposition. For example, the series BCD denotes a path starting in B, proceeding to C (across either bridge 5 or bridge 6) and then to D (across bridge 7). Similarly, $ADCBA$ denotes a path starting in A, proceeding to D, then to C, then to B (along either bridge 5 or bridge 6) and ending in A (after crossing either bridge 1 or bridge 2).

With this description of the meaning of the path notation in mind, we can make the following simple but important observation: If there is a path P that the walker may follow so as to cross each of the seven bridges once and only once, the notation for this path will be a *series of eight letters* selected from the letters A, B, C, D. It should also be noted that the arrangement of these four letters in the series of eight letters denoting the path P will be restricted in a number of ways. For example, the letters A and B must occur in juxtaposition in the series exactly twice since there are two bridges joining A to B; on the other hand, the letters A and C should never appear in juxtaposition in the desired series of eight letters since there is no bridge joining A directly to C.

Now let us concentrate our attention on a single land mass, call it X, which stands for any one of the four land masses A, B, C, or D. Note that there are either three or five bridges *leading out* of X. We shall now show that if there are three bridges leading out of X, then exactly two X's will appear in the notation for the path P. In fact, if the path P begins in X, the first letter in the path notation will be X. If the path does not begin in X, the first X will occur when the path crosses one of the bridges into X for the first time. In either case, this first X in the path notation will be followed by some other letters because the path must leave X since it eventually crosses all the bridges leading from X. A second X will then be used in the path notation for the path P when P returns to X. If P did not begin in X, this second X must be the last letter in the notation for the path P since all three bridges out of X have now been crossed. If P started in X, one bridge leading out of X remains, so the second X in the path notation for P will be followed by at least one more letter to indicate that P has crossed the last bridge. However, no more X's can occur in the path notation for P since P cannot return to X without crossing a bridge that has already been crossed. Therefore, we have shown that if there are three bridges leading out of X, exactly two X's will appear in the path notation for any path that crosses each of these three bridges once and only once. We leave it to the reader to apply similar reasoning to show that if there are five bridges leading out of X, then exactly three X's must appear in the path notation for any path that crosses each of these five bridges once and only once.

With these observations in mind, it is a very simple matter to complete Euler's solution of the Seven Bridges Problem. We have already noted that any path P that would cross each of the seven bridges in Figure 7.2 once and only once must be denoted by a series of *eight* letters. However, there are three bridges leading from each of the land masses, A, C, D and five bridges leading from B, so the conclusion of the preceding paragraph implies that the path notation for P must contain two A's, two C's, two D's and three B's—a total of *nine* letters! Therefore, no such path P can exist and the Königsberg Bridge Problem is solved!

EXERCISES

The technique used by Euler in his solution of the Königsberg Bridge Problem can be generalized to a variety of other situations. The following results are examples of conclusions that can be drawn by similar reasoning.

7.1 Prove that if there are n bridges leading out of a land mass X, then the path notation for any path P that crosses each of the n bridges once and only once must contain:

(a) $\dfrac{n+1}{2}$ X's if n is odd.

(b) $\dfrac{n}{2}$ X's if n is even and P does not begin in X.

(c) $\dfrac{n}{2} + 1$ X's if n is even and P begins in X.

7.2. Show that there is no path that will cross each of the bridges in Figure 7.3 once and only once.

7.1. THE KÖNIGSBERG BRIDGE PROBLEM

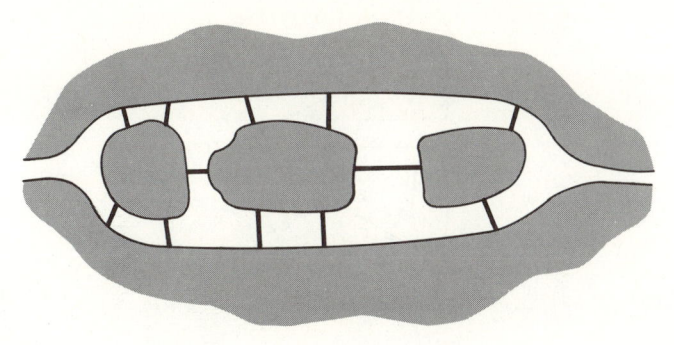

Figure 7.3

7.3. In Figure 7.3, add a new bridge joining the island on the right to the lower shore. Can a path of the desired sort be found now?

7.4. If the walker in the Königsberg Bridge Problem is allowed to leave the city and travel around the world, is it then possible for him to cross each bridge once and only once?

We have presented the preceding solution of the Königsberg Bridge Problem essentially in the form originally described by Leonhard Euler primarily for historical reasons. We shall see very soon that the problem can be disposed of more neatly when we place the problem in the context of graph theory. Nevertheless, it is worthwhile to make some observations concerning the geometric nature of this problem before we go on to the elements of graph theory.

Note that the concepts of length, angle, similarity, and shape, which are so important to Euclidean geometry are of absolutely no consequence to the formulation or solution of the above problem. That is, the sizes and shapes of the islands and bridges played no role in our considerations. Instead, the notion of *connectivity* (that is, the manner in which the land masses were interconnected by bridges) was all important. Connectivity plays a minor role in Euclidean geometry; however, we shall see that it is a central concept in graph theory.

If we strip the Seven Bridges Problem down to its essentials, we are led quite naturally to the concept of a graph and also to an alternate, more succinct proof of the fact that no path of the required sort exists. To see how this can be done, we first note that since the shapes and sizes of the land masses and bridges are not important, the land masses can be represented by points, and each bridge joining a pair of land masses can be represented by an arc joining the corresponding pair of points. The resulting replacement for Figure 7.2 is pictured in Figure 7.4, where the

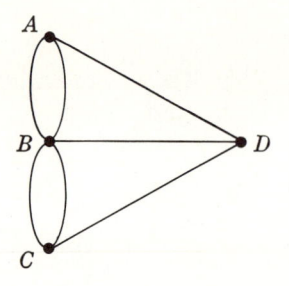

Figure 7.4

points are labeled just as the corresponding land masses are labeled in Figure 7.2. This simple geometric configuration of points and arcs connecting certain pairs of these points is an example of a graph. In terms of this graph, the Königsberg Bridge Problem can be restated as follows: Is it possible to trace the graph in Figure 7.4 starting at one of the points A, B, C, D without lifting pencil from paper in such a way that all arcs in the graph are traced once and only once?

A somewhat different argument than that given by Euler can be given to prove the impossibility of finding such a trace. Suppose a trace of the desired type did exist, and consider one of the points that is neither the starting point nor the terminal point of this trace. It must then be the case that an even number of arcs must be attached to that point, for each time we trace along an arc leading to the point we must then continue the trace along a *different* arc away from the point. That is, for every arrival at the point there is a departure since the trace neither begins nor ends at the point in question. Since no arc can be traversed more than once, we see that an even number of arcs must meet at the point. But a glance at Figure 7.4 shows that every point is met by an odd number of arcs, three at each of A, C, D and five at B. Since only one of these points can be a starting point and only one (possibly different) point can be the terminal point of the given trace, we draw the conclusion that it is impossible to trace the figure under the stated conditions.

The argument in the above paragraph can be used to solve other simple tracing problems. For example, consider the graphs in Figure 7.5. Is it possible to trace these figures without lifting pencil from paper such that no line segment is retraced? The previous argument shows that if such a trace is possible, then any point, except for possibly the starting point and terminal point, must have an even number of line segments attached to it. This leaves at most two points from which an odd number of segments can emanate. In (a) of Figure 7.5, there are four points with an odd number of segments attached, and thus no such trace can exist. However, in (b) there are no points at which an odd number of lines meet.

7.2. BASIC DEFINITIONS; EULER PATHS

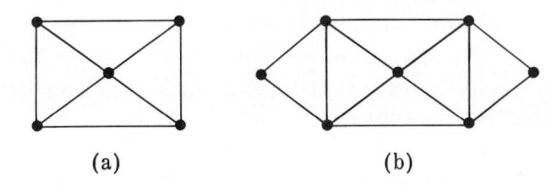

Figure 7.5

We will prove shortly that in this case a trace can be found. We invite the student to find one.

The discussion so far is an informal introduction to some of the notions of graph theory. The configurations consisting of points connected by arcs or line segments in the above discussion are examples of graphs and the conclusions we have drawn reflect the basic idea behind some elementary theorems in graph theory which we shall prove in the next section.

EXERCISES

7.5. Determine whether or not the graphs in Figure 7.6 can be traced under the conditions stated above.

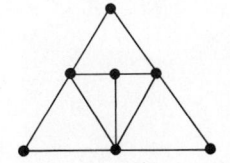

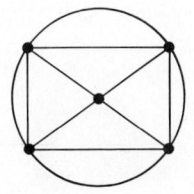

Figure 7.6

7.6. In Figure 7.7, the floor plans of two houses are given. Is it possible to walk through each door exactly once?

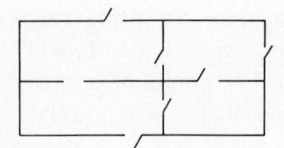

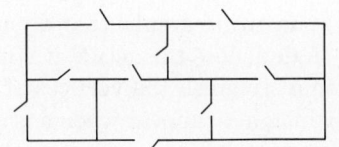

Figure 7.7

7.2. BASIC DEFINITIONS; EULER PATHS

In the preceding section, we defined a graph informally as a geometric configuration consisting of a nonempty, finite set of points in space together with some arcs joining certain pairs of these points. The particu-

282 GRAPH THEORY

lar nature of these arcs or the relative position of the points are of no consequence when such a geometric configuration is viewed as a graph; an arc only serves to indicate that its endpoints are to be regarded as connected or related to one another.

We now define the notion of a graph formally and establish some basic notation and terminology.

Definition. A *graph* is a nonempty set G consisting of a finite number of points v_i ($i = 1, \cdots, n$) in space called the *vertices* of G and a finite number of arcs e_j ($j = 1, \cdots, m$) called the *edges* of G with the following properties:

(1) Each edge of G has endpoints that are vertices of G.

(2) If an edge of G contains a vertex, then the vertex must be an endpoint of that edge.

The simple graphs considered in the last section all have the property that their edges and vertices lie in the plane. Although graphs with this property are quite important, this feature is not required by the definition of a graph. For example, the vertices and edges of a tetrahedron (see Figure 7.8(a)) or the vertices, edges, and diagonals of a cube (see Figure

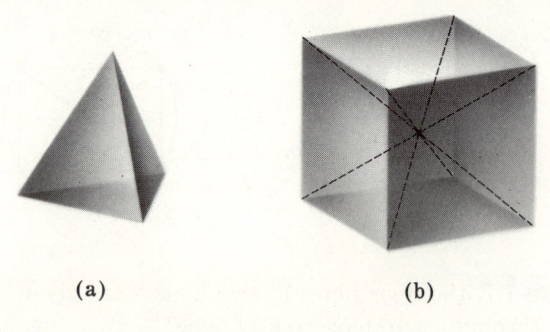

(a)   (b)

Figure 7.8

7.8(b)) constitute graphs that do not lie in a plane. In displaying a graph G, we shall denote the actual vertices in G by heavy black dots. This will serve to distinguish the vertices of G from other points of intersection of edges which may appear when a graph in space is displayed on the printed page.

Some other examples of graphs are pictured in Figure 7.9. The graph in Figure 7.9(a) is just the one considered in the alternate solution to the Königsberg Bridge Problem. The graph in Figure 7.9.(b) illustrates that not every pair of vertices of a graph need be the endpoints of an edge in the graph. Also, it is possible that the endpoints of an edge may coincide as they do for e_4 in Figure 7.9(b); such an edge is called a *loop*.

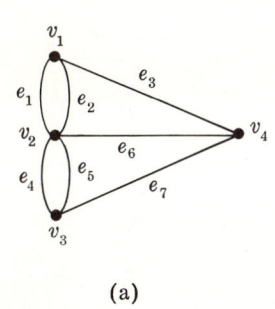

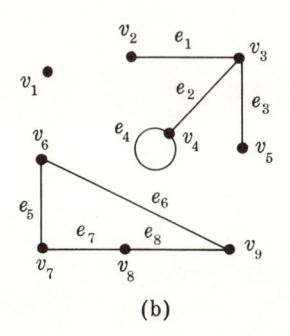

7.2 BASIC DEFINITIONS; EULER PATHS

Figure 7.9

If v is a vertex in a graph G, then we say that v has *degree n* and write $\deg(v) = n$ if v is an endpoint of precisely n edges in G. A loop at v is defined to contribute 2 to the count of the degree of v. For example, in Figure 7.9(b), $\deg(v_1) = 0$, $\deg(v_2) = \deg(v_5) = 1$, $\deg(v_6) = \deg(v_7) = \deg(v_8) = \deg(v_9) = 2$, $\deg(v_4) = \deg(v_3) = 3$. A vertex v of G is *even* if $\deg(v)$ is an even number; otherwise, v is an *odd* vertex. Thus, in Figure 7.9(b), the vertices v_1, v_6, v_7, v_8, v_9 are even while the vertices v_2, v_3, v_4, v_5 are odd.

Since the particular shape of the edges and the particular location of the vertices in a graph are not essential features of the graph, it is useful to introduce a notion of isomorphism between graphs. We want to define this notion in such a way that the statement "The graphs G_1 and G_2 are isomorphic" means that for the purposes of graph theory G_1 and G_2 may be regarded as the same graph even though G_1 and G_2 may be distinct as sets. For example, the graphs G_1, G_2 displayed in Figure 7.10 should be regarded as the same from a graph-theoretic point of view since they both have the same number of vertices and they both depict the same connectivity relations among their vertices. More specifically, we see that if we assign notation for the vertices in each of these graphs as in Figure 7.11, then two vertices v_i and v_j in G_1 are endpoints of an edge in G_1 if and only if the corresponding vertices v_i' and v_j' in G_2 are endpoints of

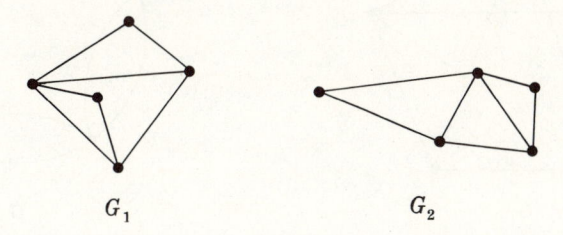

Figure 7.10

284 GRAPH THEORY

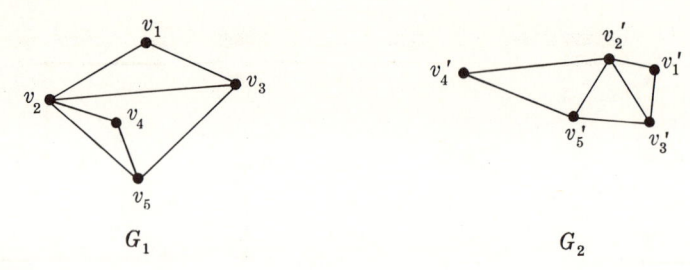

Figure 7.11

an edge in G_2. These considerations lead us quite naturally to the following definition.

Definition. Suppose that G_1 and G_2 are graphs, that $\mathcal{V}_1$ and $\mathcal{V}_2$ are the respective sets of vertices of G_1 and G_2, and that $\mathcal{E}_1$ and $\mathcal{E}_2$ are the respective sets of edges of G_1 and G_2. We say that G_1 and G_2 are *isomorphic* if there exist one-to-one correspondences

$$\alpha: \mathcal{V}_1 \to \mathcal{V}_2 \qquad \beta: \mathcal{E}_1 \to \mathcal{E}_2$$

between $\mathcal{V}_1$ and $\mathcal{V}_2$ and between $\mathcal{E}_1$ and $\mathcal{E}_2$ such that $v \in \mathcal{V}_1$ is an endpoint of $e \in \mathcal{E}_1$ if and only if $\alpha(v)$ is an endpoint of $\beta(e)$.

It is easily seen that if the graphs G_1 and G_2 are isomorphic, then:

(1) G_1 and G_2 have the same number of vertices.
(2) G_1 and G_2 have the same number of edges.
(3) If k is a nonnegative integer, then the number of vertices in G_1 with degree k must be the same as the number of vertices in G_2 with degree k.

These three conditions provide a quick, crude test for non-isomorphism of graphs; that is, if any of these conditions fails to hold for two given graphs

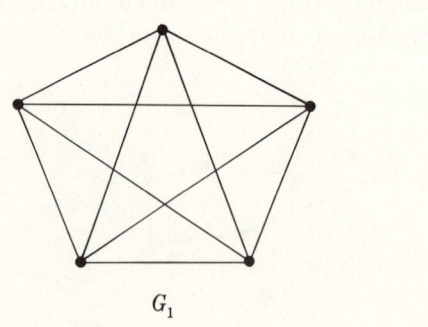

Figure 7.12

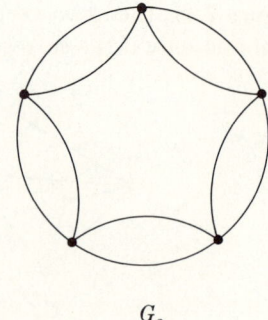

G_1 and G_2, then G_1 and G_2 are not isomorphic. However, these conditions may be satisfied by two graphs G_1 and G_2 that are not isomorphic. For example, the graphs G_1 and G_2 in Figure 7.12 satisfy these three conditions, but are not isomorphic.

Suppose that we imagine two graphs G_1 and G_2 to be constructed by using beads for vertices and rubber bands for edges. If it is possible to elastically deform G_1 into G_2, then G_1 and G_2 are evidently isomorphic. However, the converse is not true. For example, the graph in G_1 in Figure 7.13 consisting of two interlocking triangles is isomorphic to the graph G_2 in Figure 7.13 even though G_1 cannot be elastically deformed into G_2.

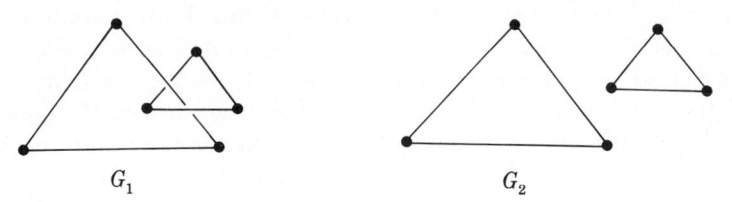

G_1 $\qquad\qquad\qquad$ G_2

Figure 7.13

As these observations suggest, it is not always an easy matter to decide whether or not two graphs are isomorphic. We shall discuss the concept of isomorphism for graphs in more detail in connection with our discussion of planar graphs in the next section.

We now wish to establish some results related to the Königsberg Bridge Problem and to the other tracing problems discussed in the last section. Our first task is to couch the simple intuitive idea of "tracing the edges of a graph without lifting the pencil from the paper and without retracing any edge" in more precise language.

Definitions. Suppose that G is a graph and that $(e^{(1)}, \cdots, e^{(k)})$ is an ordered k-tuple of edges in G such that $e^{(i)}$ and $e^{(i+1)}$ have an endpoint in common for $i = 1, \cdots, k - 1$. Then $(e^{(1)}, \cdots, e^{(k)})$ is called an *edge chain* (of length k) in G.

If $(e^{(1)}, \cdots, e^{(k)})$ is an edge chain in G and if $e^{(i)} \neq e^{(j)}$ for $i \neq j$, then $(e^{(1)}, \cdots, e^{(k)})$ is called a *path* (or *trace*) in G. If every edge of G is included in the path $(e^{(1)}, \cdots, e^{(k)})$, this path is called an *Euler path* in G.

The graph G is *connected* if for each pair of vertices in G, there is an edge chain in G including edges with these vertices as endpoints.

Thus, an edge chain $(e^{(1)}, \cdots, e^{(k)})$ in a graph G is simply a succession of k edges in G. This edge chain is a path if no edges of G are repeated in the chain. An Euler path in G is simply a path that uses all of the edges in G. Finally, a graph G is connected if every pair of vertices can be joined by an edge chain in G.

Note that the Königsberg Bridge Problem amounts to the question of whether or not the graph in Figure 7.9(a) has an Euler path. Of course, we we have already shown that it does not.

The following result will be useful in obtaining a necessary and sufficient condition for a graph to have an Euler path.

Proposition 7.1. Every graph G has an even number (possibly zero) of odd vertices.

PROOF. Remove any loops that may occur in G to obtain a graph G' with the same vertices as G, but which is loop-free. Note that G' has the same number of odd vertices as G since each loop at a vertex contributes 2 to the degree of that vertex. If n_k is the number of vertices in G' of degree k for each positive integer k, then the total number N of odd vertices is $N = n_1 + n_3 + n_5 + n_7 + \cdots$. If E is the number of edges in G', then

$$2E = n_1 + 2n_2 + 3n_3 + 4n_4 + \cdots$$

since each edge of G' contains two vertices. Thus,

$$2E - N = 2n_2 + 2n_3 + 4n_4 + 4n_5 + 6n_6 + 6n_7 + \cdots$$

so that $2E - N$ is an even number. Since $2E$ is certainly an even number, we see that N must also be an even number. Therefore, G' and hence G must contain an even number of odd vertices.

We shall now show that the number of even and odd vertices in a graph G is closely related to the nature of the paths in G (compare the alternate solution of the Königsberg Bridge Problem).

Proposition 7.2. If G is a connected graph, then there is an Euler path in G if and only if G contains either zero or two odd vertices.

PROOF. We may assume that G contains no loops since the removal of all loops in G would not affect the existence of an Euler path or the even or odd character of any vertex.

Suppose that there is an Euler path $(e^{(1)}, \cdots, e^{(k)})$ in G and suppose that this path starts at the vertex v_1 and ends at v_2. (We allow the possibility that $v_1 = v_2$.) If v is any vertex in G that is distinct from v_1 and v_2, then v must be an even vertex. For if v is the endpoint of some edge $e^{(q)}$ in the Euler path, then v is also the endpoint of either $e^{(q-1)}$ or $e^{(q+1)}$ by definition of edge chain. Hence, the edges in the Euler path that have v as one endpoint must appear in pairs in this path. Since all of the edges of G are

7.2. BASIC DEFINITIONS; EULER PATHS

included in the Euler path, the vertex v must be even. Therefore, only v_1 and v_2 may be odd vertices. This shows that G has at most two odd vertices. In view of Proposition 7.1, it follows that G has either 0 or 2 odd vertices.

Now suppose that G has no odd vertices. We can construct an Euler path for G as follows. Choose any vertex v_1 and let $e^{(1,1)}$ be any edge in G with v_1 as one endpoint. If v_2 is the other endpoint of $e^{(1,1)}$, then $v_2 \neq v_1$ since G contains no loops by assumption. Since v_2 is also an even vertex, there is an edge $e^{(1,2)} \neq e^{(1,1)}$ with v_2 as one endpoint. If v_3 is the other endpoint of $e^{(1,2)}$, then $(e^{(1,1)}, e^{(1,2)})$ is a path joining v_1 to v_3 of length 2. If there are other edges distinct from $e^{(1,1)}$ and $e^{(1,2)}$ with v_3 as one endpoint (which will be the case unless $v_3 = v_1$ and v_1 is a vertex of degree 2), we can select such an edge $e^{(1,3)}$. If v_4 is the other endpoint of $e^{(1,3)}$, then $(e^{(1,1)}, e^{(1,2)}, e^{(1,3)})$ is a path of length 3 joining v_1 to v_4. Proceeding in this way, this path extension procedure must eventually terminate after k steps when there are no remaining unused edges with v_1 as one endpoint. The final vertex of this path must be the initial vertex v_1 since all vertices in G are even. (Figure 7.14 shows a path that terminates after six steps.)

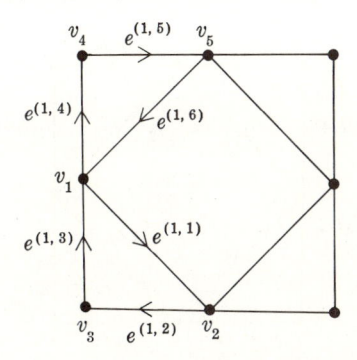

Figure 7.14

At this point, either all edges in G have occurred in the path just constructed, in which case $(e^{(1,1)}, \cdots, e^{(1,k)})$ is the desired Euler path, or else there is an edge $e^{(1,j)}$ in this path having a vertex v_1' that is the endpoint of an unused edge. In the latter case, we may repeat the above procedure starting at v_1' and only using edges not used previously to obtain a second path $(e^{(2,1)}, \cdots, e^{(2,k')})$ that will terminate at v_1' after k' steps when there are no more unused edges with v_1' as one endpoint. (Figure 7.15 shows such a second path for the graph in Figure 7.14 starting and ending in v_2.)

The two paths may now be combined to yield a path

$$(e^{(1,1)}, \cdots, e^{(1,j)}, e^{(2,1)}, \cdots, e^{(2,k')}, e^{(1,j+1)}, \cdots, e^{(1,k)})$$

starting and terminating at v_1. If this path exhausts the edges of G, then it is

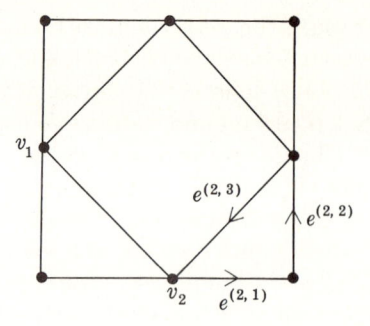

Figure 7.15

the desired Euler path. If not, we choose a vertex v_1'' attached to a yet unused edge and repeat the procedure once more to obtain a path containing still more edges of G. Since G has only a finite number of edges, this augmentation procedure must eventually terminate and yield an Euler path of G. (For example, the resulting Euler path for the graph in Figure 7.16 is $(e^{(1,1)}, e^{(2,1)}, e^{(2,2)}, e^{(3,1)}, e^{(3,2)}, e^{(3,3)}, e^{(2,3)}, e^{(1,2)}, e^{(1,3)}, e^{(1,4)}, e^{(1,5)}, e^{(1,6)})$.)

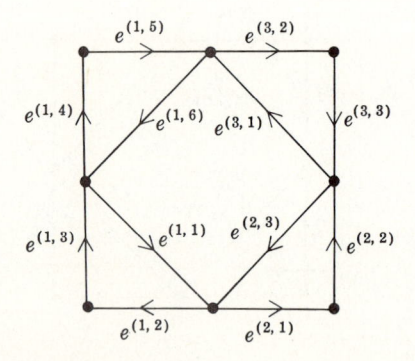

Figure 7.16

The construction of an Euler path for a graph G with precisely two odd vertices can be performed in a manner quite similar to the above construction except that v_1 should be selected to be one of the odd vertices. Alternatively, one could augment the graph G by introducing an additional edge having the two odd vertices as endpoints. The new graph G' thus obtained would have no odd vertices and the Euler trace for G', which we know exists by the previous argument, would yield an Euler path for G starting at one odd vertex and terminating at the other odd vertex. The details of this final part of the proof are left to the reader.

An Euler path in a graph G is an edge chain in G that contains each edge of G once and only once. Proposition 7.2 provides a complete answer to the question of whether or not a given graph G has an Euler path. It would be natural to attempt to establish a corresponding theory for edge chains that "pass through" each vertex of a graph once and only once. Such edge chains are referred to as Hamiltonian lines. More precisely, a *Hamiltonian line* in a graph G is an edge chain $(e^{(1)}, \cdots, e^{(k)})$ in G such that each vertex in G is an endpoint of precisely two edges in the chain. Thus, if we start at a vertex v and travel along a Hamiltonian line, then we must return to v after passing through every other vertex of the graph on the way. Of course, it is clear that no Hamiltonian line can exist in a graph that is not connected or that has any vertices of degree one.

Despite the apparent similarity between the existence problems for Hamiltonian lines and Euler paths, these problems are essentially quite different. Although some results have been obtained concerning Hamiltonian lines, no general criterion assuring the existence of a Hamiltonian line analogous to Proposition 7.2 for Euler paths is known.

Hamiltonian lines are named after W. R. Hamilton (1805–1865) who

William Hamilton
The Granger Collection

invented a puzzle, often referred to as Hamilton's Game, based on the existence of a Hamiltonian line for the graph consisting of the edges and vertices of a dodecahedron, a polyhedron having 12 pentagonal faces and 20 vertices (see Figure 7.17). This puzzle made use of a solid dodecahedron whose twenty vertices were labeled with the names of various important cities in the world. The idea of the game was to find a route following the edges of the dodecahedron that would pass through each "city" once and only once.

A solution to Hamilton's Game is most conveniently displayed by using the graph in Figure 7.18 which is isomorphic to the graph of edges

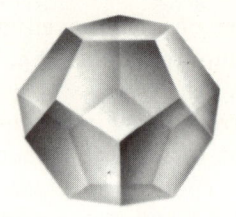

Figure 7.17

and vertices of a dodecahedron. (In the next section, we shall discuss a method of exhibiting this isomorphism.) The heavy lines in Figure 7.18 indicate a Hamiltonian line that solves the problem.

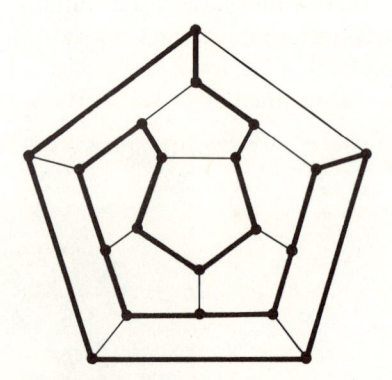

Figure 7.18

EXERCISES

7.7. Determine which of the graphs in Figure 7.19 have Euler paths. Find an Euler path for any graphs possessing such a path.

7.8. Show that it is not possible to draw an unbroken curve that *crosses* each edge in the graph of Figure 7.20 once and only once. Crossings should *not* occur at vertices. Show that one vertex and one edge can be added to this graph in such a way that it is possible to draw such a curve for the new graph.

7.9. (The Highway Inspection Problem.) A highway inspector wishes to check all of the roads in the map in Figure 7.21. Can he do this without traveling any road more than once? If so, find the suitable route for him to travel.

7.10. (The Traveling Salesman Problem.) A traveling salesman based in Champaign-Urbana wishes to travel to each city in the map in Figure 7.21 and then return home, but he does not want to go

7.2. BASIC DEFINITIONS; EULER PATHS   291

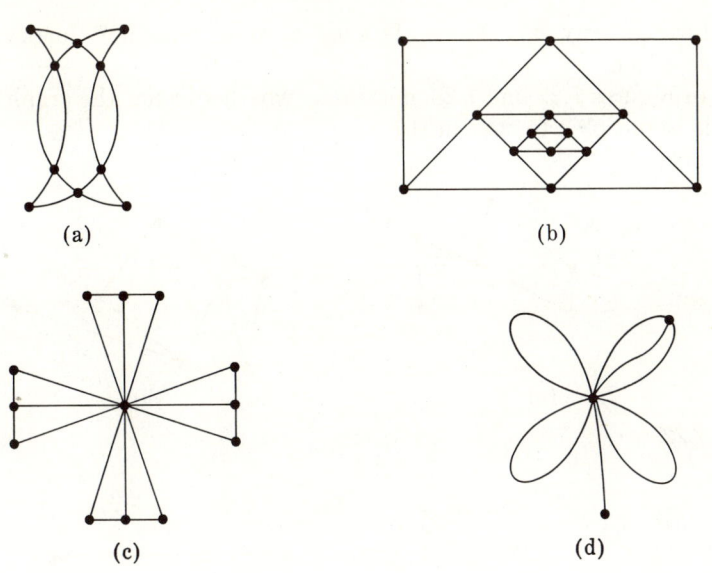

Figure 7.19

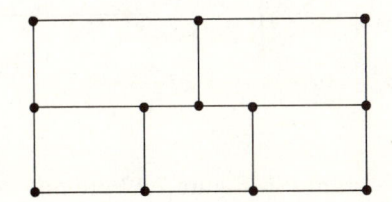

Figure 7.20

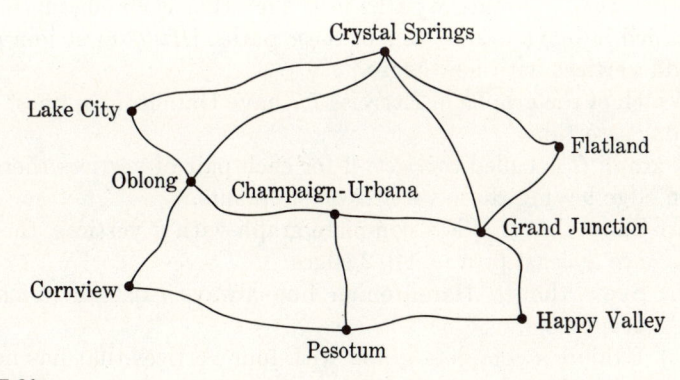

Figure 7.21

through any city twice. Is such a route possible? Justify your assertion.

7.11. In Figures 7.22 and 7.23, determine whether or not the graph in (a) is isomorphic to that in (b).

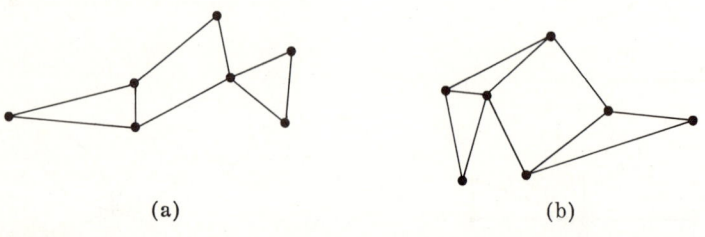

(a) (b)

Figure 7.22

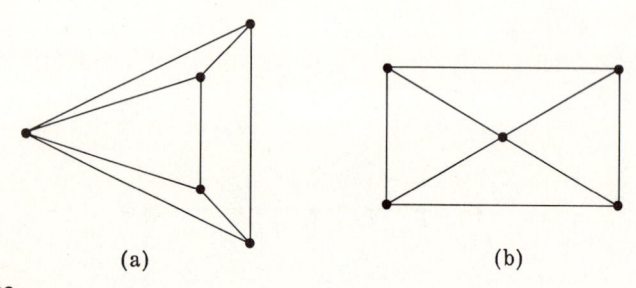

(a) (b)

Figure 7.23

7.12. Prove that the graphs in Figure 7.12 are not isomorphic.
7.13. Show that if an Euler path exists for a graph G and if G is isomorphic to a graph G', then G' also has an Euler path.
7.14. If G is a connected graph with an even number $2k$ of odd vertices, prove that there are k paths in G such that each edge in G is contained in one and only one of these paths. [*Hint:* First join pairs of odd vertices with new edges.]
7.15. Which of the graphs in Exercise 7.7 have Hamiltonian lines? Justify your assertions.
7.16. A graph G is called *complete* if for each pair of vertices there exists an edge having those vertices as endpoints.
  (a) Show that if G is a complete graph with n vertices, then there are at least $[n(n-1)]/2$ edges.
  (b) Show that a Hamiltonian line always exists in a complete graph.
  (c) Exhibit a complete graph with four vertices that has no Euler path. Can you generalize?

7.3. PLANAR GRAPHS AND EULER'S FORMULA

We begin this section with the definition of an important special class of graphs.

Definition. A graph G is *planar* if it is isomorphic to a graph G' such that:

(a) All of the edges and vertices of G' lie in a plane.

(b) If two edges of G' intersect at a point, that point must be a vertex of G'.

Of course, any graph G in a plane whose edges intersect only at vertices of G is planar. A simple example of a planar graph that does not lie in a plane is provided by the graph G consisting of the vertices and edges of a cube. In this case, the isomorphic graph G' with properties (a) and (b) is depicted in Figure 7.24(b). An appropriate isomorphism between

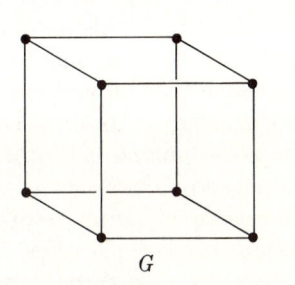

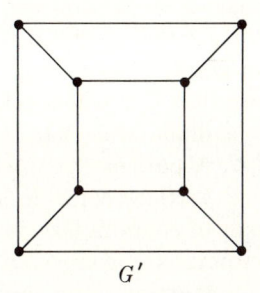

Figure 7.24

G and G' in Figure 7.24 is defined by corresponding the edges and vertices of the top and bottom faces of the cube G to those of the inner and outer squares of G', respectively.

The graph G in Figure 7.25 lies in a plane but its diagonal edges intersect at a point that is not a vertex of G. Nevertheless, G is a planar

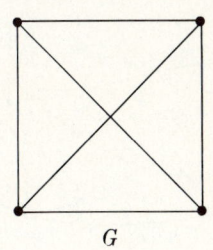

Figure 7.25

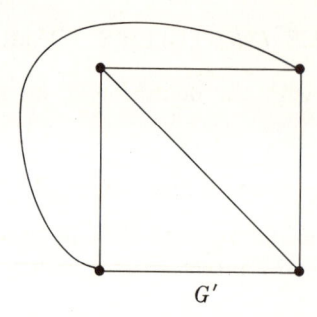

Figure 7.26

graph since G is isomorphic to the graph G' in Figure 7.26 and G' satisfies conditions (a) and (b) above. For a more complicated graph G, it is not always a simple matter to determine by inspection whether or not there is a graph G' isomorphic to G such that G' has properties (a) and (b). However, as we shall soon see, a remarkable result known as Euler's Formula will provide a useful tool for dealing with this as well as other problems in graph theory. First, we need to introduce some terminology.

Suppose that G is a connected graph such that all vertices and edges in G lie in a plane and that two distinct edges of G intersect only at vertices of G. A path $(e^{(1)}, \cdots, e^{(k)})$ in G is a *simple circuit* if its initial and terminal vertices coincide and if for each vertex v in G there is at most one pair of edges in this path with v as an endpoint. In other words, a simple circuit is a connected succession of distinct edges in G that returns to its starting point and that does not pass through any vertex more than once. A simple circuit $(e^{(1)}, \cdots, e^{(k)})$ in G encloses a region in the plane. If this region contains no other simple circuits in G, then this region is called an *interior face* of G and $(e^{(1)}, \cdots, e^{(k)})$ is called a *minimal circuit* in G. For example, in the graph G in Figure 7.27, the path (e_1, e_4, e_6, e_2) is a simple circuit. However, the region enclosed by this circuit is not an interior face since it contains the simple circuits (e_1, e_3, e_2) and (e_3, e_4, e_6). On the other hand, both of these latter simple circuits are minimal.

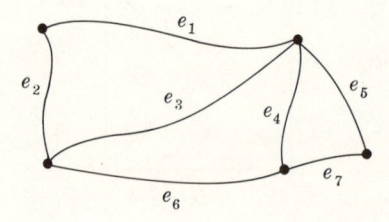

Figure 7.27

7.3. PLANAR GRAPHS AND EULER'S FORMULA

The region of the plane consisting of all points that are not contained in any interior face of G is called the *exterior face* of G. A region in the plane that is either an exterior or interior face of G is simply called a *face* of G. In Figure 7.28, F_1, F_2, F_3 are interior faces of G while F_4 is the exterior

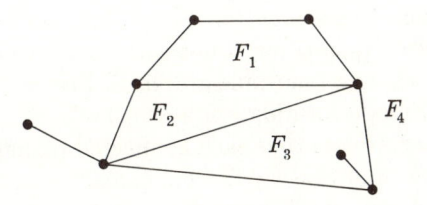

Figure 7.28

face of G. We shall denote the number of vertices of the graph G by V_G, the number of edges of G by E_G, and the number of faces of G (including the exterior face) by F_G.

We are now in a position to state and prove the main result of this section.

Proposition 7.3. *(Euler's Formula)* If G is a connected graph whose vertices and edges lie in a plane and whose edges intersect only at vertices of G, then

$$V_G - E_G + F_G = 2.$$

PROOF. First of all, suppose that $F_G = 1$, that is, suppose that there are no interior faces in G. Then there are no simple circuits in G. If $E_G = 1$, then it is obvious that $V_G = 2$ and the stated formula is valid. If $E_G > 1$, it is possible to remove a single edge and a single vertex from G to obtain a connected graph G' with no simple circuits. Consequently, mathematical induction may be applied to establish the fact that $V_G = E_G + 1$. (The reader should supply the details.) We conclude that $V_G - E_G + F_G = (E_G + 1) - E_G + 1 = 2$, and the formula is proved in the case $F_G = 1$.

Proceeding by mathematical induction, let us assume that Euler's formula holds for all graphs under consideration having k faces and let G be a connected planar graph for which $F_G = k + 1$. If we remove an edge from a minimal circuit in G, then we obtain a graph G' that is still connected and has k faces (since each such edge borders exactly two faces of G), one less edge than G, and the same number of vertices as G. Consequently, since the inductive hypothesis implies that $V_{G'} - E_{G'} + F_{G'} = 2$, it follows that $V_G - E_G + F_G = V_{G'} - (E_{G'} + 1) + (F_{G'} + 1) = 2$. That is, Euler's formula is valid for G. Therefore, the proposition follows from the Principle of Mathematical Induction.

If G is a connected planar graph, then the definition of planarity implies that there is an isomorphic graph G' in a plane whose edges intersect only at vertices in G'. The graph G' is by no means uniquely determined by G. However, if G'' is another graph in a plane that is isomorphic to G such that the edges of G'' intersect only at vertices of G'', then G' and G'' are isomorphic. Consequently, $V_{G'} = V_{G''}$ and $E_{G'} = E_{G''}$. Even though minimal circuits in G' need not correspond to minimal circuits in G'' (for example, see the isomorphic graphs in Figure 7.10), it is true that $F_{G'} = F_{G''}$ since Euler's formula applies to both G' and G''. In view of these considerations, we see that each connected planar graph G uniquely determines the three numbers V_G, E_G, F_G, namely, the numbers of vertices, edges, and faces of any graph isomorphic to G that lies in a plane with edges intersecting only at vertices.

In elementary geometry, we ordinarily consider the concepts of face, vertex, and edge in connection with the study of convex polyhedra. Recall that a polyhedron is a three-dimensional solid bounded by planar surfaces. A polyhedron is *convex* if it contains the entire line segment joining any pair of its points. Figure 7.29 shows an example of a polyhedron that is

Figure 7.29

not a convex polyhedron. Of course, the vertices and edges of a polyhedron constitute a graph in space. Moreover, this graph can be shown to be planar if the polyhedron is convex. Although we shall not prove this fact formally, it is a simple matter to present the intuitive basis for a formal proof. If we imagine a "skeleton" of the given convex polyhedron P constructed by replacing the edges of P by thin metal rods and if we appropriately place a light source s at a point slightly above one face f of P as in Figure 7.30, then the "shadow" G' on a plane below P and parallel to f is a graph in a plane that is isomorphic to the graph G consisting of the vertices and edges of P. Of course, the light source should be placed so that the shadows of distinct edges and vertices are also distinct. The exterior face of the resulting planar graph corresponds to the face f in Figure 7.30 over which the light is located.

In their geometrical investigations, the ancient Greeks discovered five so-called regular polyhedra, that is, convex polyhedra with congruent

7.3. PLANAR GRAPHS AND EULER'S FORMULA 297

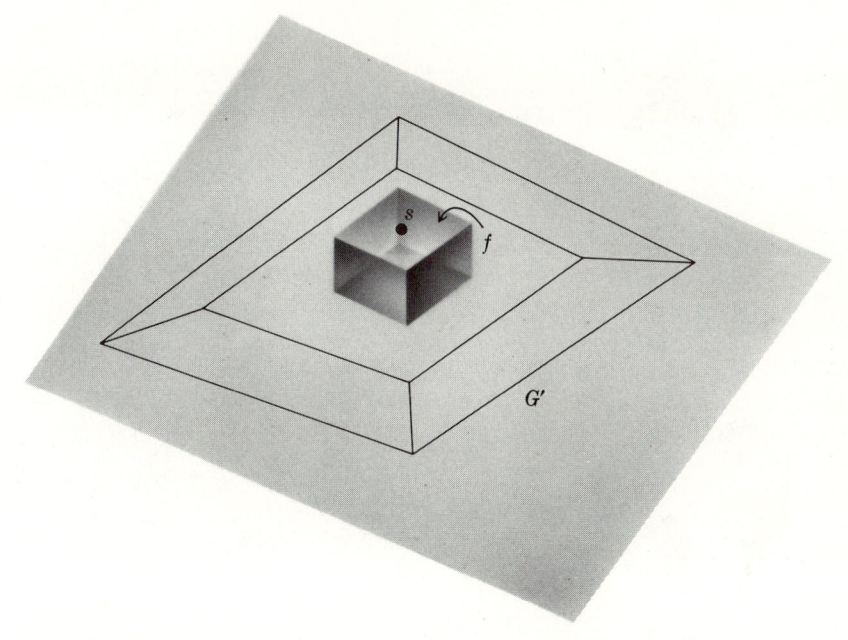

Figure 7.30

faces. These polyhedra, which are frequently called the *Platonic bodies*, are pictured below on the left-hand side of Figure 7.31; the right-hand side of Figure 7.31 displays graphs contained in a plane that are isomorphic to the corresponding graphs of vertices and edges of these regular polyhedra.

The reader should attempt to visualize how each of the graphs on the right result from the convex polyhedron on the left by the "shadow isomorphism" discussed earlier.

The Greek geometers searched for other examples of regular polyhedra, but their efforts were fruitless. Finally, Theaetetus (414–368 B.C.) proved that only five regular polyhedra exist.

We shall now prove this result as an application of Euler's formula.

Proposition 7.4. The tetrahedron, cube, octahedron, dodecahedron, and icosahedron are the only regular polyhedra.

PROOF. Suppose that P is a regular polyhedron. Then each face is bounded by the same number n of edges and each vertex is met by the same number r of edges; moreover, $n \geq 3$ and $r \geq 3$. Since each edge is on the boundary of two faces, we have $nF = 2E$. Also, since each edge has two

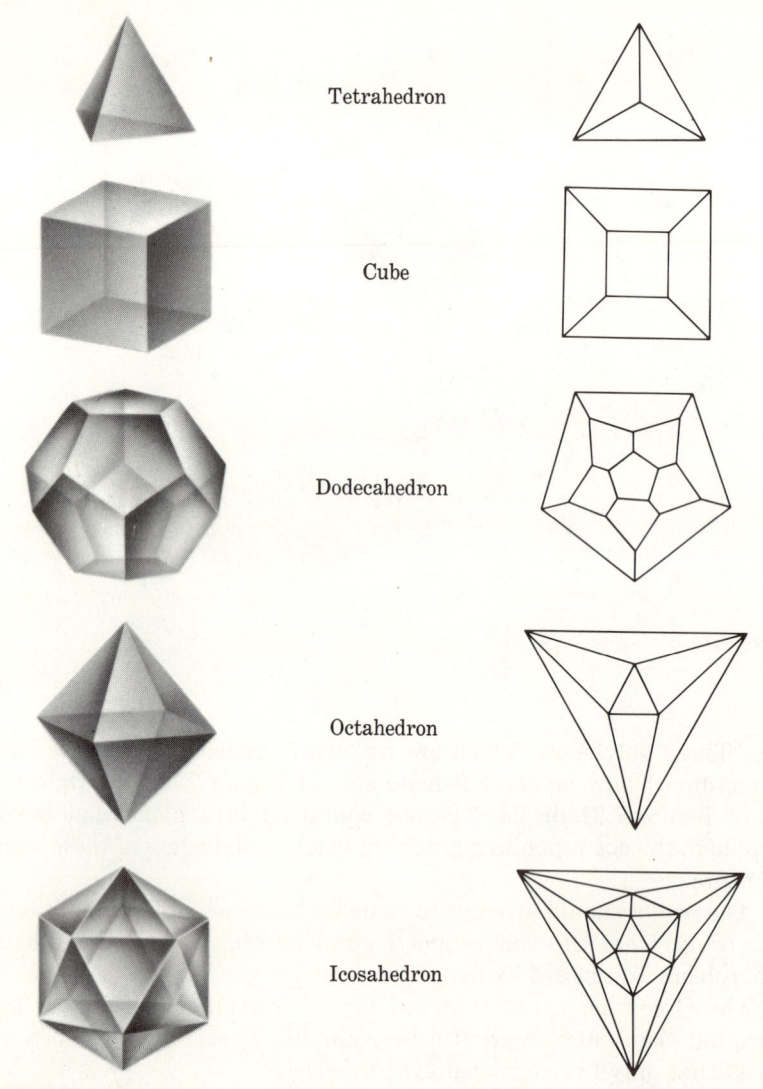

Figure 7.31

- Tetrahedron
- Cube
- Dodecahedron
- Octahedron
- Icosahedron

endpoints, we have $rV = 2E$. Therefore, by Euler's formula,

$$2 = V - E + F = \frac{2E}{r} - E + \frac{2E}{n},$$

which can be rewritten in the form

$$\frac{1}{E} = \frac{1}{r} - \frac{1}{2} + \frac{1}{n}. \tag{*}$$

Since E is a positive integer, (*) shows that both r and n cannot be ≥ 4 simultaneously. Consequently, since $r \geq 3$ and $n \geq 3$, only the following possibilities are allowed by (*):

$$n = 3 \text{ and } r = 3, 4, \text{ or } 5$$
$$r = 3 \text{ and } n = 3, 4, \text{ or } 5.$$

In Table 7.1 we show that each of the five distinct cases for n and r yield one of the five regular polyhedra in Figure 7.31. Thus, we have shown that P must be one of the five regular polyhedra mentioned in the proposition and the proof is complete.

Table 7.1

Case	$\dfrac{1}{E} = \dfrac{1}{r} - \dfrac{1}{2} + \dfrac{1}{n}$	$V = \dfrac{2E}{r}, F = \dfrac{2E}{n}$	P
$n = 3, r = 3$	$\dfrac{1}{E} = \dfrac{1}{6}$	$V = 4, F = 4$	Tetrahedron
$n = 4, r = 3$	$\dfrac{1}{E} = \dfrac{1}{12}$	$V = 8, F = 6$	Cube
$n = 5, r = 3$	$\dfrac{1}{E} = \dfrac{1}{30}$	$V = 20, F = 12$	Dodecahedron
$n = 3, r = 4$	$\dfrac{1}{E} = \dfrac{1}{12}$	$V = 6, F = 8$	Octahedron
$n = 3, r = 5$	$\dfrac{1}{E} = \dfrac{1}{30}$	$V = 12, F = 20$	Icosahedron

We were able to apply Euler's formula in the preceding proposition because the graph consisting of the edges and vertices of a regular polyhedron is planar. On the other hand, we shall now see that Euler's formula may also be used to prove that certain graphs are not planar. We shall discuss the techniques involved in the context of a well-known puzzle known as the Utilities Problem. The solution of the problem hinges on whether or not a certain graph is planar.

Suppose that each of n houses is to be connected to n utilities (for example, electricity, water, gas, cable TV, and so on). Is it possible to do this without having one utility line pass over another?

If the ith house is denoted by h_i, and the ith utility is denoted by u_i, and the path joining h_i to u_j is denoted by $e^{(ij)}$, then the problem reduces to deciding whether or not the graph G_n, called the *utilities graph of order n*, with vertices $\{h_i, u_j: i, j = 1, 2, \cdots, n\}$ and edges $\{e^{(ij)}: i, j = 1, 2,$

..., n} is planar. The utilities graphs of orders 2 and 3 are shown in Figure 7.32.

It is a very simple matter to show that G_2 is a planar graph; in fact, the graph in Figure 7.33 is obviously isomorphic to G_2. It is not so easy to decide about the planarity of G_3. The planar graph in Figure 7.34 comes

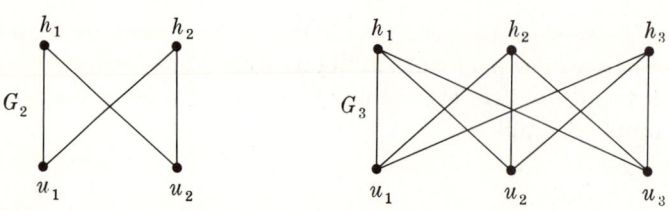

Figure 7.32

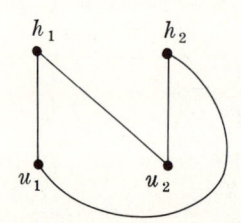

Figure 7.33

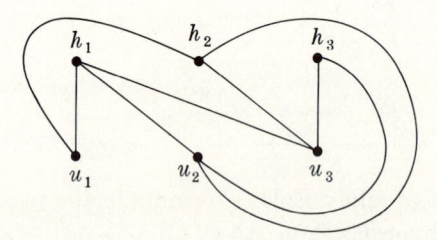

Figure 7.34

to within a single edge (joining h_3 to u_1) of being isomorphic to G_3. However, by making use of Euler's formula, we can establish the following result.

Proposition 7.5. If $n \geq 3$, then the utilities graph G_n of order n is not a planar graph.

PROOF. Since G_3 is contained in G_n for all $n \geq 3$, it suffices to show that G_3 is not planar. Suppose, to the contrary, that G_3 is planar. Then we can and shall assume that G_3 is a graph in a plane whose edges intersect at vertices only. If $i \neq j$, then neither $\{h_i, h_j\}$ nor $\{u_i, u_j\}$ can be the pair of endpoints of an edge in G_3 since neither two houses nor two utilities are ever to be connected by a single edge. It follows that each face of G_3 has *at least* four edges. Since each edge borders exactly two faces, it follows that $4F \leq 2E$.

For the graph G_3, we have $V = 6$ and $E = 9$. Consequently, Euler's formula implies that $2 = V - E + F = 6 - 9 + F$ so that $F = 5$. But then $20 = 4F \leq 2E = 18$, which is a contradiction. Therefore, G_3 is not a planar graph.

Any graph G whose edges and vertices lie on a sphere S and whose edges intersect only at vertices in G is necessarily planar. This can be verified by making use of "shadow isomorphism" similar to that used earlier in verifying the same conclusion for convex polyhedra. We place the light source s at a point on the sphere S that is not a vertex or on an edge in G. If P is the plane tangent to the sphere S at the point diametrically opposite to s, then the shadow G' cast by G on P (see Figure 7.35)

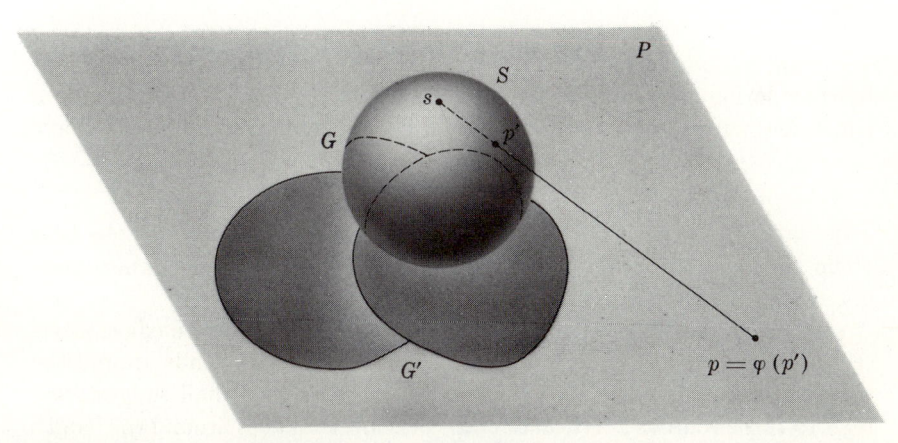

Figure 7.35

is a planar graph isomorphic to G. More precisely, if each point $p' \neq s$ of S is mapped onto the point p in the plane P that lies on the line through s and p' (see Figure 7.35), the resulting mapping φ is a one-to-one correspondence between the set $S' = \{p' \in S: p' \neq s\}$ and P. If e_i ($i = 1, \cdots, m$) and v_j ($j = 1, \cdots, n$) are the edges and vertices of G respectively, then $\varphi(e_i)$ ($i = 1, \cdots, m$) and $\varphi(v_j)$ ($j = 1, \cdots, n$) are the corresponding edges and vertices of G'. The face of G containing s corresponds to the exterior face of G'. In particular, it follows that if V_G, E_G, F_G

302 GRAPH THEORY

are the numbers of vertices, edges, and faces for G, then

$$V_G - E_G + F_G = 2.$$

If graphs are drawn on surfaces other than a plane or sphere, then the notion of face can be defined in a manner analogous to planar graphs except we deal with regions lying in the surface. However, Euler's formula is generally no longer valid and requires modification. We shall consider the situation for graphs on a torus, which is a doughnut-shaped surface as in Figure 7.36.

Figure 7.36

That Euler's formula is not correct for graphs on a torus can be seen by considering the graph G in Figure 7.37. This graph has one vertex, two edges, and with respect to the surface of the torus, one face. Thus, we have

$$V_G - E_G + F_G = 1 - 2 + 1 = 0.$$

This happens because the edge e_1 goes around the torus through the hole while the edge e_2 goes around the length of the torus. Together, they determine only one face rather than three.

Suppose that G is a connected graph on the torus whose edges intersect only at vertices. If G contains a simple circuit that winds around the torus through the hole (as e_1 does in Figure 7.37), we shall suggestively refer to this circuit as a *vertical circuit*. A circuit that goes around the length

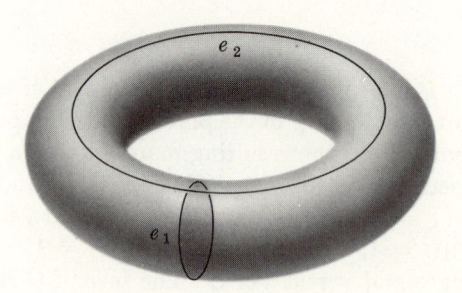

Figure 7.37

of the torus (as e_2 does in Figure 7.37) will be called a *horizontal circuit*. We shall now show that if G contains at least one vertical and one horizontal circuit, then the correct version of Euler's formula is

$$V_G - E_G + F_G = 0.$$

Imagine the torus on which G is situated to be made of rubber. Choose a vertical circuit in G and cut the rubber torus along the edges in this circuit. This amounts to repeating each of the edges and vertices in the circuit. The result is a new graph G' for which $V_{G'} - E_{G'} + F_{G'} = V_G - E_G + F_G$ since the number of new edges is equal to the number of new vertices. The cut rubber torus on which G' is situated can now be elastically deformed into a spherical surface with two holes in it as indicated in the succession of figures below (see Figure 7.38). This deformation

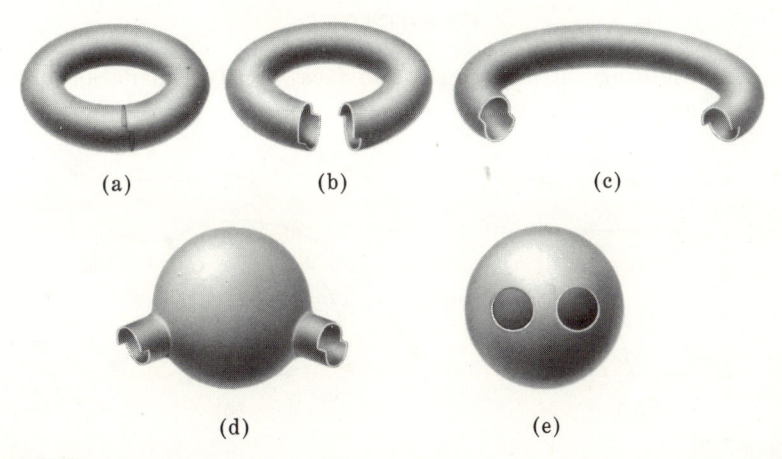

Figure 7.38

transforms G' into a graph G'' on the sphere with two holes in it. If we regard these two holes as new faces on the sphere, then G'' is a connected graph on the sphere since G contains a horizontal circuit. Thus, we have $V_{G''} - E_{G''} + F_{G''} = 2$ by Euler's formula. Therefore, if we count only those faces that correspond to faces in G (that is, if we do not count the two "faces" created by the holes in the sphere), we obtain $V_G - E_G + F_G = 0$. The above cut and deformation procedure can be made precise to obtain the following result.

Proposition 7.6. If G is a connected graph on a torus whose edges intersect only at vertices of G and if G contains a horizontal and a vertical circuit, then $V_G - E_G + F_G = 0$.

304 GRAPH THEORY

Similarly, if G is such a graph on an "anchor ring" surface of the sort pictured in Figure 7.39 and if G contains three circuits of the type suggested by the curves in Figure 7.39, then $V_G - E_G + F_G = -2$. For after making two cuts of the surface at 1 and 2, the resulting surface can be deformed into the surface of a sphere with four holes in it.

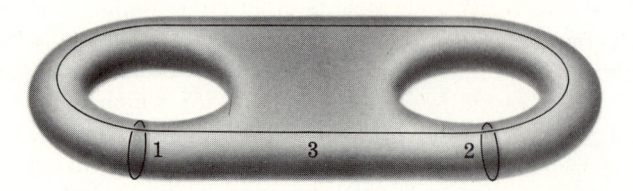

Figure 7.39

EXERCISES

7.17. By counting vertices, edges, and faces, verify Euler's formula for the graphs in Figure 7.40.

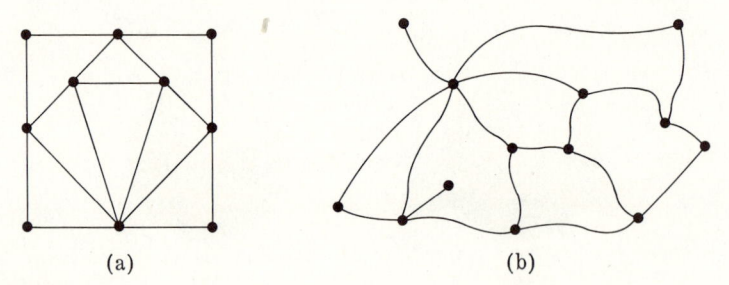

Figure 7.40

7.18. In Section 7.2, we saw that the graph consisting of the edges and vertices of a dodecahedron has a Hamiltonian line. Can the same statement be made for the graphs associated with the other regular polyhedra?

7.19. Suppose G is a planar graph with no loops and with at most one edge joining two vertices in G. Prove the following:
(a) $3F_G \leq 2E_G$.
(b) $E_G \leq 3V_G - 6$.
(c) There is a vertex v in G such that $\deg(v) \leq 5$.

7.20. A certain connected planar graph is such that each face is bounded by five edges and each vertex has the same degree d. Prove that $d < 4$. If we wished to draw such a graph for the case $d = 3$, how many faces would we need?

7.4. THE FOUR COLOR PROBLEM AND THE FIVE COLOR THEOREM

7.21. The *minimal complete graph* C_n *of order* n is the graph consisting of n vertices with no loops such that every pair of distinct vertices is joined by precisely one edge. Figure 7.41 displays the complete graphs of order 2, 3, 4, and 5.

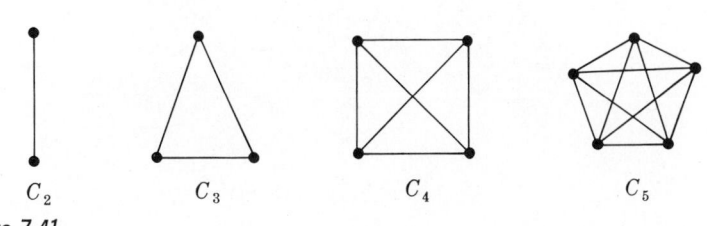

C_2 C_3 C_4 C_5

Figure 7.41

(a) Prove that the minimal complete graph of order 5 is not planar.
(b) Determine all values of n for which C_n is not a planar graph. [*Remark:* It is obvious that if a graph G contains a subgraph that is not planar, then G is not planar. In particular, if a graph G contains the utilities graph G_3 or the minimal complete graph C_5 of order 5 as a subgraph, then G is not planar. On the other hand, it can be shown that if a graph G is not planar, then (after possibly replacing certain vertices of degree 2 and the two adjacent edges in G by a single edge), the graph G must contain a subgraph that is isomorphic to either G_3 or C_5. See R. G. Busacker and T. L. Saaty, *Finite Graphs and Networks*. New York: McGraw-Hill, Inc., 1965, p. 70.]

7.4. THE FOUR COLOR PROBLEM AND THE FIVE COLOR THEOREM

Ordinarily, when a geographical map of a certain region (for example, continent, country, state) is printed, the various subregions (for example, countries, states, counties) are colored in such a way that no two subregions with a portion of boundary in common are colored the same. Experience seems to indicate that any such map may be colored without making use of more than four colors. It is a simple matter to provide examples of such maps that require four colors. For example, the map in Figure 7.42 can clearly be properly colored with four colors as indicated, but no smaller number of colors will do.

In view of these facts, a natural problem presents itself. Suppose that we regard a map M in the plane to be *properly colored* if the various regions (including the "outside" or infinite region) are colored so that no two regions having a portion of boundary in common are colored the same.

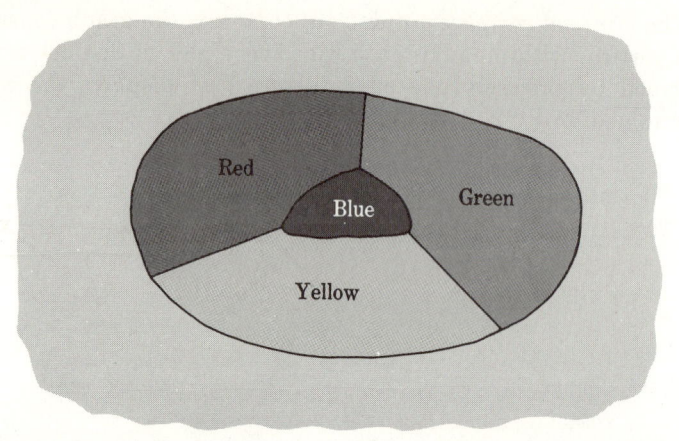

Figure 7.42

Then, is it possible to prove that four colors are always sufficient to properly color any map in the plane, or does there exist a map that actually requires five colors for its proper coloration? This problem is known as the *Four Color Problem*. Despite the apparent simplicity of its statement, it remains unsolved to this day. No one has found a proof that four colors are always adequate, yet no map has been drawn that actually requires more than four colors to properly color it.

The vigorous development of graph theory during the last hundred years is due in part to the very serious efforts made to solve this problem on the part of many gifted mathematicians. Although their efforts to resolve this seemingly childish and certainly impractical problem have not been successful, the mathematical discoveries in graph theory and other areas that these efforts spawned have found numerous significant and useful applications in mathematics, science, and technology.

Although the Four Color Problem was known and was studied by some mathematicians including A. De Morgan and A. F. Möbius as early as 1850, it was first proposed to the general mathematics community through a communication in the *Proceedings of the London Mathematical Society* in 1878 by the eminent algebraist A. Cayley (1801–1895). A proof that four colors would always suffice was published the following year by A. B. Kempe. However, an error in this proof was discovered by P. J. Heawood in 1890; nevertheless, Heawood was able to use a reduction procedure similar to that employed by Kempe to prove that *five* colors will always suffice. In the search for an example of a map that actually requires five colors, computers have been used to examine maps with less than a fixed number N of regions. So far, it has been shown that no map requiring five colors can be found among maps with less than 40 countries.

7.4. THE FOUR COLOR PROBLEM AND THE FIVE COLOR THEOREM

Arthur Cayley
Library of Congress

For the purposes of our discussion of coloring problems, we shall regard the geographic maps in question to be connected graphs in a plane such that there are no loops, all edges border on two distinct faces, and each vertex is met by at least three edges. The borders of the regions in the map meet at the vertices of the graph, and the edges of the graph are simply the portions of the border between two such vertices. We shall refer to connected planar graphs of the above sort as *maps*. Because of these restrictions, the planar graphs pictured in Figure 7.43 are *not* maps.

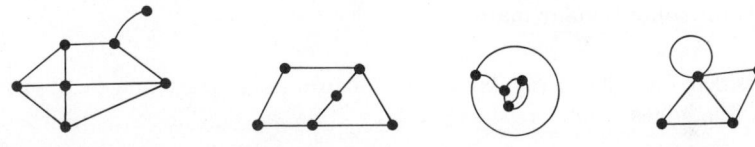

Figure 7.43

We have restricted the concept of map in the above manner for the sake of convenience in exposition and so that the resultant concept agrees with our intuitive feeling as to what a map should be. These restrictions are superficial and are not imposed to avoid deep or subtle difficulties in connection with the coloration problem.

A map that can be properly colored by using at most n colors will be called *n-colorable*. Our primary objective in this section is to establish the fact that every map in the plane is 5-colorable. We now proceed to this task by proving a sequence of propositions that culminate in the Five Color Theorem. The first step is to show that we may concentrate on a special type of map known as a regular map.

308 GRAPH THEORY

Definition. A map is called *regular* if every vertex in the map has degree 3.

Proposition 7.7. If n is a positive integer such that every regular map is n-colorable, then every map is n-colorable.

PROOF. Suppose that M is a map containing at least one vertex of degree greater than 3. We can modify M to obtain a regular map M' by replacing each vertex in M whose degree exceeds three with a small circle as shown in Figure 7.44. By hypothesis, M' can be colored with at most n colors in

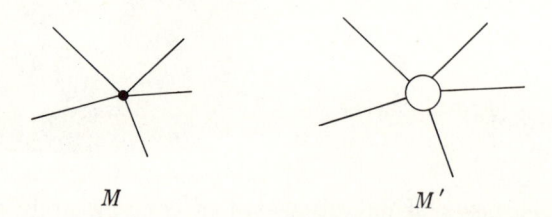

M $\qquad$ M'

Figure 7.44

such a way that no two faces having an edge in common are colored alike. If M' is so colored, we then shrink the circles back to the original vertices while extending the coloring of the faces bordering these circles. The result is a proper coloring of M using at most n colors.

Euler's formula permits us to draw the following important conclusion concerning regular maps.

Proposition 7.8. Each regular map M contains at least one face with at most five edges.

PROOF. Suppose M is a regular map and that V, E, F denote the respective numbers of vertices, edges, and faces of M. Define F_n to be the number of faces in M with precisely n edges. Then $F_1 = 0$ since M does not have loops. Hence, the number F of faces of M is

$$F = F_2 + F_3 + \cdots + F_k + \cdots.$$

Also, since each edge is common to two faces, it follows that

$$2E = 2F_2 + 3F_3 + \cdots + kF_k + \cdots.$$

Finally, since each vertex is the endpoint of precisely three edges, we have

$$3V = 2F_2 + 3F_3 + \cdots + kF_k + \cdots.$$

7.4. THE FOUR COLOR PROBLEM AND THE FIVE COLOR THEOREM

As a consequence of Euler's formula, we obtain

$$2 = V - E + F = \frac{2F_2 + 3F_3 + \cdots + kF_k + \cdots}{3}$$
$$- \frac{2F_2 + 3F_3 + \cdots + kF_k + \cdots}{2}$$
$$+ (F_2 + F_3 + \cdots + F_k + \cdots),$$

which may be written in the form

$$12 = 4F_2 + 3F_3 + 2F_4 + F_5 + (-1)F_7 + \cdots + (6 - n)F_n + \cdots.$$

Since the number on the left-hand side of the preceding equation is positive and since $F_k \geq 0$ for all k, it cannot be the case that $F_2 = F_3 = F_4 = F_5 = 0$. But then, by definition of F_n, there must be at least one face in M with at most five edges.

Note that the preceding proof shows that if $F_2 = F_3 = F_4 = 0$, then there must be at least twelve faces with five edges.

Proposition 7.9. If M is any regular map, then there is a regular map M' with a smaller number of faces than M such that *if M' is 5-colorable*, then M is also 5-colorable.

PROOF. Since M is a regular map, M has at least one face f with at most five edges. Thus, the portion of the map M immediately surrounding f must be typified by one of the configurations described in Figure 7.45. If f is described by either of the configurations in (a) or (b) of Figure 7.45, remove from M the edge e_1 and the two vertices that are endpoints of e_1 to define a regular map M' having one less face than M.

If f is described by either of the configurations (c) or (d) in Figure 7.45, choose two distinct faces f_1 and f_2 that border f such that f_1 and f_2 have no edges in common. Of course, it may happen that two nonadjacent faces bordering f "wrap around" the intermediate face or faces and have an edge in common with one another (see f' and f'' in Figure 7.46(a)) or that two distinct edges of f belong to the same face (see f' in Figure 7.46(b)). However, in either of these cases, this will "isolate" some pair of faces f_1 and f_2 bordering f such that f_1 and f_2 have no edge in common. In this case, remove from M the edges of f_1 and f_2 bordering f and the vertices that are endpoints of these edges to define a regular map M'. In all cases, the map M' will have fewer faces than the given map M.

Suppose that M' may be properly colored with at most five colors and suppose that such a coloration of M' has been carried out. Now each face of M corresponds to either a face of M' or a portion of a face of M'.

310   GRAPH THEORY

Figure 7.45

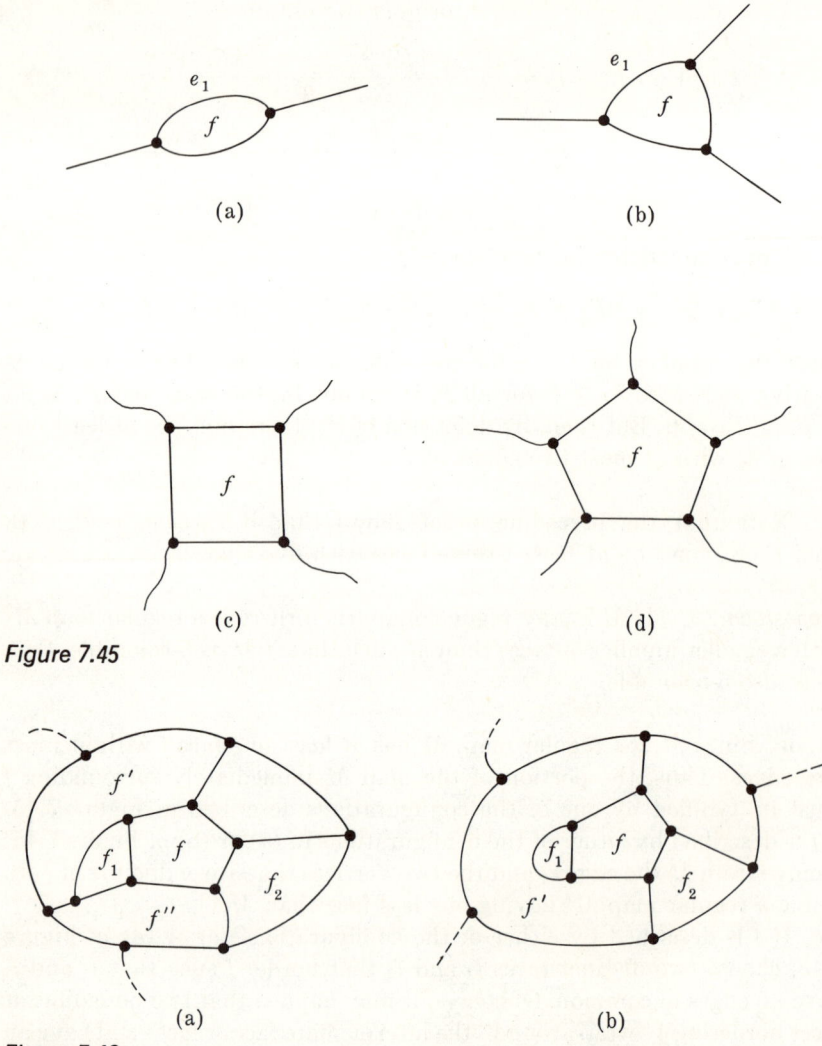

Figure 7.46

If we color each face of M, *except f*, with the color of the corresponding face of M', then at most four colors are used to color the faces bordering f. This leaves at least one color for f so that f can be colored differently from its neighboring faces. Thus, M can be properly colored in five colors.

We may now combine the conclusions of Propositions 7.7 and 7.9 to obtain Heawood's result.

7.4. THE FOUR COLOR PROBLEM AND THE FIVE COLOR THEOREM

Proposition 7.10. (The Five Color Theorem) Every map in the plane can be properly colored with at most five colors.

PROOF. According to Proposition 7.7, it is sufficient to show that any regular map is 5-colorable. Suppose, contrary to this statement, that the set $\mathfrak{M}$ of all regular maps that cannot be properly colored in five colors is not empty. Choose a map M_0 in $\mathfrak{M}$ such that the number of faces in M_0 is less than or equal to the number of faces of any other map in $\mathfrak{M}$. (How do we know that such a map M_0 can be found in $\mathfrak{M}$?) By Proposition 7.9, there is a regular map M_0' with fewer faces than M_0 such that *if M_0' is 5-colorable, then M_0 is 5-colorable.* But $M_0' \notin \mathfrak{M}$ since it has fewer faces than M_0. Consequently, M_0' is 5-colorable by definition of $\mathfrak{M}$. Therefore, by Proposition 7.9, M_0 can be properly colored with at most five colors, contrary to the choice of M_0. This contradiction implies that $\mathfrak{M}$ must be empty. Hence, every map can be properly colored with at most five colors.

As we have already remarked, it is known that every map with less than 40 faces can be properly colored with at most four colors. The following less ambitious result along these lines can be derived from the techniques developed in this section.

Proposition 7.11. If a map M has fewer than 12 faces, then it can be properly colored with at most four colors.

PROOF. Since each vertex is of degree at least 3 and since each edge contains at most two vertices, it is true that $3V \leq 2E$. Therefore, by Euler's formula $2 = V - E + F \leq -\frac{1}{3}E + F$, that is, $6 + E \leq 3F$. Hence, if $F \leq 11$, it follows that $E \leq 27$.

Now if every face in M has five or more edges, then $5F \leq 2E$ so that

$$2 = V - E + F \leq \tfrac{2}{3}E - E + \tfrac{2}{5}E = \tfrac{1}{15}E$$

by Euler's formula. But this would imply that $30 \leq E$, contrary to the conclusion of the preceding paragraph. Therefore, M must contain a face with four or fewer edges.

The argument used in the proof of Proposition 7.9 now shows that there is a map M' with fewer faces than M such that if M' can be properly colored with at most four colors, the same is true of M. The argument used in Proposition 7.10, applied to the class $\mathfrak{M}$ of all maps with fewer than 12 faces that cannot be properly colored with at most four colors, then shows that $\mathfrak{M}$ is empty. Thus, every map M with fewer than twelve faces can be colored with at most four colors.

Of course, geographical maps are sometimes displayed on spheres as in the case of a world globe. However, the shadow isomorphism discussed in the last section shows that any map on a sphere is isomorphic to a map in the plane so that the coloration problem for spherical maps is equivalent to the coloration problem for maps on a plane.

With the Four Color Problem still unsolved for plane and spherical maps, it would seem hopeless to even attempt to solve the corresponding problem for maps on other more complicated surfaces such as the torus (see Figure 7.36) or the "anchor ring" surface (see Figure 7.39). Surprisingly enough however, the coloration problem for several such surfaces has been completely solved. The following result makes use of the techniques developed in connection with the Five Color Theorem to solve the coloration problem for the torus.

Proposition 7.12. Every map on a torus can be colored with at most seven colors and there are maps on the torus that cannot be colored with fewer than seven colors.

PROOF. As in the case of maps in the plane, we need only consider regular maps M. If V, E, and F denote the number of vertices, edges, and faces in M respectively, then $V - E + F \geq 0$. (In fact, equality holds by Proposition 7.6 if M includes a vertical and horizontal circuit; otherwise, $V - E + F$ is either 1 or 2.) The proof of Proposition 7.8 is easily modified to this situation to show that M must contain a face with six or fewer edges. Considerations entirely similar to those in the proof of Proposition 7.9 then show that to each regular map M on the torus there corresponds a regular map M' with fewer faces than M such that if M' is 7-colorable, then M is 7-colorable. From this fact we conclude as in the proof of Proposition 7.10 that M can be colored with at most seven colors.

To show that seven colors may be required for a map on a torus, we need only consider the map in Figure 7.47. This completes the proof of the proposition.

It can also be shown that every map on the anchor ring surface in Figure 7.39 can be properly colored with at most eight colors and that eight colors are actually required for some such maps.

The coloration problem has also been discussed for the curious surface called the *Möbius strip*. A model of this surface can be constructed by making a half-twist in a narrow piece of paper or tape and then gluing the two ends together. (See Figure 7.48.)

There are several reasons why this surface is deserving of serious interest. We shall only discuss one such reason here since our main concern is the coloration problem.

7.4. THE FOUR COLOR PROBLEM AND THE FIVE COLOR THEOREM 313

Figure 7.47

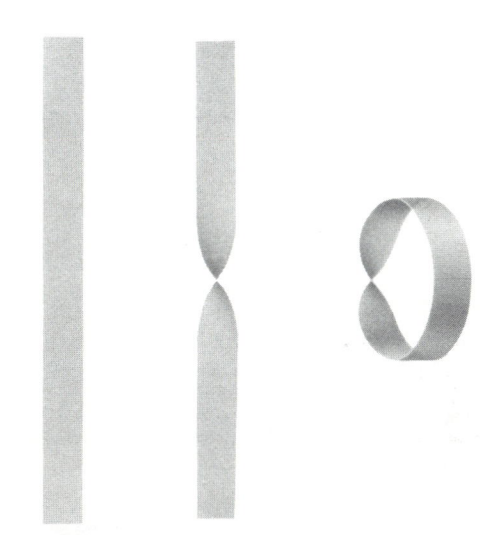

Figure 7.48

The Möbius strip is a simple example of a so-called *nonorientable* surface. Roughly speaking, a surface is *orientable* if it is possible to assign a perpendicular direction to the surface S at each point of S in such a way that this direction varies continuously along curves on the surface. If such an assignment is impossible, the surface S is said to be nonorientable. For example, the surfaces of a sphere or a torus are orientable, and Figure 7.49 depicts one of the two possible assignments of directions for these surfaces, namely, the outward direction. On the other hand, the Möbius strip is nonorientable since a continuous assignment of direction on the surface is impossible. For example, in Figure 7.50 if we begin at the point

314   GRAPH THEORY

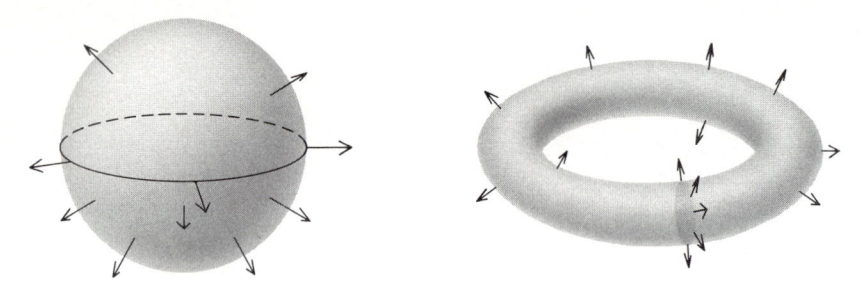

Figure 7.49

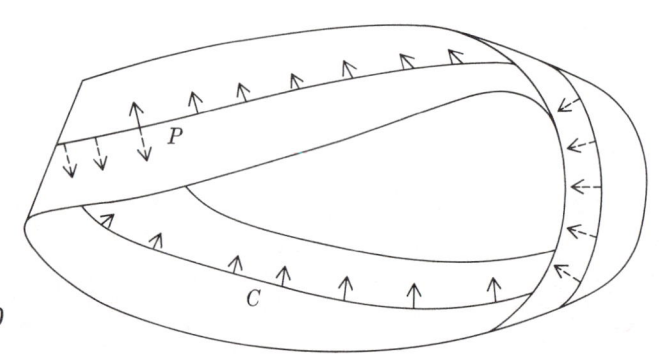

Figure 7.50

P and continuously assign a perpendicular direction along the curve C at the middle of the strip, then when we return to P along C, we will be forced to assign the direction opposite to the initial direction. This situation can be described more colorfully as follows: If a fly lands at the point P and crawls along C until he returns to P, he will be upside-down with respect to his original landing position!

Figure 7.51

7.4. THE FOUR COLOR PROBLEM AND THE FIVE COLOR THEOREM

Returning to the coloration problem, we remark that it has been proved that six colors will always suffice to properly color any map on the Möbius strip. Moreover, the map displayed in Figure 7.51 shows that six colors may be required. Thus, the coloration problem has been completely solved for the Möbius strip.

Despite the fact that the coloration problem has been solved for all surfaces, excepting the plane and sphere, for which a serious effort has been made, these two exceptions remain and will probably continue to perplex and inspire mathematicians for some time to come.

EXERCISES

7.22. Show that the faces of each of the five regular polyhedra (see Figure 7.31) can be properly colored with four colors. Can any of these polyhedra be properly colored with fewer colors?

7.23. Use the procedure in Proposition 7.7 to associate a regular map M' with the map M in Figure 7.52.

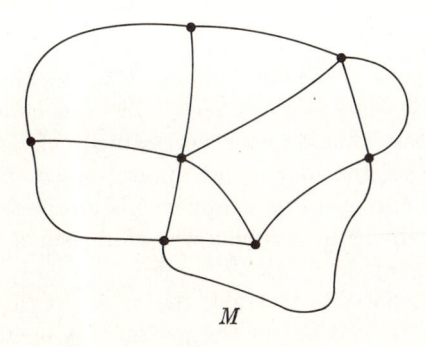

Figure 7.52

7.24. Show that the map M in Figure 7.52 can be properly colored with three colors but not two colors.

7.25. There is a theorem that states that if every vertex of a map in a plane is even, then the map is 2-colorable. Convince yourself of this fact by considering some examples.

7.26. Is it possible to construct a regular map M in the plane with fourteen faces such that M has 3 faces with two edges, 4 faces with four edges, 3 faces with five edges, 1 face with six edges, 2 faces with seven edges and 1 face with eight edges?

7.27. Suppose a map in the plane has the property that every face has three edges and the degree of every vertex is 5. Prove that there must be exactly 20 faces.

APPENDIX: THE CASE OF THE COLORED CUBES

As another indication of how graph theory may be used, we now discuss briefly the problem of the colored cubes. A commercial puzzle, which is now in fairly general circulation, consists of four cubes stacked in a single column in a clear celophane packet (see Figure 7.53). Each face of each

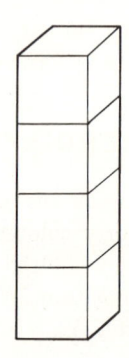

Figure 7.53

cube is colored with one of the four colors red, white, blue, green, and each of the cubes has at least one face of each color. The cubes are colored and stacked in such a way that on each vertical side of the stack appearing in the packet, each of the four colors appears once and only once. The puzzle challenges one to open the wrapper, mix up the four cubes and then attempt to stack them again in such a way that each of the colors appears on each vertical side of the stack.

Of course, it is possible to stack the cubes in the required fashion since they were so stacked before the packet was opened. Nevertheless, if one merely proceeds by trial and error, it ordinarily requires a considerable amount of time to stack the cubes properly again because of the large number of possible stackings and the small number of proper stackings for the cubes. As a matter of fact, it is a simple matter to compute an upper bound for the number of all possible stackings of the cubes as follows: The first cube may be placed in essentially three different ways, while the second, third, and fourth cubes can then each be placed in 24 different positions. It follows that an upper bound for the number of all possible positions of the cubes is

$$3 \cdot 24 \cdot 24 \cdot 24 = 41{,}472.$$

Of course, since all six faces of each cube are colored with one of four colors, some colors will be repeated on a given cube so that the actual number of *distinct* stackings of the cubes will be smaller than 41,472. However, it will still be quite a large number, and the number of proper stackings is very small in comparison.

On the face of it, this problem seems to have little to do with graphs. Of course, the vertices and edges of the cubes involved constitute a graph, but it is obvious that this graph bears no connection to the problem of finding a proper stacking. What is needed is a graph that reflects the basic information that is relevent to the solution of the problem. To find such a graph, suppose that we unwrap the packet of properly stacked colored cubes and then place the cubes in a line on a table without rotating them in any direction (see Figure 7.54). Suppose the vertical faces of the four

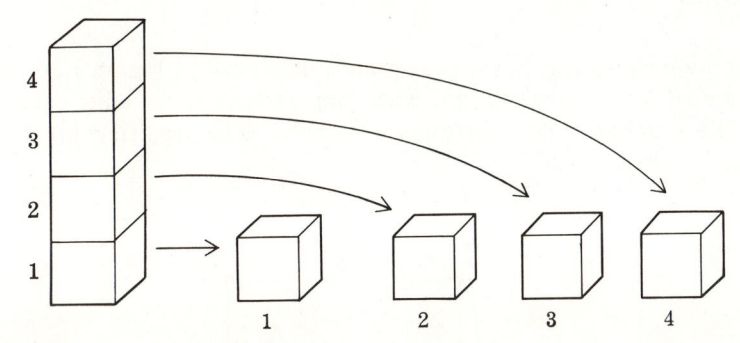

Figure 7.54

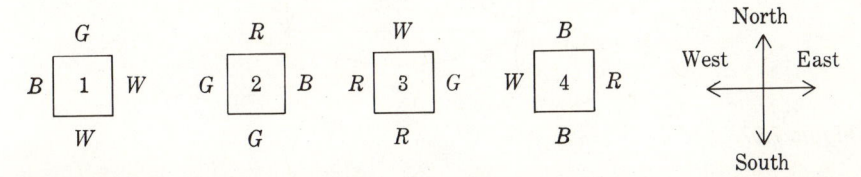

Figure 7.55

cubes have the colorations as indicated in Figure 7.55 where R, W, B, G denote the colors red, white, blue, green, respectively. The compass directions refer to the orientation of the cubes as we look at them. Suppose also the top and bottom of each cube is colored as indicated below:

	1	2	3	4
Top	R	G	B	G
Bottom	B	W	W	B

The basic information concerning the individual cubes can be put in terms of four graphs. Each of these graphs has four vertices, each vertex corresponding to one of the four colors red (R), white (W), blue (B), green

318 GRAPH THEORY

(G). In each of these graphs, we join two vertices by an edge if and only if the two colors represented by the vertices are the colors of one of the three pairs of opposite faces for the cube in question. For the cubes under consideration, the resulting graphs are pictured in Figure 7.56. All of the

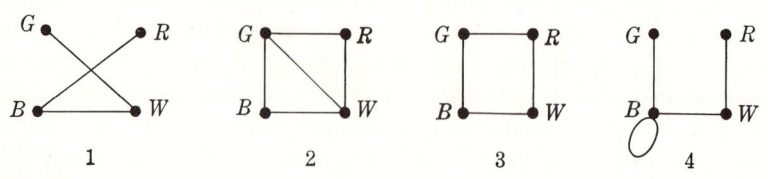

Figure 7.56

information contained in the four graphs displayed in Figure 7.56 can be compressed into a single graph with four vertices if we label each edge with the number of the corresponding cube. The resulting graph G is pictured in Figure 7.57.

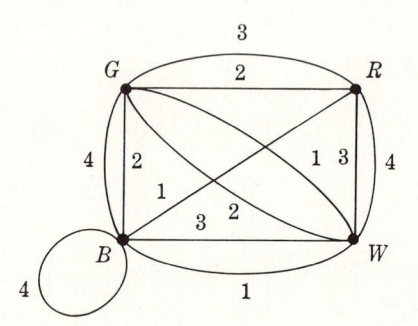

Figure 7.57

The stacking of the cubes as described in Figure 7.55 is related to two subgraphs of G corresponding to the East-West faces of the cubes and the North-South faces of the cubes. The graph G_1 in Figure 7.58 is the graph for the coloration of the East-West face pairs, and G_2 in Figure 7.58 corresponds to the coloration of the North-South face pairs. Each of these

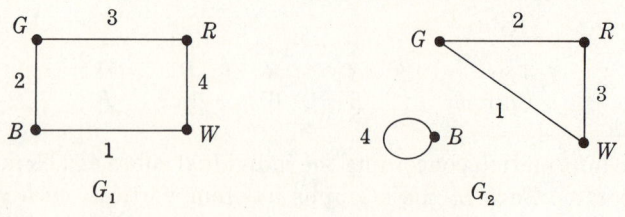

Figure 7.58

graphs consists of vertices and edges of G. The graph G_1 includes each vertex of G since each of the four colors must appear once on both the East and West side of the stack. Moreover, each edge in G_1 is labeled with a different number since the East-West pair of faces for each cube must be represented in G_1, and each vertex of G_1 is of degree 2 since each color appears once on an East face and once on a West face. The same conclusions can be drawn concerning the graph G_2. Finally, we note that G_1 and G_2 have no edges of G in common since G_1 corresponds to East-West face pairs while G_2 corresponds to North-South face pairs. Thus, we have shown that to the given proper stacking of the cubes, there correspond two subgraphs G_1 and G_2 of the graph G such that:

(a) Both G_1 and G_2 contain all the vertices of G.
(b) Both G_1 and G_2 contain an edge labeled i for each $i = 1, 2, 3, 4$.
(c) Every vertex in G_1 and G_2 is of degree 2.
(d) G_1 and G_2 have no edges of G in common.

On the other hand, if G_1' and G_2' are *any* two subgraphs of G which have the properties (a), (b), (c), (d) stated above for G_1 and G_2, then G_1' and G_2' can be used to describe another (possibly different) stacking of the given cubes. For example, if $G_1' = G_1$ and G_2' is the graph pictured in Figure 7.59, then G_1' and G_2' clearly satisfy (a), (b), (c), (d). G_1' and G_2'

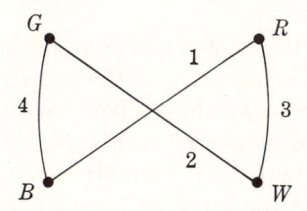

Figure 7.59

correspond to the stacking of the given cubes indicated in Figure 7.60. Figure 7.60 describes a stacking of the given set of four cubes that satisfies the requirements of the puzzle and yet is distinct from the original stacking in the packet.

Let us now summarize our conclusions. Suppose that we have opened the puzzle and that we have mixed up the cubes. We now proceed to

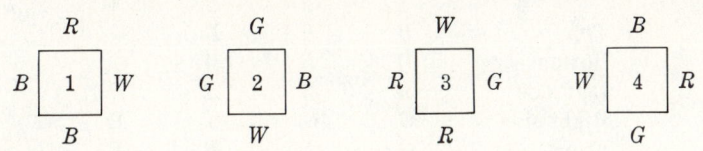

Figure 7.60

construct the graph G associated with this particular set of cubes as follows: Given four vertices labeled G, R, B, W, respectively, select one of the cubes and join two of the vertices with an edge labeled 1 if and only if the two vertices represent the colors on a pair of opposite faces of this cube. Repeat this procedure for a second cube labeling the new edges with a 2. Proceed in a similar fashion with the third and fourth cubes, being careful to keep track of the order in which the cubes were selected. The result will be a graph G of the sort in Figure 7.57. Next we find two subgraphs G_1 and G_2 of G with properties (a), (b), (c), (d). These subgraphs will then prescribe a proper stacking of the four cubes just as in the case of G_1' and G_2' above.

The appropriate subgraphs leading to a solution of the case of the colored cubes can generally be found by trial and error without too much difficulty. However, it is possible to systematize the process as follows. The color pairings for the three sets of opposing faces can be listed for each cube as in the following chart.

	(a)	(b)	(c)
1	RB	WG	WB
2	RG	WG	BG
3	WB	RW	RG
4	RW	BB	BG

This chart corresponds to the coloration given for the cubes considered above. A subgraph of the desired type then corresponds to a selection of one term from each row in the above chart such that when the letters of these terms are combined there are two R's, two W's, two B's and two G's. For example, the terms 1(a), 2(b), 3(b), 4(c) have this property and these pairings correspond to the graph G_2' in Figure 7.59. With a little practice, the selection of such terms can be performed without exhausting all 81 possible cases, but usually it involves no less labor than determining the desired subgraphs by inspection.

EXERCISE

7.28. Suppose the four cubes have the following colorations:

	1	2	3	4
Top	B	R	B	G
Bottom	W	B	W	G
Left side	R	W	R	B
Right side	G	G	G	W
Front	G	R	R	R
Back	W	W	R	B

Draw the graph for these cubes and find appropriate subgraphs leading to a solution. Describe the corresponding color arrangement of the stack. (The coloration scheme in the above table is the one used in the commercial version of this puzzle called "Instant Insanity" and is sold by Parker Brothers, Inc.)

REMARKS AND REFERENCES

We included an introduction to the theory of graphs in this book for several reasons. First of all, even though graph theory is not a standard topic in secondary school mathematics, it provides the mathematical basis for the solution of significant problems and interesting puzzles that secondary school students find exciting. The teacher who is familiar with the graph-theoretic analysis of problems and puzzles such as those considered in this chapter will be in a position to direct a student's natural curiosity toward a worthwhile mathematical experience. Secondly, graph theory is rich in intuitive and geometric content; nevertheless, its methods, conclusions, and emphasis are in sharp contrast to those of Euclidean geometry. This contrast should help one to appreciate the nature and objectives of Euclidean geometry and also the breadth of modern geometry. Thirdly, the portion of the theory of graphs that we have developed in this chapter is elementary in character and is quite accessible to secondary school students. For this reason, it is an excellent source of enrichment material for independent reading projects and mathematics club talks. Finally, the wealth of applications that graph theory has found in mathematics, the physical, biological, and social sciences, economics, and engineering suggest that the subject will find its proper place in the standard mathematics program in the near future.

The reader should find the following references to be useful sources for additional reading on graph theory. The book by Tietze [5] has a pleasant, nontechnical discussion of neighboring domains, map coloring, and other popular problems and it requires very little mathematical background to be enjoyed. The paperback book by Ore [4] is a very well written introduction to graph theory and lies well within the reach of a good secondary school student. It also contains a discussion of several aspects of graph theory not mentioned in this chapter. A slightly higher level account of the subject is given by Busacker and Saaty [2]. Their book also outlines myriad applications that reveal the scope of the subject. The book by Liu [3] is another good source for further reading on graph theory and its applications. Information concerning the historical development of the subject as well as discussions of other puzzles and problems can be found in Ball and Coxeter [1].

1. Ball, W. W. R. and H. S. M. Coxeter, *Mathematical Recreations and Essays*, New York: The Macmillan Company, 1962.
2. Busacker, R. G. and T. L. Saaty, *Finite Graphs and Networks*, New York: McGraw-Hill Book Company, Inc., 1965.
3. Liu, C. L., *Introduction to Combinatorial Mathematics*, New York: McGraw-Hill Book Company, Inc., 1968.
4. Ore, Oystein, *Graphs and Their Uses*, New York: Random House, 1963.
5. Tietze, Heinrich, *Famous Problems of Mathematics*, New York: Graylock Press, 1965.

Chapter 8

THE GEOMETRY OF COMPLEX NUMBERS

Although the basic arithmetic properties of the system of complex numbers were fairly well known and understood by the end of the sixteenth century, the geometrical aspects of this system were not investigated until the nineteenth century. The first major forays into the geometric theory of complex numbers were made by C. Wessel in 1797, by J. Argand in 1806, and by K. F. Gauss in 1831. B. Riemann (1826–1866) was largely

Bernhard Riemann
David Smith Collection

responsible for carrying forward these initial efforts to the state of a mature theory.

The first systematic study of transformations in the geometry of complex numbers was carried out by A. Möbius (1790–1860) around the year 1850. His investigation was in large part the motivation for Felix Klein's *Erlangen Programm* of 1872 in which he proposed the study of

Augustus F. Möbius
David Smith Collection

geometry from the point of view of transformations with emphasis on those quantities that remain invariant under specified classes of transformations.

This chapter is devoted to a study of the geometric aspects of the complex number system. We first discuss the two main geometric representations of the complex number system, namely the complex plane and the Riemann sphere, and then examine an important class of transformations of the complex numbers. Some applications of the theory are developed in the final section of the chapter. We shall assume that the reader is familiar with the material in Section 2.4 dealing with the arithmetic of complex numbers.

8.1. THE COMPLEX PLANE

In Section 2.4 we defined a complex number to be an ordered pair of real numbers. This immediately suggests one geometric interpretation of the set **C** of complex numbers. If we introduce a Cartesian coordinate system on a plane as in Figure 8.1 and if the complex number $z = x + iy$ is viewed as the ordered pair (x,y) of real numbers, then we may represent z by the point in the plane having Cartesian coordinates x and y, respectively. For example, the complex number $-1 + 2i$ is represented by the point in the plane two units above and one unit to the left of the origin.

On the other hand, since every point in the Cartesian plane is determined by its ordered pair of coordinates and since these coordinates are real numbers, it follows that each point in the Cartesian plane determines a unique complex number and this complex number is represented by the given point. Thus, the mapping that assigns to each complex number

8.1. THE COMPLEX PLANE

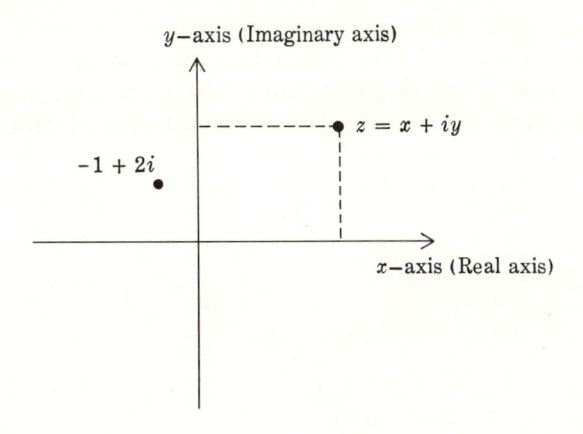

Figure 8.1

$z = x + iy$ the point in the Cartesian plane with first and second coordinates equal to the real part $\text{Re}(z) = x$ and imaginary part $\text{Im}(z) = y$ of z, respectively, is a one-to-one correspondence between the set **C** of complex numbers and the set of points in the Cartesian plane. Note that the imaginary numbers in **C** (that is, $z \in \mathbf{C}$ such that $\text{Re}(z) = 0$) correspond precisely to the points of the y-axis and the real numbers in **C** (that is, $z \in \mathbf{C}$ such that $\text{Im}(z) = 0$) correspond exactly to the points of the x-axis in the Cartesian plane. For these reasons, the y-axis is called the *imaginary axis* while the x-axis is called the *real axis*. The Cartesian plane used to represent the complex numbers in this way is called the *complex plane*.

The absolute value $|z| = \sqrt{x^2 + y^2}$ of the complex number $z = x + iy$ is simply the distance from the point z in the complex plane to the origin. The complex conjugate $\bar{z} = x - iy$ of $z = x + iy$ is the reflection of z in the real axis. (See Figure 8.2.) Note that $|z| = |\bar{z}|$.

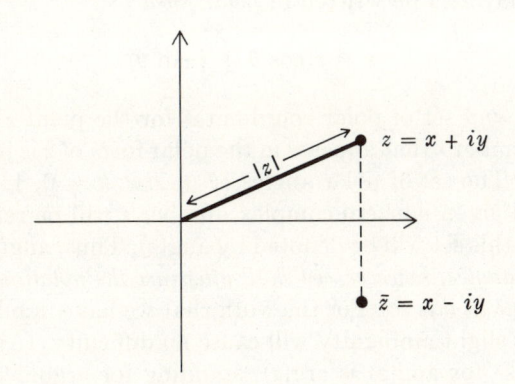

Figure 8.2

326 THE GEOMETRY OF COMPLEX NUMBERS

It is often useful to represent the points in the complex plane by their polar coordinates instead of the rectangular coordinates described above. Recall that a point P distinct from the origin in the Cartesian plane has polar coordinates (r,θ) if r is the distance of the point P from the origin and θ is the angle between the positive x-axis and the ray through P issuing from the origin. (See Figure 8.3.) Thus, the polar coordinates (r,θ)

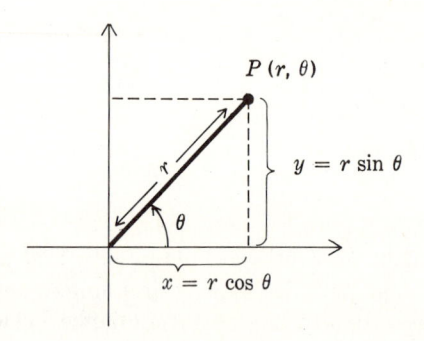

Figure 8.3

and the rectangular coordinates (x,y) of P are related by the equations:

$$x = r \cos \theta, \qquad r = \sqrt{x^2 + y^2}$$
$$y = r \sin \theta, \qquad \tan \theta = \frac{y}{x}.$$

Of course, the polar coordinates $(r, \theta \pm 2k\pi)$ for any nonnegative integer k describe the same point P in the plane as the polar coordinates (r,θ).

In view of the above relations, we see that a nonzero complex number $z = x + iy$ may also be written in *polar form*

$$z = r(\cos \theta + i \sin \theta),$$

where (r,θ) is *any* set of polar coordinates for the point z in the complex plane. The number r that appears in the polar form of z is just the absolute value $|z|$ of z. The set of polar angles $\{\theta \pm 2k\pi : k = 0, 1, 2, \cdots\}$ corresponding to a given nonzero complex number z will be referred to as the *angle of z* and this set will be denoted by ang(z). Thus, ang(z) *is actually an infinite set of angles; however, we shall often use the notation* ang(z) *for any particular angle in this set.* For the work that we have in mind for complex numbers, this slight ambiguity will cause no difficulty. (Another notation in common use for ang(z) is arg(z), standing for argument of z, but we shall not use this notation here.)

Example 8.1. If $z = -1 - i$, then $r = |z| = \sqrt{2}$. Also, since $\tan \theta = 1$ and z lies in the third quadrant, we may take $\theta = 5\pi/4$. (See Figure 8.4.) Thus, we have

$$-1 - i = \sqrt{2}\left(\cos\frac{5\pi}{4} + i \sin \frac{5\pi}{4}\right).$$

Any of the angles in

$$\text{ang}(-1 - i) = \left\{\frac{5\pi}{4} \pm 2k\pi \colon k = 0, 1, 2, \cdots\right\}$$

can be selected for θ. That is,

$$-1 - i = \sqrt{2}\left(\cos\left(\frac{5\pi}{4} \pm 2k\pi\right) + i \sin\left(\frac{5\pi}{4} \pm 2k\pi\right)\right)$$

for any nonnegative integer k.

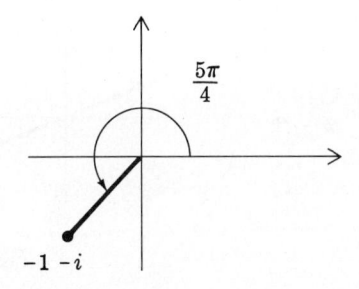

Figure 8.4

EXERCISES

8.1. Plot the complex numbers $5i$, $-2 + 2i$ and $\sqrt{3} - i$ in the complex plane and write them in polar form.
8.2. Write the complex number having absolute value 4 and polar angle $\pi/3$ in the form $x + iy$.
8.3. If z has polar form $r(\cos \theta + i \sin \theta)$, show that
 (a) $\bar{z} = r(\cos \theta - i \sin \theta)$,
 (b) $-z = r(\cos (\theta + \pi) + i \sin (\theta + \pi))$.
8.4. Locate all points z in the complex plane that satisfy
 (a) $\text{Re}(z) = \text{Im}(z)$,
 (b) $z = \bar{z}$,
 (c) $z = |z|$.

328 THE GEOMETRY OF COMPLEX NUMBERS

The complex plane is more than just a simple geometric visualization of the set of complex numbers. As we shall soon see, the arithmetic operations on **C** correspond to elementary geometric constructions, and this interplay of arithmetic and geometry provides valuable insight into the nature of the complex number system. In fact, it was not until the geometry of complex numbers was developed in the nineteenth century that their importance was truly appreciated.

We shall now proceed to the geometric interpretation of the basic arithmetic operations for complex numbers.

Addition and Subtraction

If $z_1 = x_1 + iy_1$ and $z_2 = x_2 + iy_2$ are given complex numbers, the sum $z_1 + z_2$ is defined to be the complex number $(x_1 + x_2) + i(y_1 + y_2)$. Thus, if z_1 and z_2 are represented as points in the complex plane, and if $0, z_1, z_2$ are not collinear, then the point of the complex plane representing $z_1 + z_2$ is the fourth vertex of the parallelogram with vertices $0, z_1, z_2$. (See Figure 8.5(a).) If $0, z_1, z_2$ are collinear, the parallelogram in question

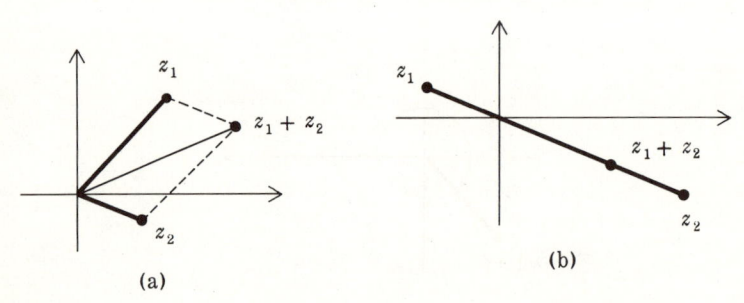

Figure 8.5

is degenerate; in this case, the point of the complex plane corresponding to $z_1 + z_2$ is the endpoint of the directed line segment that is the sum of the two directed line segments joining 0 to z_1 and 0 to z_2 respectively. (See Figure 8.5(b).) These assertions follow directly from the definition of the rectangular coordinates of a point; that is, the x-coordinate of a point P in the plane is precisely the *directed distance* to P from the y-axis while the y-coordinate of P is the directed distance to P from the x-axis.

Since the sum of the lengths of two sides of a triangle always exceeds the length of the third side, this interpretation of addition yields the triangle inequality

$$|z_1 + z_2| \leq |z_1| + |z_2|.$$

The difference $z_1 - z_2$ is the complex number $(x_1 - x_2) + i(y_1 - y_2)$, which is the same as $z_1 + (-z_2)$. Geometrically, $-z_2$ is the reflection of

z_2 with respect to the origin. Hence, subtraction has the geometric interpretation indicated in Figure 8.6. Notice that $|z_1 - z_2|$ is the usual distance between the points z_1 and z_2 in the complex plane.

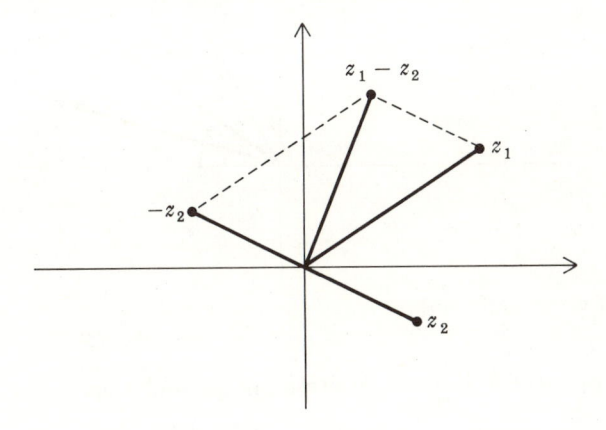

Figure 8.6

Multiplication and Division

If $z_1 = x_1 + iy_1$ and $z_2 = x_2 + iy_2$, then the product $z_1 z_2$ is defined by $z_1 z_2 = (x_1 x_2 - y_1 y_2) + i(x_1 y_2 + x_2 y_1)$. In order to geometrically interpret multiplication, it is useful to use the polar form of complex numbers. If z_1 and z_2 are nonzero complex numbers having polar forms

$$z_1 = r_1(\cos \theta_1 + i \sin \theta_1),$$
$$z_2 = r_2(\cos \theta_2 + i \sin \theta_2),$$

then

$$\begin{aligned} z_1 z_2 &= r_1 r_2 [(\cos \theta_1 \cos \theta_2 - \sin \theta_1 \sin \theta_2) \\ &\quad + i(\cos \theta_1 \sin \theta_2 + \sin \theta_1 \cos \theta_2)] \\ &= r_1 r_2 [\cos(\theta_1 + \theta_2) + i \sin(\theta_1 + \theta_2)]. \end{aligned}$$

The last equality was obtained by applying the addition formulas for the sine and cosine functions. Thus, we see that the point of the complex plane representing $z_1 z_2$ lies on the ray $\theta = \theta_1 + \theta_2$ at a distance $r_1 r_2$ from the origin. That is, $|z_1 z_2| = |z_1| |z_2|$ and $\text{ang}(z_1 z_2) = \text{ang}(z_1) + \text{ang}(z_2)$. (See Figure 8.7.) Thus, when we multiply two complex numbers, we multiply their absolute values and add their angles. For example, since i has polar coordinates $(1, \pi/2)$, the product iz is obtained from the point z by a rotation of that point through $\pi/2$ in the counter-clockwise direction.

A straightforward induction argument can be used to extend the above observations to the product of any finite number of complex num-

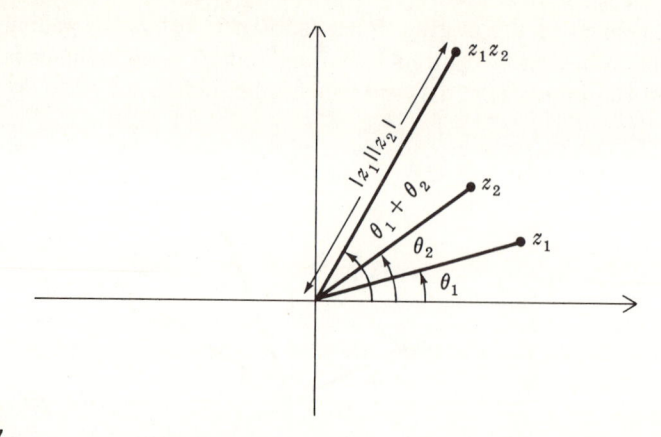

Figure 8.7

bers. In particular, if n is a positive integer and z has polar form

$$z = r(\cos\theta + i\sin\theta),$$

then

$$z^n = r^n(\cos n\theta + i \sin n\theta).$$

That is, $|z^n| = |z|^n$ and $\text{ang}(z^n) = n\,\text{ang}(z)$.

For example, if $z = 1 + i$, then $r = |z| = \sqrt{2}$ and we may take $\text{ang}(z) = \pi/4$. Thus, $(1+i)^{14}$ has absolute value $(\sqrt{2})^{14} = 128$ and angle $7\pi/2$, or equivalently, $3\pi/2$. Hence,

$$(1+i)^{14} = 128\left(\cos\frac{3\pi}{2} + i\sin\frac{3\pi}{2}\right) = -128i.$$

Division of z_1 by z_2 can be similarly interpreted. Since

$$\frac{1}{z_2} = \frac{1}{r_2(\cos\theta_2 + i\sin\theta_2)}$$

$$= \frac{1}{r_2(\cos\theta_2 + i\sin\theta_2)} \cdot \frac{\cos\theta_2 - i\sin\theta_2}{\cos\theta_2 - i\sin\theta_2}$$

$$= \frac{1}{r_2}(\cos\theta_2 - i\sin\theta_2)$$

$$= \frac{1}{r_2}(\cos(-\theta_2) + i\sin(-\theta_2)),$$

we see as above that the polar form of z_1/z_2 is

$$\frac{z_1}{z_2} = \frac{r_1}{r_2}(\cos(\theta_1 - \theta_2) + i\sin(\theta_1 - \theta_2)).$$

EXERCISES

8.5. Compute $\dfrac{10}{(1-i)(2-i)(3-i)}$.

8.6. Compute $\dfrac{1+i}{1-i}$ directly and also by using polar forms.

8.7. Compute $(\sqrt{3} - i)^6$ by using the polar form of $\sqrt{3} - i$.

8.8. If n is a positive integer, what are the possible values of $(i)^n$ and of $(i)^n - (1/i)^n$?

8.9. Show that $z\bar{z} = |z|^2$ by using the polar form of z. (Note that $\operatorname{ang}(\bar{z}) = -\operatorname{ang}(z)$.)

8.10. If $z^2 = (\bar{z})^2$, what can be said about z?

8.11. Show that $|\operatorname{Re}(z)| + |\operatorname{Im}(z)| \leq \sqrt{2}\,|z|$.

8.12. If z_1 and z_2 are given nonzero complex numbers, let the complex number z be such that the triangle with vertices $0, z_1, z$ is similar to the triangle with vertices $0, 1, z_2$. (See Figure 8.8.) Show that $z = z_1 z_2$.

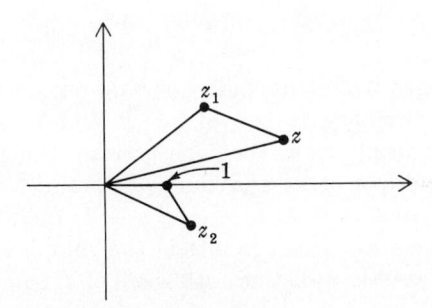

Figure 8.8

8.13. If z_1, z_2, z_3 are three distinct points of the complex plane, find a geometric interpretation of the angle of the complex number

$$\frac{z_2 - z_1}{z_3 - z_1}.$$

Roots of Complex Numbers

If z is a given complex number and n is a positive integer, then any complex number w satisfying $w^n = z$ is called an nth *root* of z. If $z = 0$, then the only solution to $w^n = 0$ is $w = 0$, so that 0 is the only nth root

332 THE GEOMETRY OF COMPLEX NUMBERS

of 0. We shall now show how to determine the nth roots of nonzero complex numbers.

Suppose $z \neq 0$ is given. Then z has polar form

$$z = r(\cos \theta + i \sin \theta),$$

where $r > 0$ and θ may be chosen so that $0 \leq \theta < 2\pi$. We wish to find all complex numbers w such that $w^n = z$. If w has polar form,

$$w = \rho(\cos \varphi + i \sin \varphi), \tag{1}$$

where $0 \leq \varphi < 2\pi$, then we have

$$\rho^n(\cos n\varphi + i \sin n\varphi) = r(\cos \theta + i \sin \theta). \tag{2}$$

Since $\rho^n = |w^n| = |z| = r$ and ρ is a positive real number, we see that $\rho = \sqrt[n]{r}$, the unique positive nth root of the positive real number r. Also, by equating the real and imaginary parts of (2), we have

$$\cos n\varphi = \cos \theta, \quad \sin n\varphi = \sin \theta.$$

This implies that $n\varphi = \theta + 2k\pi$, where k is an integer. Thus, a complex number w with polar form (1) will be an nth root of z if and only if

$$\rho = \sqrt[n]{r}, \quad \varphi = \frac{\theta}{n} + \frac{2k\pi}{n},$$

where k is an integer. Different values of k do not necessarily determine different complex numbers. In fact, if $k = k_1$ and $k = k_2$ are values of k that differ by a multiple of n, then the corresponding angles φ_1 and φ_2 will differ by a multiple of 2π and thus will produce the same complex number. However, if $k = 0, 1, 2, \cdots, n - 1$, then the corresponding angles will determine n distinct points in the complex plane. Hence, we see that there are exactly n distinct nth roots of z, namely, $w_0, w_1, \cdots, w_{n-1}$, where

$$w_k = \sqrt[n]{r}\left(\cos\left(\frac{\theta}{n} + \frac{2k\pi}{n}\right) + i \sin\left(\frac{\theta}{n} + \frac{2k\pi}{n}\right)\right) \tag{3}$$

for $k = 0, 1, 2, \cdots, n - 1$. Formula (3) is usually referred to as *de Moivre's formula* for the nth roots of a nonzero complex number z. The *set* of nth roots of z given by (3) will be denoted by $z^{1/n}$.

Example 8.2. Compute the 5th roots of $-4 + 4i$. The polar form of this number is

$$4(-1 + i) = 4\sqrt{2}\left(\cos \frac{3\pi}{4} + i \sin \frac{3\pi}{4}\right).$$

Consequently, $(-4 + 4i)^{1/5} = \{w_0, w_1, w_2, w_3, w_4\}$, where

$$w_k = \sqrt{2}\left(\cos\left(\frac{3\pi}{20} + \frac{2k\pi}{5}\right) + i\sin\left(\frac{3\pi}{20} + \frac{2k\pi}{5}\right)\right)$$

for $k = 0, 1, 2, 3, 4$. (See Figure 8.9.)

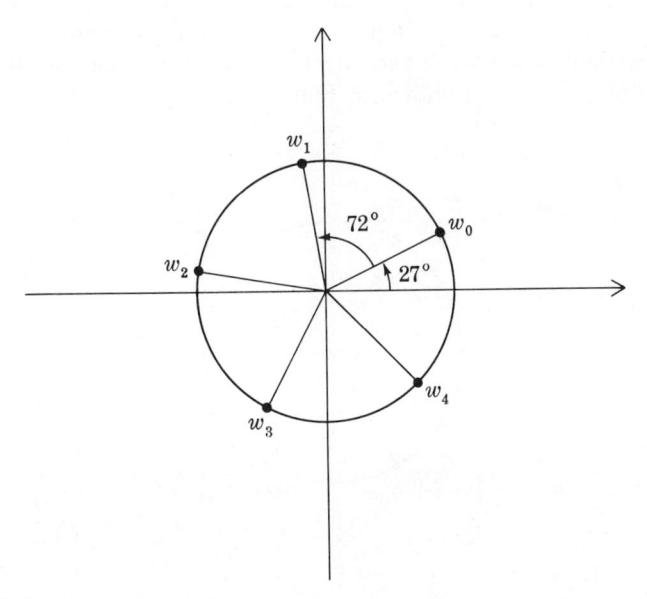

Figure 8.9

In general, the points of $z^{1/n}$ are located on the circle of radius $\sqrt[n]{r}$ centered at 0 and these points divide this circle into n equal arcs, each subtending a central angle of $2\pi/n$. If we locate w_0 on the circle having angle θ/n, then the remaining points $w_1, w_2, \cdots, w_{n-1}$ are located by starting at w_0 and going around the circle, measuring off an angle of $2\pi/n$ at a time. The resulting picture resembles a wheel with n equally spaced spokes.

Example 8.3. Complex numbers z satisfying $z^n = 1$ are called *nth roots of unity*. These n points are just the vertices of the regular polygon of n sides inscribed in the circle of radius 1 centered at the origin. Since 1 has polar form $\cos 0 + i \sin 0$, we see that

$$1^{1/n} = \left\{\cos\frac{2k\pi}{n} + i\sin\frac{2k\pi}{n} : k = 0, 1, \cdots, n-1\right\}.$$

The nth roots of unity are neatly related as follows. If we set

$$\omega = \cos\frac{2\pi}{n} + i\sin\frac{2\pi}{n},$$

then

$$1^{1/n} = \{1,\omega,\omega^2, \cdots ,\omega^{n-1}\}.$$

This is because the nth roots of unity all have absolute value 1 and thus raising ω to the kth power is equivalent to multiplying ang(ω) by k.

In particular, if $n = 6$, then $2\pi/6 = \pi/3$, or 60 degrees, and the six 6th roots of unity are as shown in Figure 8.10.

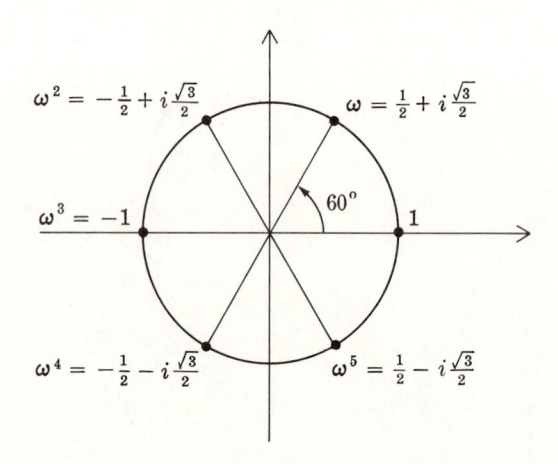

Figure 8.10

EXERCISES

8.14. Find all solutions of the following:
  (a) $z^9 + z = 0$.
  (b) $z^6 + 2z^3 + 2 = 0$.

8.15. Find the following complex numbers and describe their location in the complex plane:
  (a) The cube roots of -16.
  (b) The 8th roots of unity.
  (c) $(i)^{1/4}$.
  (d) $(1 + i\sqrt{3})^{1/2}$.

8.16. Derive the identities:
$$\cos 3\theta = 4\cos^3\theta - 3\cos\theta$$
$$\sin 3\theta = -4\sin^3\theta + 3\sin\theta.$$

[*Hint:* Consider $(\cos\theta + i\sin\theta)^3$.]

8.17. (a) Show that the sum of the nth roots of unity is 0.
[Hint: $z^n - 1 = (z - 1)(z^{n-1} + \cdots + z + 1)$.]
(b) Solve $z^4 + z^3 + z^2 + z + 1 = 0$.

8.18. If n and m are positive integers, we define $(z^{1/n})^m$ to be the set

$$(z^{1/n})^m = \{w^m : w \in z^{1/n}\}.$$

If n and m have no prime factors in common, then $(z^m)^{1/n} = (z^{1/n})^m$. Use this fact to compute $(i)^{2/3}$ in two different ways. However, show that $(i^4)^{1/6} \neq (i^{1/6})^4$.

Simple curves and regions in the plane can be easily described in terms of complex numbers. For example, since the circle C centered at a point a with radius r is the set of all points in the complex plane that are a distance r from a, we see that C can be described analytically as

$$C = \{z \in \mathbf{C} : |z - a| = r\}.$$

Similarly, since an ellipse E can be described as the set of all points P such that the sum of the distance from P to two fixed points a, b (called the foci of E) is a constant r, E has the following analytic description

$$E = \{z \in \mathbf{C} : |z - a| + |z - b| = r\}.$$

Other familiar loci from plane analytic geometry can also be described analytically by using similar ideas.

Example 8.4. The set $\{z : |z - i| = |z - 2|\}$ can be seen to be a straight line in two different ways.

First, if we substitute $z = x + iy$ and square both sides of the given equation, we obtain

$$x^2 + (y - 1)^2 = (x - 2)^2 + y^2$$

which can be simplified to

$$4x - 2y = 3.$$

Thus, the given set is a straight line.

Alternatively, we can observe that if z satisfies $|z - i| = |z - 2|$, then the distance $|z - i|$ from z to i must equal the distance $|z - 2|$ from z to 2. That is, z must lie on the line that is the perpendicular bisector of the line segment joining i to 2. Conversely, each point z on this line is equidistant from i and 2. Hence, we see that the set is the straight line shown in Figure 8.11. Similar reasoning shows that the set $\{z : |z - i| > |z - 2|\}$ is the half-plane shaded in Figure 8.11.

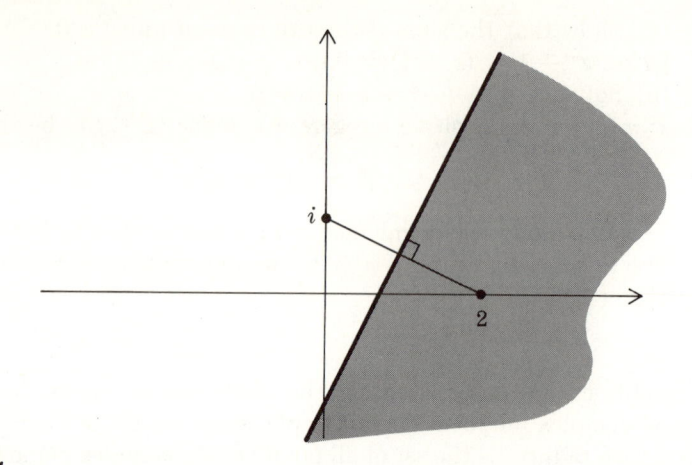

Figure 8.11

EXERCISES

8.19. Describe geometrically the sets in the complex plane determined by the following conditions:
   (a) $|z + i| \leq 3$.
   (b) $1 < |z - 1 - i| < 2$.
   (c) $|z + i| = |z - 3i|$.
   (d) $|z - 2| - |z + 2| = 1$.
   (e) $|z|^2 = 2\,\mathrm{Re}(z)$.

8.20. Show that the equation $\mathrm{Re}(z^2) = c$, where $c > 0$, represents a hyperbola.

8.21. Show that the equation $\mathrm{Re}(1/z) = c$, where $c > 0$, represents a circle.

8.22. Show that the equation $\left|\dfrac{z - a}{z - b}\right| = \lambda$, where $\lambda > 0$ and a, b are fixed complex numbers, represents a circle or a straight line.

8.2. THE RIEMANN SPHERE

Besides the complex plane, there is another interesting and useful geometric representation of the set **C** of complex numbers that is obtained as follows. We place a sphere of diameter 1 on the complex plane so that it rests at the origin. (See Figure 8.12.) We shall suggestively refer to the point on the sphere that is coincident with the origin as the *South Pole* of the sphere, while the diametrically opposite point N will be referred to as the *North Pole*. Given any complex number z, the straight line joining

8.2. THE RIEMANN SPHERE

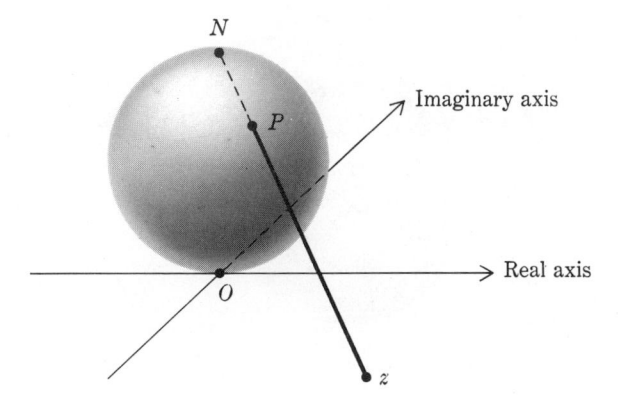

Figure 8.12

the North Pole N to the point of the complex plane representing z will intersect the sphere at exactly one point P other than N. (See Figure 8.12.) On the other hand, given any point of the sphere other than N, the extension of the straight line through the given point and N will intersect the complex plane at exactly one point. Thus, this procedure establishes a one-to-one correspondence between the complex plane and the set of points on the sphere excepting N. (The reader who has studied Chapter 7 on graph theory will recognize this procedure as being essentially the same as the one used to establish isomorphisms between plane and spherical maps.) This one-to-one correspondence is generally referred to as *stereographic projection*. Thus, by means of stereographic projection, it is possible to interpret each complex number z geometrically as a point other than N on the sphere; conversely, each point of the sphere other than N determines a unique complex number. We shall refer to the set of points on the sphere as the *Riemann sphere*. Thus, by means of stereographic projection, the Riemann sphere less the point N provides a second geometric interpretation of the set of complex numbers.

Each circle on the Riemann sphere is the intersection of the sphere and a plane in space. For example, the "equator" of the Riemann sphere is the intersection of the sphere with the plane one-half unit above and parallel to the complex plane. If L is a straight line in the complex plane, then under stereographic projection, L corresponds to the circle K obtained by intersecting the sphere with plane π determined by the line L and the North Pole N. (See Figure 8.13.)

Not every point on the circle K corresponds to a point on L since N is not the image of any point in the plane under stereographic projection. Notice, however, that if we look at points on L that are a great distance out on L, the corresponding points on K are quite close to N. That is, if

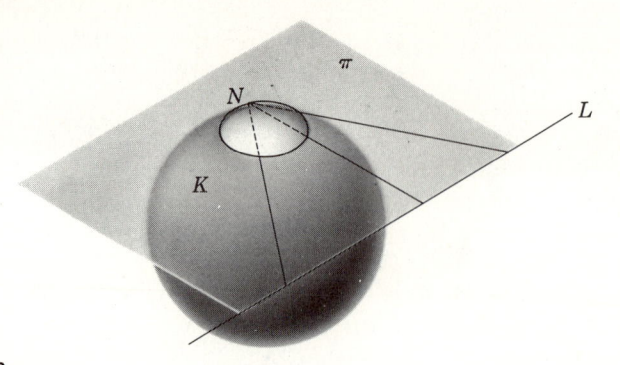

Figure 8.13

we think of points "tending to infinity" along L, then the corresponding points on K come closer and closer to the North Pole N of the sphere.

The preceding observation provides one motivation for adjoining to the complex plane an extra point that will correspond to the North Pole of the Riemann sphere. Of course, there is no room in the plane itself for such a point, but any object not in **C** can serve as this adjunct to **C**. This extra "ideal" point will be denoted by ∞ and referred to as the *point at infinity*. The one-to-one correspondence between the complex plane and the Riemann sphere excepting N can then be extended to include ∞ and N, and we shall say that ∞ and N correspond to one another under stereographic projection.

When complex numbers are viewed as points in the complex plane, the attachment of the point at infinity might appear somewhat artificial since there is no convenient visualization of this point. However, the representation of **C** by means of the Riemann sphere makes this adjunct seem quite natural since the North Pole is a perfectly acceptable geometric object distinguished only by its position on the sphere. Moreover, we shall see later that the use of the North Pole of the Riemann sphere as the point at infinity will be consistent and convenient from the point of view of the arithmetic and geometry of complex numbers.

It is possible to define certain arithmetic relations between ∞ and complex numbers as follows: for $z \in$ **C**, we define

$$z \pm \infty = \infty, \qquad \frac{z}{\infty} = 0$$

and if $z \neq 0$, we define

$$z \cdot \infty = \infty, \qquad \frac{z}{0} = \infty.$$

These conventions are consistent with the arithmetic properties of **C** and

8.2. THE RIEMANN SPHERE

are in accord with our intuitive feelings concerning ∞. However, the expressions ∞/∞, $0 \cdot \infty$, $\infty - \infty$ are left undefined since any attempt to define them leads to inconsistencies with the usual arithmetic properties of complex numbers.

The complex plane together with the point at infinity is called the *extended complex plane* and will be denoted by **C***. The set **C*** with the arithmetic operations of **C** extended in the above manner will be referred to as the *extended complex number system*. Since the extended complex plane is in one-to-one correspondence with the Riemann sphere through stereographic projection, we shall also use the notation **C*** for the Riemann sphere as well.

We shall see that both the Riemann sphere and the extended complex plane will be useful in our study of the geometry of complex numbers. As a matter of fact, it will be useful to consider both interpretations simultaneously in some contexts. Consequently, it is necessary for us to examine stereographic projection in some detail and to derive analytic relationships between the coordinates of a point in the complex plane and the space coordinates of the stereographic projection of that point on the Riemann sphere. We shall now proceed to this task.

First of all, we introduce a coordinate t-axis, perpendicular to the complex plane and passing through the origin. (See Figure 8.14.) This

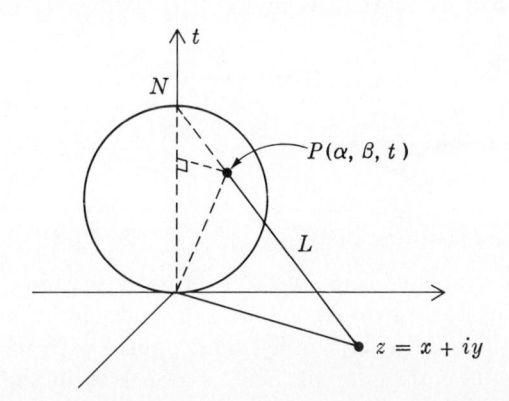

Figure 8.14

axis, together with the real and imaginary axes in the complex plane provide a Cartesian coordinate system in space. Note that the North Pole N has coordinates $(0,0,1)$ since the diameter of the sphere is 1.

Suppose that $z = x + iy$ is a given point in the complex plane and suppose that (α,β,t) are the coordinates of the point P on the Riemann sphere corresponding to z by stereographic projection. Since the sphere

has center $(0,0,\frac{1}{2})$ and radius $\frac{1}{2}$, the coordinates of P must satisfy
$$\alpha^2 + \beta^2 + (t - \tfrac{1}{2})^2 = (\tfrac{1}{2})^2$$
or equivalently,
$$\alpha^2 + \beta^2 + t^2 - t = 0. \tag{4}$$

Also, since P lies on the straight line L through z and N, the numbers x and y are related to α, β, and t by
$$\alpha = x(1 - t), \qquad \beta = y(1 - t). \tag{5}$$

This may be seen by considering similar triangles or by using the equations of the line L. Substituting the expressions for α and β in (5) into Equation (4), we obtain the following quadratic equation in t:
$$t^2(x^2 + y^2 + 1) - t(2x^2 + 2y^2 + 1) + (x^2 + y^2) = 0.$$

If we set $r^2 = x^2 + y^2 = |z|^2$, then the solutions of this quadratic equation are given by
$$t = \frac{2r^2 + 1 \pm 1}{2(1 + r^2)}.$$

The plus sign corresponds to the North Pole N. Thus, the coordinates of P are obtained by taking the minus sign. This gives us
$$t = \frac{r^2}{1 + r^2},$$
and substitution into (5) yields
$$\alpha = \frac{x}{1 + r^2}, \qquad \beta = \frac{y}{1 + r^2}.$$

Thus, we have expressed α, β, and t in terms of x, y, and r. On the other hand, by solving these expressions for x, y, and r in terms of α, β, and t, we see that if the point P on the Riemann sphere with coordinates (α,β,t) is given, then the corresponding point $z = x + iy$ in the complex plane is determined by
$$x = \frac{\alpha}{1 - t}, \qquad y = \frac{\beta}{1 - t},$$
and also
$$r^2 = \frac{t}{1 - t}.$$

Summarizing these results, we have the following proposition.

8.2. THE RIEMANN SPHERE

Proposition 8.1. If (α, β, t) are the coordinates of a point P other than N on the Riemann sphere, and if $z = x + iy$ is the point on the complex plane corresponding to P by stereographic projection, then

$$\alpha = \frac{x}{1 + r^2}, \qquad \beta = \frac{y}{1 + r^2}, \qquad t = \frac{r^2}{1 + r^2},$$

and

$$x = \frac{\alpha}{1 - t}, \qquad y = \frac{\beta}{1 - t}, \qquad r^2 = \frac{t}{1 - t},$$

where $r^2 = x^2 + y^2 = |z|^2$.

We shall now derive some important properties of stereographic projection. It was noted earlier that each circle on the Riemann sphere containing N was the image of a straight line in the complex plane. If we think of a line L in the extended complex plane as passing through the point at infinity, then the North Pole N is even included in the image of L. Also, every straight line in the extended plane is mapped onto a circle passing through N.

It is also fairly evident that circles on the sphere that are parallel to the equator of the sphere (that is, intersections of the sphere with planes parallel to the complex plane) correspond to circles in the complex plane centered at the origin under stereographic projection. In particular, the equator of the sphere is mapped onto the unit circle $\{z : |z| = 1\}$ in the complex plane. The fact that other circles on the sphere also correspond to circles in the plane is not as easily seen and it is best to verify this assertion analytically. This is done in the proof of the following proposition which summarizes the relationship between circles on the sphere and lines and circles in the plane.

Proposition 8.2. (a) The image of a straight line in the extended complex plane under stereographic projection is a circle on the Riemann sphere that passes through N, and conversely, each such circle on the sphere is the image of a straight line in the extended plane.

(b) The image of a circle in the complex plane under stereographic projection is a circle on the Riemann sphere that does not pass through N, and conversely, each such circle on the sphere is the image of a circle in the plane.

PROOF. We shall prove the first assertion in (b). Each circle K in the plane is given by an equation of the form

$$(x^2 + y^2) + Ax + By + C = 0.$$

By Proposition 8.1, the image (α,β,t) of a point $(x,y,0)$ lying on this circle must satisfy

$$\frac{\alpha^2 + \beta^2}{(1-t)^2} + A\frac{\alpha}{1-t} + B\frac{\beta}{1-t} + C = 0,$$

which is equivalent to

$$(1-C)t + A\alpha + B\beta + C = 0, \tag{6}$$

since $\alpha^2 + \beta^2 = t - t^2$. This implies that (α,β,t) lies on a plane in space, and hence, we conclude that the image of K is a circle on the sphere. Furthermore, since $\alpha = 0$, $\beta = 0$, $t = 1$ does not satisfy (6), this circle does not pass through N. The converse assertion is proved by reversing the above argument.

The assertion (a) can also be proved in a similar fashion by using the fact that each line in the plane is given by the equation $Ax + By + C = 0$. We leave the details to the reader.

If we adopt the point of view that a straight line in the extended complex plane is a circle of infinite radius that passes through the point at infinity, then the preceding proposition can be stated more succinctly, namely, *stereographic projection is circle preserving*.

Two intersecting circles in the plane or on the sphere are *orthogonal* if the tangents to the circles at each point of intersection are orthogonal (that is, perpendicular). Similarly, a line is orthogonal to a circle in the plane if the line is orthogonal to the tangent lines of the circle at the points of intersection. Note that it is impossible for two circles to be orthogonal at just a single point; in fact, if two circles intersect with orthogonal tangent lines at one point, then there must be a second point of intersection with orthogonal tangent lines at that point also.

The following proposition states that stereographic projection is "orthogonality preserving."

Proposition 8.3. If K_1 is a line or circle in the complex plane and if K_2 is a line or circle that intersects K_1, then K_1 and K_2 are orthogonal if and only if their images K_1' and K_2' under stereographic projection are orthogonal circles on the Riemann sphere.

PROOF. Since circles can be replaced by their tangent lines, if necessary, it suffices to consider the case when K_1 and K_2 are straight lines. Then the images K_1' and K_2' are circles that pass through the North Pole N of the Riemann sphere. The circles K_1' and K_2' will be orthogonal if and only if their tangent lines at N are orthogonal. We wish to show this is the case if and only if K_1 and K_2 are perpendicular. To do this, suppose K_1 and K_2 are given by the equations

$$K_1: \quad A_1 x + B_1 y = C_1,$$
$$K_2: \quad A_2 x + B_2 y = C_2.$$

By applying Proposition 8.1, we see that the circles K_1' and K_2' are the intersections of the Riemann sphere with the following planes in space:

$$\pi_1: \quad A_1 \alpha + B_1 \beta = C_1(1 - t),$$
$$\pi_2: \quad A_2 \alpha + B_2 \beta = C_2(1 - t).$$

Since the plane $t = 1$ is the tangent plane to the Riemann sphere at the point N, the tangent lines to K_1 and K_2 at N are given by

$$L_1': \quad A_1 \alpha + B_1 \beta = 0, \quad t = 1$$
$$L_2': \quad A_2 \alpha + B_2 \beta = 0, \quad t = 1$$

respectively. Now K_1 and K_2 are orthogonal if and only if $A_1 A_2 + B_1 B_2 = 0$ (that is, if and only if their slopes are negative reciprocals), which occurs if and only if the tangent lines L_1' and L_2' are orthogonal. This completes the proof.

Note that the above proof can be extended slightly to show that stereographic projection is angle preserving. That is, lines and/or circles in the plane intersect at an angle θ if and only if the images on the Riemann sphere also intersect at the angle θ. This is because the coefficients A_1, A_2, B_1, B_2 in the preceding proof determine the angle at which two lines intersect. Thus, any angle, not just 90 degrees, is preserved under stereographic projection.

EXERCISES

8.23. Use the relations in Proposition 8.1 to show that the unit circle in the complex plane corresponds to the equator of the sphere under stereographic projection. Show that the exterior of the unit circle corresponds to the northern hemisphere and the interior of the unit circle corresponds to the southern hemisphere of the Riemann sphere.

8.24. Show that if z_1 and z_2 are complex numbers that correspond to diametrically opposite points on the Riemann sphere, then $z_1 \bar{z}_2 = -1$.

8.25. Suppose a circle on the Riemann sphere is obtained by intersecting the sphere with a plane perpendicular to the complex plane and a distance $\frac{1}{4}$ from the origin. What is the radius of the corresponding circle in the plane?

8.26. Suppose a hemisphere of radius 1 is placed on the complex plane at the origin and points on the plane are projected onto the hemisphere relative to the point (0,0,1) shown in Figure 8.15. Find the relations between z and the coordinates of P similar to those of Proposition 8.1.

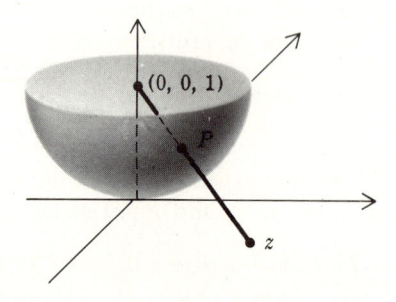

Figure 8.15

8.3. BILINEAR TRANSFORMATIONS

In this section, we shall examine the basic geometric properties of an important class of functions defined on the extended complex number system $\mathbf{C}^*$. Our discussion will begin with the study of several very simple types of functions that will serve as "building blocks" for the class we wish to consider. Because of their simple and basic nature, these special functions will be referred to as the *elementary transformations*. Each of these elementary transformations is defined in terms of an arithmetic operation on the complex number system, but the geometric interpretations of these operations show that they are the basic transformations studied in Euclidean geometry. As a consequence, the study of Euclidean geometry from the transformational point of view could be developed in terms of complex numbers. Some geometric applications of the material in this section are presented in Section 8.4 to indicate the flavor of this approach.

(1) Translations

Given a fixed complex number c, define the function $T_c: \mathbf{C}^* \to \mathbf{C}^*$ as follows: $T_c(z) = z + c$ for all $z \in \mathbf{C}^*$, that is,

$$T_c(z) = \begin{cases} z + c, & \text{if } z \in \mathbf{C} \\ \infty, & \text{if } z = \infty. \end{cases}$$

It is easy to see that T_c is a one-to-one function from $\mathbf{C}^*$ onto $\mathbf{C}^*$; moreover, the restriction of T_c to the set $\mathbf{C}$ of complex numbers is also a one-to-one function from $\mathbf{C}$ onto $\mathbf{C}$. In view of our geometric interpretation for addition in the complex plane (see Section 8.1), we see that the point $T_c(z)$ is obtained from z by translating z through a distance $|c|$ in the direction ang(c). (See Figure 8.16.) Thus, if A is a subset of the complex plane, then $T_c(A)$ is obtained by translating each point of A by the directed line segment joining 0 to c. For these reasons, the function T_c is called *translation by c*.

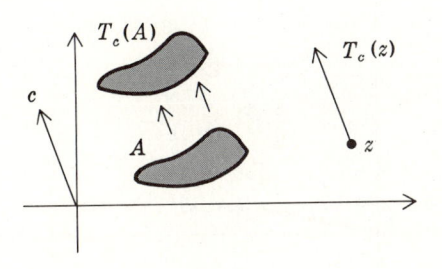

Figure 8.16

(2) Rotation-Dilations

Given a nonzero complex number b, define $H_b: \mathbf{C}^* \to \mathbf{C}^*$ by $H_b(z) = bz$ for all $z \in \mathbf{C}^*$, that is,

$$H_b(z) = \begin{cases} bz, & \text{if } z \in \mathbf{C} \\ \infty, & \text{if } z = \infty. \end{cases}$$

Since $b \neq 0$, H_b is a one-to-one function from $\mathbf{C}^*$ onto $\mathbf{C}^*$ and the restriction of H_b to $\mathbf{C}$ is also a one-to-one function from $\mathbf{C}$ onto $\mathbf{C}$. According to the geometric interpretation of multiplication in the complex plane, the point $H_b(z)$ is obtained from the point z in the complex plane as follows. Rotate the ray L from 0 through z by ang(b) to obtain a ray L' from 0. Then $H_b(z)$ is the point on L' whose distance from 0 is $|b|$ times the distance from 0 to z. (See Figure 8.17.) Thus, the image $H_b(A)$ of a

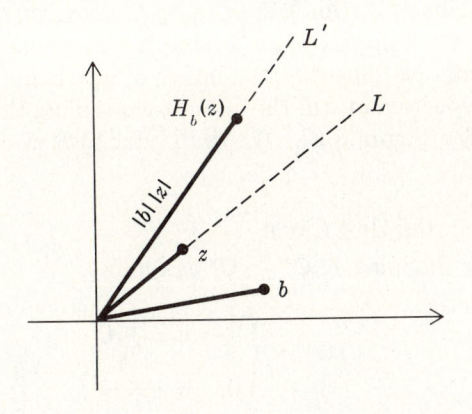

Figure 8.17

set A in the complex plane is obtained from A by rotating A around the origin through the angle of b to obtain a set A', and then dilating (that is, "stretching" or "shrinking") the directed line segments joining 0 to each point of A' by the factor $|b|$ to determine the points of $H_b(A)$. (See Figure 8.18.) For this reason, it is suggestive to refer to the function H_b

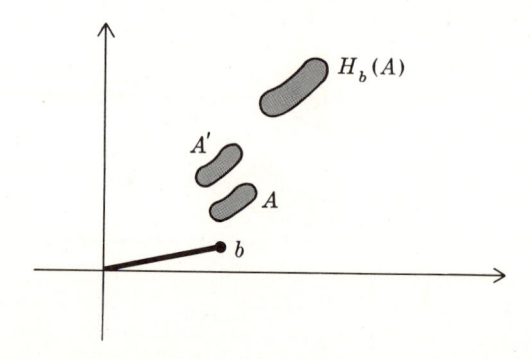

Figure 8.18

as *rotation-dilation by b*. (It should be pointed out that some authors refer to H_b as a *homology* with center 0, ratio $|b|$ and angle ang(b).) Note that if $|b| = 1$, then the point $H_b(z)$ is obtained from the point z in the complex plane by rotating z through ang(b). In this case, H_b is referred to more simply as *rotation by ang(b)*.

(3) **Reflection in the Real Line**

Define the function $R: \mathbf{C}^* \to \mathbf{C}^*$ by

$$R(z) = \begin{cases} \bar{z}, & \text{if } z \in \mathbf{C} \\ \infty, & \text{if } z = \infty. \end{cases}$$

R is a one-to-one function from $\mathbf{C}^*$ onto $\mathbf{C}^*$ and the restriction of R to $\mathbf{C}$ is a one-to-one function from $\mathbf{C}$ onto $\mathbf{C}$. Note that the image $R(z)$ of any point z in the complex plane is simply the reflection of z in the real axis. It is equally obvious that the image of any point z on the Riemann sphere is the reflection of z in the plane π containing the real axis and the North Pole. (See Figure 8.19.) We shall refer to R as *reflection in the real line*.

(4) **Inversion in the Unit Circle**

Define the mapping $I: \mathbf{C}^* \to \mathbf{C}^*$ as follows:

$$I(z) = \begin{cases} \dfrac{1}{z}, & \text{if } z \in \mathbf{C} \\ 0, & \text{if } z = \infty. \end{cases}$$

8.3. BILINEAR TRANSFORMATIONS

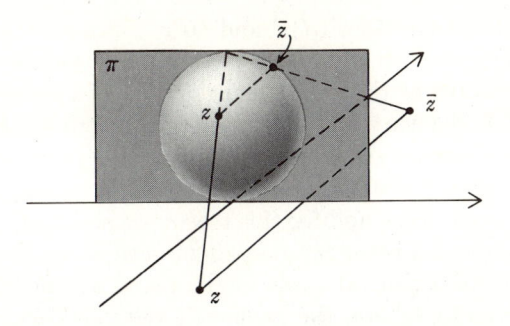

Figure 8.19

Note that this definition implies that $I(0) = \infty$ since we have defined $\frac{1}{0}$ to be ∞. Also, observe that I is a one-to-one mapping from $\mathbf{C}^*$ onto $\mathbf{C}^*$; however, the restriction of I to $\mathbf{C}$ is not a mapping from $\mathbf{C}$ onto $\mathbf{C}$ since $I(0) \notin \mathbf{C}$.

We shall now describe the geometric relationship between the points z and $I(z)$ for any point $z \neq 0$ of the complex plane. First of all, we note that z and $I(z)$ lie on the same ray issuing from 0 in the complex plane since $\text{ang}(z) = \text{ang}(1/\bar{z})$. Furthermore, $|I(z)| = 1/|z|$. Thus, if z has polar form $z = r(\cos \theta + i \sin \theta)$, then $I(z)$ has polar form

$$I(z) = \frac{1}{r}(\cos \theta + i \sin \theta).$$

We see from this that if $|z| = 1$, then $I(z) = z$ so that each point on the unit circle is left fixed by I.

If $|z| \neq 1$, then the image $I(z)$ of z can be determined by a simple geometric construction as follows. Suppose $|z| > 1$ and let L be the ray from the origin through z. (See Figure 8.20.) If T is a line tangent to the unit circle that passes through z, then the line perpendicular to L passing through the point P of tangency for T intersects L at the point Q as shown in Figure 8.20. It is easily shown that Q is the point $I(z) = 1/\bar{z}$ by using

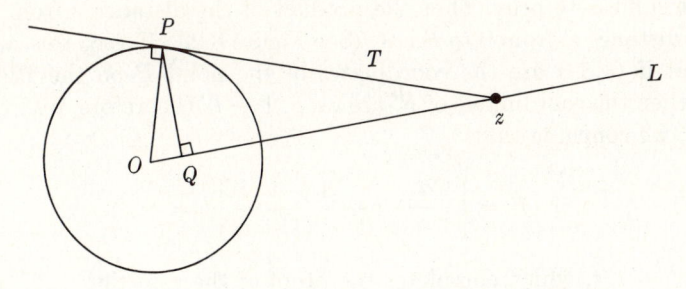

Figure 8.20

the similarity of the triangles OQP and OPz. Since $|OP| = 1$, this similarity implies $|OQ| = 1/|z|$. Since we have already noted that $\text{ang}(z) = \text{ang}(1/\bar{z})$, the validity of the above construction of $I(z)$ is verified when $|z| > 1$. If $|z| < 1$, the construction can be reversed to yield the point $1/\bar{z}$ outside the unit circle by starting with z inside the unit circle.

The geometric construction described above is usually referred to as inversion in the unit circle, and for this reason we shall call the mapping I of the extended complex plane *inversion in the unit circle*. Geometric inversion with respect to a general circle in the plane will be discussed later.

On the Riemann sphere, the geometric relation between the points representing z and $I(z)$ is very simple to describe. In fact, the point representing $I(z)$ on the Riemann sphere is simply the reflection in the plane of the equator of the point representing z. (See Figure 8.21.) To

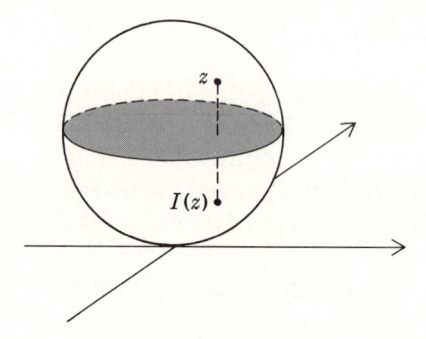

Figure 8.21

verify this assertion, we first note that the plane passing through the origin O, the North Pole N, and the point P on the Riemann sphere that represents z also contains the point representing $I(z)$ on the Riemann sphere since $\text{ang}(z) = \text{ang}(1/\bar{z})$. Therefore, if P' is the point on the Riemann sphere obtained by reflecting P in the plane of the equator, and if z and z' are the stereographic projections of P and P' on the complex plane, it suffices to prove that the product of the distance r from 0 to z and the distance r' from 0 to z' is 1. (See Figure 8.22.) To see this, we first note that if (α, β, t) are the coordinates of the point P on the Riemann sphere, then the coordinates of P' are $(\alpha, \beta, 1 - t)$. Therefore, by Proposition 8.1, we conclude that

$$r \cdot r' = \left[\frac{t}{1-t} \cdot \frac{1-t}{1-(1-t)} \right]^{1/2} = 1.$$

Hence, $z' = 1/\bar{z}$, which completes the proof of the assertion.

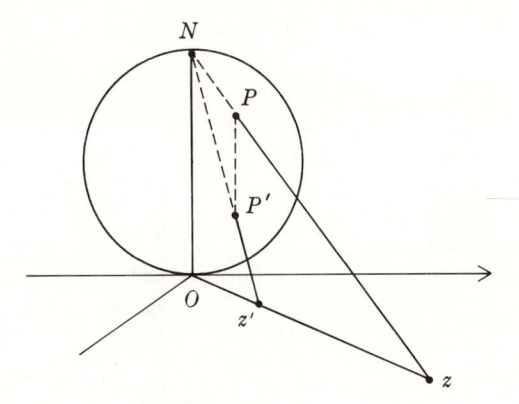

Figure 8.22

We have now completed the list of those functions that we shall refer to as the *elementary transformations* in the sequel. These elementary transformations may be composed to produce new functions mapping $\mathbf{C}^*$ onto $\mathbf{C}^*$. For example, if we follow inversion I in the unit circle by reflection R in the real line, we obtain the function $R \circ I$ defined by

$$(R \circ I)(z) = \begin{cases} \dfrac{1}{z}, & \text{if } z \in \mathbf{C} \\ 0, & \text{if } z = \infty. \end{cases}$$

It is easy to verify that the order of composition is not important in this case, that is, $R \circ I = I \circ R$. As another example of a composition of elementary transformations, consider the function defined by following the rotation-dilation H_b by the translation T_c. In this case, note that

$$(T_c \circ H_b)(z) = \begin{cases} bz + c, & \text{if } z \in \mathbf{C} \\ \infty, & \text{if } z = \infty. \end{cases}$$

The order of composition is important here; in fact, since $bz + c = b(z + c/b)$, we see that $T_c \circ H_b = H_b \circ T_{c/b}$.

Example 8.5. If f is defined on $\mathbf{C}^*$ by

$$f(z) = iz + 1 + i, \quad z \in \mathbf{C}^*,$$

then f can be interpreted geometrically as a rotation through $\pi/2$ followed by a translation through a distance of $|1 + i| = \sqrt{2}$ in the direction $\text{ang}(1 + i) = \pi/4$. That is, $f = T_{1+i} \circ H_i$. The effect of f on a region in the plane is illustrated in Figure 8.23.

350   THE GEOMETRY OF COMPLEX NUMBERS

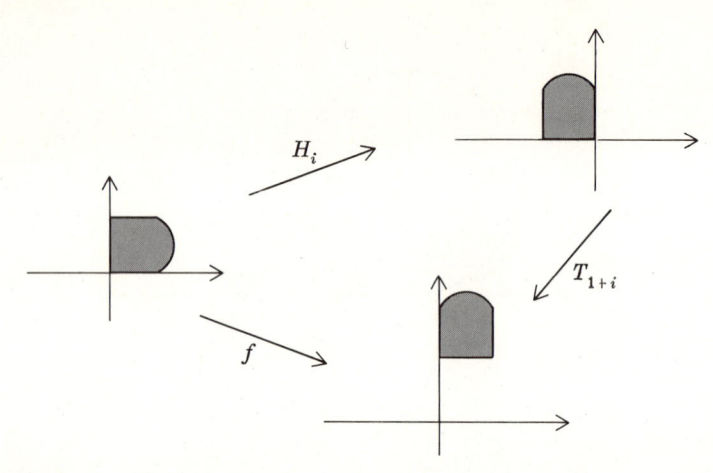

Figure 8.23

The following result identifies all of the functions from $\mathbf{C}^*$ onto $\mathbf{C}^*$ that can be written as the composite of a finite number of elementary transformations.

Proposition 8.4. A function $f: \mathbf{C}^* \to \mathbf{C}^*$ is the composite of a finite number of elementary transformations if and only if there exist complex numbers a, b, c, d with $ad - bc \neq 0$ such that either

$$f(z) = \begin{cases} \dfrac{az+b}{cz+d}, & \text{if } z \in \mathbf{C} \\ \dfrac{a}{c}, & \text{if } z = \infty \end{cases} \quad (A)$$

or

$$f(z) = \begin{cases} \dfrac{a\bar{z}+b}{c\bar{z}+d}, & \text{if } z \in \mathbf{C} \\ \dfrac{a}{c}, & \text{if } z = \infty. \end{cases} \quad (B)$$

PROOF. First of all, it is clear that such complex numbers a, b, c, d can be found for each of the elementary transformations. (For example, inversion I in the unit circle is of type (B) since we can take $a = 0, b = 1, c = 1, d = 0$.)

If $f: \mathbf{C}^* \to \mathbf{C}^*$ is of the form (A) and if $c = 0$, then $d \neq 0$ and $a \neq 0$ since $ad - bc \neq 0$, and we can write

$$f(z) = \frac{a}{d}z + \frac{b}{d} = (T_{b/d} \circ H_{a/d})(z), \quad z \in \mathbf{C}^*.$$

If $c \neq 0$, then for $z \in \mathbf{C}$,

$$\frac{az+b}{cz+d} = \frac{bc-ad}{c}\left[\frac{1}{cz+d}\right] + \frac{a}{c},$$

which implies that f can be expressed as

$$f = T_{a/c} \circ H_s \circ R \circ I \circ T_d \circ H_c,$$

where $s = (bc - ad)/c$. Thus, any function of the form (A) is a composite of a finite number of elementary transformations.

If f is of the form (B), then since

$$\frac{a\bar{z}+b}{c\bar{z}+d} = \left[\overline{\frac{\bar{a}z+\bar{b}}{\bar{c}z+\bar{d}}}\right]$$

it follows that f can be written as a composite of a function of the form (A) and reflection R in the real line. Hence, any function of the form (B) is the composite of a finite number of elementary transformations.

In view of the above conclusions, the proof of the proposition will be complete if we can show that the composition of any two functions of the form (A) or (B) is again a function of one of these two forms. Moreover, since any function f of the form (B) can be written as $f = R \circ g$ where g is of the form (A), it is even sufficient to prove that the composite of two functions of the form (A) is again of the form (A).

With this in mind, suppose f_1 and f_2 are defined as follows:

$$f_1(z) = \begin{cases} \dfrac{a_1 z + b_1}{c_1 z + d_1}, & \text{if } z \in \mathbf{C} \\ \dfrac{a_1}{c_1}, & \text{if } z = \infty \end{cases}$$

$$f_2(z) = \begin{cases} \dfrac{a_2 z + b_2}{c_2 z + d_2}, & \text{if } z \in \mathbf{C} \\ \dfrac{a_2}{c_2}, & \text{if } z = \infty \end{cases}$$

where $a_1 d_1 - b_1 c_1 \neq 0$ and $a_2 d_2 - b_2 c_2 \neq 0$. Then for $z \in \mathbf{C}$, we have

$$(f_2 \circ f_1)(z) = \frac{a_2\left[\dfrac{a_1 z + b_1}{c_1 z + d_1}\right] + b_2}{c_2\left[\dfrac{a_1 z + b_1}{c_1 z + d_1}\right] + d_2} = \frac{Az + B}{Cz + D},$$

where $A = a_2 a_1 + b_2 c_1$, $B = a_2 b_1 + b_2 d_1$, $C = c_2 a_1 + d_2 c_1$, $D = c_2 b_1 + d_2 d_1$. Also, $(f_2 \circ f_1)(\infty) = A/C$ and $AD - BC = (a_2 d_2 - b_2 c_2)(a_1 d_1 - b_1 c_1) \neq 0$. Therefore, $f_2 \circ f_1$ is of the form (A) and the proof is complete.

The constants a, b, c, d mentioned in Proposition 8.4 are only determined up to a nonzero common multiple; that is, the constants a, b, c, d may be replaced with $a' = \lambda a$, $b' = \lambda b$, $c' = \lambda c$, $d' = \lambda d$ for any nonzero complex number λ without altering the function f in (A) or (B). The expression $ad - bc$ occurring in the statement of Proposition 8.4 is called the *determinant of f*, and it is also only determined up to a nonzero multiple by the function f. The significance of the determinant of f lies in the following fact. A function f that can be written in the form

$$f(z) = \frac{az+b}{cz+d}, \quad z \in \mathbf{C}$$

is nonconstant on $\mathbf{C}$ if and only if $ad - bc \neq 0$. Thus, the condition that the determinant be nonzero is imposed to avoid the degenerate case of a constant function.

A function $f\colon \mathbf{C}^* \to \mathbf{C}^*$ that is of the form (A) in Proposition 8.4 is called a *bilinear transformation*. (These transformations are also referred to as Möbius transformations or linear fractional transformations in some books.) This terminology permits the following reformulation of Proposition 8.4. A function $f\colon \mathbf{C}^* \to \mathbf{C}^*$ is the composite of a finite number of elementary transformations if and only if f is either a bilinear transformation or the composite of a bilinear transformation with reflection in the real line.

In the sequel, the explicit designation of the value of a bilinear transformation at the point at infinity will usually be suppressed. That is, if f is defined by

$$f(z) = \frac{az+b}{cz+d}, \quad ad - bc \neq 0$$

for $z \in \mathbf{C}$, then it will be implicitly understood that $f(\infty) = a/c$.

A number of conclusions concerning bilinear transformations that can be gleaned from Proposition 8.4 or its proof are listed in the following result.

Proposition 8.5. A bilinear transformation f is a one-to-one function from $\mathbf{C}^*$ onto $\mathbf{C}^*$ and the inverse function f^{-1} of f is also a bilinear transformation. The composite of two bilinear transformations is a bilinear transformation.

We leave the details of the verification of this result to the reader as an exercise.

EXERCISES

8.27. Describe the image of the unit disc $\{z: |z| \leq 1\}$ under the transformations:
(a) $f(z) = i(z + 2)$.
(b) $g(z) = 3iz - i$.
(c) $h(z) = \dfrac{2}{z} - 2$.

8.28. Write the following bilinear transformations as composites of elementary transformations.
(a) $f(z) = (1 + 2i)z - 7$.
(b) $g(z) = \dfrac{iz - 1 + i}{z + 1}$.

What is the image of ∞ in each of these cases?

8.29. Find the bilinear transformations that correspond to the following geometric operations:
(a) Rotation by $\pi/4$ around the point $3 - i$.
(b) Reflection in the line $\text{Re}(z) = \text{Im}(z)$.

8.30. Show that the inverse f^{-1} of a bilinear transformation

$$f(z) = \frac{az + b}{cz + d}$$

is given by

$$f^{-1}(z) = \frac{-dz + b}{cz - a}.$$

8.31. Prove Proposition 8.5.

8.32. A bilinear transformation $f: \mathbf{C}^* \to \mathbf{C}^*$ is said to be *real* if there exist real numbers a, b, c, d such that

$$f(z) = \frac{az + b}{cz + d}$$

for $z \in \mathbf{C}$. Prove that:
(a) The composite of two real bilinear transformations is a real bilinear transformation.
(b) The inverse of a real bilinear transformation is a real bilinear transformation.
(c) The image of the real axis (including ∞) under a real bilinear transformation is the real axis (including ∞).

8.33. Suppose T denotes the set of all one-to-one functions from $\mathbf{C}^*$ onto $\mathbf{C}^*$. A nonempty subset G of T is called a *complex transformation group* if
(a) $f \in G$ implies $f^{-1} \in G$.
(b) $f \in G, g \in G$ imply $f \circ g \in G$.

Prove that if G is a complex transformation group, then G contains the identity map on $\mathbf{C}^*$ and, for any f, g in G, there is an h in G such that $f \circ h = g$.

8.34. Which of the following sets of mappings on $\mathbf{C}^*$ are complex transformation groups?
(a) The set of all bilinear transformations.
(b) The set of all functions $g \colon \mathbf{C}^* \to \mathbf{C}^*$ such that $g = f \circ R$ where f is a bilinear transformation and R is reflection in the real line.
(c) The union of the sets in (a) and (b).
(d) The set of all real bilinear transformations.
(e) The set of all functions $h \colon \mathbf{C}^* \to \mathbf{C}^*$ such that $h = f \circ R$ where f is a real bilinear transformation and R is reflection in the real line.
(f) The union of the sets in (d) and (e).
(g) The set of all translations T_c for $c \in \mathbf{C}$.
(h) The set of all rotation-dilations H_b for $b \neq 0$, $b \in \mathbf{C}$.
(i) The set of all rotations H_b for $b \neq 0$, $|b| = 1$.

We shall now turn to an analysis of the basic geometric properties of bilinear transformations. The following result shows that bilinear transformations are circle preserving mappings of the extended complex plane. This property led A. Möbius, who was the first to systematically analyze bilinear transformations in 1853, to refer to these transformations as *Kreisverwandtschaften* (that is, having an affinity for circles).

Proposition 8.6. If K is a circle or a straight line (including ∞) in the extended complex plane $\mathbf{C}^*$ and if f is a bilinear transformation, then $f(K)$ is a circle or a straight line in $\mathbf{C}^*$.

[*Note:* We are *not* claiming that the image of a circle is a circle and that the image of a line is a line, but rather that the image of a circle is a circle *or* a line and that the image of a line is a circle *or* a line.]

PROOF. By virtue of Proposition 8.4, f can be written as the composite of elementary transformations. Since it is clear that translations, rotation-dilations, and reflections in the real line map circles into circles and straight lines into straight lines, it suffices to prove the assertion of the proposition for the special case that f is inversion I in the unit circle.

The circle or straight line K in the extended complex plane corresponds to a circle K' on the Riemann sphere under stereographic projection since stereographic projection is circle preserving (see Proposition 8.2). Since inversion I in the unit circle corresponds to reflection in the plane of the equator on the Riemann sphere, we know that $I(K')$ is a circle on the Riemann sphere. Therefore, by again applying the circle preserving property of stereographic projection, we conclude that $I(K)$ is a circle or a straight line in the extended complex plane. This completes the proof of the proposition.

The technique used in the proof of the preceding proposition together with the orthogonality preserving property of stereographic projection can be used to establish the corresponding property for bilinear transformations of the extended complex plane. We shall omit the details of the verification.

Proposition 8.7. If K_1 and K_2 are circles or straight lines in the extended complex plane and if f is a bilinear transformation, then K_1 and K_2 are orthogonal if and only if $f(K_1)$ and $f(K_2)$ are orthogonal.

The preceding pair of propositions together with the fact that each line or circle in the extended complex plane is completely determined by three points can be used to find the images of lines and circles under bilinear transformations. In order to find the image of a particular line or circle under a bilinear transformation, it suffices to find the images of any three points on that line or circle. This technique is illustrated in the following example.

Example 8.6. Suppose the bilinear transformation f is defined by

$$f(z) = \frac{z-i}{z+i}.$$

Let us determine the images of the real and imaginary axes and the unit circle under f.

Let L_1 and L_2 denote the imaginary and real axes, respectively. The images of the points $0, i, \infty$ lying on L_1 in the extended complex plane are easily computed to be $f(0) = -1$, $f(i) = 0$, $f(\infty) = 1$. Thus, since $f(L_1)$ is either a line or a circle, and since $f(L_1)$ must contain the points $-1, 0, 1$, we see that $f(L_1)$ is the real axis. The image $f(L_2)$ of the real axis can be similarly found by noting that the points $f(0) = -1, f(1) = -i$, $f(\infty) = 1$ determine the unit circle; consequently, $f(L_2)$ is the unit circle.

The image of the unit circle under f must be a line since $f(-i) = \infty$, and it must be orthogonal to both the real axis $f(L_1)$ and the unit circle $f(L_2)$ since the unit circle is orthogonal to the real and imaginary axes. Thus, f must map the unit circle onto the imaginary axis.

In the geometric analysis of bilinear transformations, an important role is played by those points that remain fixed under a given transformation. More explicitly, if $f: \mathbf{C}^* \to \mathbf{C}^*$ is a given function and if $z_0 \in \mathbf{C}^*$ satisfies $f(z_0) = z_0$, then z_0 is called a *fixed point* for f. For example, every point in $\mathbf{C}^*$ is a fixed point for the identity transformation $e(z) = z$ on $\mathbf{C}^*$. On the other hand, a point $z_0 \in \mathbf{C}^*$ is a fixed point for reflection R in the real line if and only if either z_0 is a real number or $z_0 = \infty$. The following result describes the set of fixed points for a bilinear transformation.

Proposition 8.8. Suppose

$$f(z) = \frac{az + b}{cz + d}, \quad z \in \mathbf{C}$$

is a bilinear transformation distinct from the identity transformation.

(a) If $c \neq 0$, then the point at infinity is not a fixed point for f. There are at least one and at most two fixed points for f in $\mathbf{C}$.

(b) If $c = 0$, then the point at infinity is the only fixed point for f if and only if $a = d$. If $a \neq d$, then $b/(d - a)$ is also a fixed point for f.

PROOF. If $c \neq 0$, then $f(\infty) = a/c \neq \infty$, so that ∞ is not a fixed point for f. Also, in this case, a point $z_0 \in C$ is a fixed point for f if and only if z_0 satisfies the equation

$$z = \frac{az + b}{cz + d}$$

or equivalently

$$cz^2 + (d - a)z - b = 0.$$

This quadratic equation has solutions

$$z = \frac{-(d - a) \pm [(d - a)^2 + 4bc]^{1/2}}{2c}.$$

Thus, f will have one or two fixed points according as the expression $(d - a)^2 + 4bc$ is either zero or nonzero.

Now suppose $c = 0$. Then $d \neq 0$ and $a \neq 0$ since $ad - bc \neq 0$; hence, f is given by $f(z) = (a/d)z + (b/d)$ for $z \in \mathbf{C}$ and $f(\infty) = \infty$. Thus, ∞ is a fixed point for f. Any other fixed point for f must satisfy

$$(d - a)z - b = 0.$$

This equation possesses solutions if and only if either $a \neq d$ or $d - a = b = 0$. But the latter case cannot occur since this would force f to be the identity transformation, which is ruled out by hypothesis. Thus, ∞ is the only fixed point for f if and only if $a = d$ in which case $b \neq 0$. Also, if $a \neq d$, then a second fixed point for f is $z = b/(d - a)$.

8.3. BILINEAR TRANSFORMATIONS

Corollary 1. A bilinear transformation f that is not the identity transformation is a translation if and only if the point at infinity is the only fixed point for f.

Corollary 2. If f and g are bilinear transformations and if z_1, z_2, z_3 are three distinct points of $\mathbf{C}^*$ for which $f(z_k) = g(z_k)$, $k = 1, 2, 3$, then $f(z) = g(z)$ for all $z \in \mathbf{C}^*$.

PROOF. Since inverses and composites of bilinear transformations are also bilinear transformations, we see that $h = g^{-1} \circ f$ is a bilinear transformation. Since $h(z_k) = g^{-1}(f(z_k)) = z_k$ for $k = 1, 2, 3$, the points z_1, z_2, z_3 are all fixed points for h. However, Proposition 8.8 shows that the only bilinear transformation capable of having more than two fixed points is the identity transformation. Hence, $h(z) = z$ for all $z \in \mathbf{C}^*$, which implies that $f(z) = g(z)$ for all $z \in \mathbf{C}^*$.

The second corollary states that if a bilinear transformation is determined at three points, then it is determined everywhere. This may appear quite remarkable at first sight, but the fact is that a bilinear transformation is defined by four constants a, b, c, d and these constants need only be determined up to proportionality (see the remarks following Proposition 8.4). The information provided by knowing the images of three points is adequate for this determination.

Proposition 8.9. Given three distinct points z_1, z_2, z_3 and three distinct image points w_1, w_2, w_3 of $\mathbf{C}^*$, there is one and only one bilinear transformation f such that $w_k = f(z_k)$ for $k = 1, 2, 3$.

PROOF. We shall first assume that none of the six given points is the point at infinity. In this case, we define the bilinear transformations f_1 and f_2 as follows:

$$f_1(z) = \frac{z - z_1}{z - z_3} \cdot \frac{z_2 - z_3}{z_2 - z_1},$$

$$f_2(z) = \frac{z - w_1}{z - w_3} \cdot \frac{w_2 - w_3}{w_2 - w_1}.$$

Then, $f_1(z_1) = f_2(w_1) = 0$, $f_1(z_2) = f_2(w_2) = 1$, and $f_1(z_3) = f_2(w_3) = \infty$. The bilinear transformation $f = f_2^{-1} \circ f_1$ thus has the desired property since

$$f(z_k) = f_2^{-1}(f_1(z_k)) = w_k$$

for $k = 1, 2, 3$. We have already noted in Corollary 2 above that there can be at most one bilinear transformation with this property, and thus f

is the one and only bilinear transformation satisfying $f(z_k) = w_k$ for $k = 1, 2, 3$.

If one of the z_k's or one of the w_k's happens to be the point at infinity, the argument is the same except that in the definition of f_1 and f_2 those terms involving ∞ are omitted. For example, if $z_1 = \infty$, then f_1 is defined to be

$$f_1(z) = \frac{z_2 - z_3}{z - z_3}.$$

We again have $f_1(z_1) = 0$ and the proof proceeds as before.

In order to construct the bilinear transformation f mentioned in the above proposition, it is not actually necessary to construct f_1 and f_2, then compute f_2^{-1} and then compute $f = f_2^{-1} \circ f_1$. For if $z \in \mathbf{C}$, $w = f(z) \in \mathbf{C}$, then $w = f_2^{-1} \circ f_1(z)$, so that $f_2(w) = f_1(z)$. That is, for all z and w in $\mathbf{C}$, we have

$$\frac{w - w_1}{w - w_3} \cdot \frac{w_2 - w_3}{w_2 - w_1} = \frac{z - z_1}{z - z_3} \cdot \frac{z_2 - z_3}{z_2 - z_1}. \tag{7}$$

Thus, in order to compute f, it is only necessary to substitute the given values of z_i, w_i, $i = 1, 2, 3$ in (7) and then solve for w in terms of z. The following example illustrates this technique.

Example 8.7. Suppose $z_1 = 0$, $z_2 = 1$, $z_3 = i$, and $w_1 = 1 + i$, $w_2 = 2$, $w_3 = \infty$. Then (7) takes the form:

$$\frac{w - (1 + i)}{2 - (1 + i)} = \frac{z - 0}{z - i} \cdot \frac{1 - i}{1 - 0}.$$

This can be solved for w to yield the bilinear transformation f defined by

$$w = f(z) = \frac{(1 - i)z + (1 - i)}{z - i}.$$

It is an easy matter to verify that $f(0) = 1 + i$, $f(1) = 2$, $f(i) = \infty$.

Since each line and circle is determined by three points, the fact that a bilinear transformation can always be found that maps three prescribed points into any other three prescribed points relates nicely to the fact that bilinear transformations are circle preserving. This enables us to map a given line or circle into any other line or circle by means of a bilinear transformation.

Example 8.8. Suppose we wished to find a transformation that mapped the unit circle onto the imaginary axis. We choose the points $1, i, -1$, to determine the unit circle and the points $0, i, \infty$ to determine the imaginary axis. The bilinear transformation f satisfying $f(1) = 0$, $f(i) = i$, and $f(-1) = \infty$ will then perform the desired task and f is readily found to be given by

$$f(z) = \frac{z-1}{z+1}.$$

This is not the only bilinear transformation that maps the unit circle onto the imaginary axis, for any other choice of three points on the circle or the line would determine another possibly different bilinear transformation that achieves the same end.

The preceding series of results leads quite naturally to a certain symmetry preserving property of bilinear transformations that we shall now describe. If L is a given straight line and if z_1 and z_2 are given points in the complex plane, then it is reasonable to say that z_1 and z_2 are *symmetric with respect to* L if L is the perpendicular bisector of the line segment joining z_1 and z_2. Note that if z_1 and z_2 are symmetric with respect to L, then every circle through z_1 and z_2 is orthogonal to L. (See Figure 8.24.)

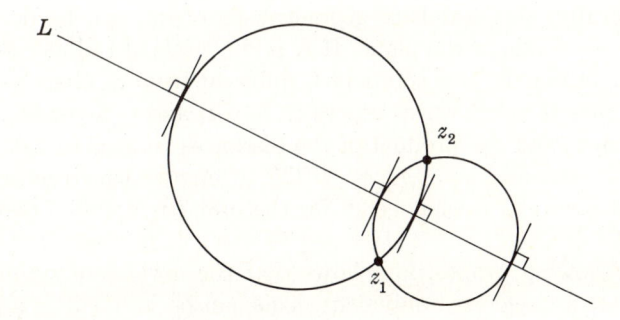

Figure 8.24

With these observations in mind, we formulate the following definition. If K is a given line or circle and if z_1 and z_2 are two points in $\mathbf{C}^*$ not on K, we say that z_1 and z_2 are *symmetric with respect to* K if every circle or line passing through z_1 and z_2 is orthogonal to K. Note that if K is a circle with center z_0, then z_0 and the point at infinity are symmetric with respect to K.

Since bilinear transformations are both circle preserving and orthogonality preserving, it is not surprising that they are also symmetry preserving.

Proposition 8.10. If z_1 and z_2 are symmetric with respect to a line or circle K and if f is a bilinear transformation, then $f(z_1)$ and $f(z_2)$ are symmetric with respect to $f(K)$.

PROOF. According to Proposition 8.6, $f(K)$ is a circle or a straight line. Moreover, if K' is any circle or straight line through $f(z_1)$ and $f(z_2)$, then $K' = f(f^{-1}(K'))$ and $f^{-1}(K')$ is a circle or a straight line through z_1 and z_2. Hence, $f^{-1}(K')$ is orthogonal to K since z_1 and z_2 are symmetric with respect to K. Consequently, by Proposition 8.7, $K' = f(f^{-1}(K'))$ is orthogonal to $f(K)$. It follows that $f(z_1)$ and $f(z_2)$ are symmetric with respect to $f(K)$.

At first glance, it is not clear that for a given circle K and a given point z_1 not on K, there is a point z_2 such that z_1 and z_2 are symmetric with respect to K. However, if we map K onto a straight line L by means of a bilinear transformation f, then it is clear that there exists a point w such that $f(z_1)$ and w are symmetric with respect to L. It then follows from Proposition 8.10 that z_1 and $z_2 = f^{-1}(w)$ are symmetric with respect to $K = f^{-1}(L)$. We see in addition that z_2 is uniquely determined since w is unique and f is a one-to-one function.

We previously discussed inversion with respect to the unit circle. More generally, the notion of geometric inversion can be defined with respect to any circle in the plane. If K is the circle of radius r centered at z_0 and if z_1 and z_2 are two points of **C** different from z_0, then z_1 and z_2 are said to be *inverse points* with respect to K if z_1 and z_2 lie on the same ray issuing from z_0 and the product of the distances from z_0 to z_1 and z_0 to z_2 is r^2, that is, $|z_0 - z_1| |z_0 - z_2| = r^2$. The geometric construction relating z_1 and z_2 is identical to that given for the unit circle and is illustrated in Figure 8.25.

The following proposition shows that the notions of symmetry and inversion for a circle are equivalent. This can be verified by a geometric argument, but the following proof using bilinear transformations is particularly elegant.

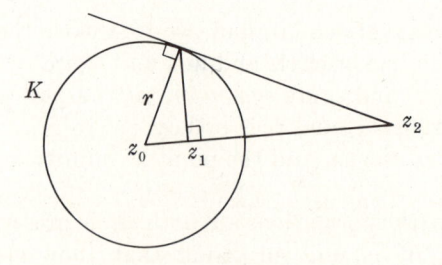

Figure 8.25

Proposition 8.11. Suppose K is a circle in the complex plane with center z_0 and radius r. Then two points z_1 and z_2 in the complex plane are symmetric with respect to K if and only if z_1 and z_2 are inverse points with respect to K.

PROOF. Since translations and rotations preserve distances and orthogonality, we can assume without loss of generality that K is centered at the origin and z_1, z_2 lie on the positive real axis. (See Figure 8.26.) The bilinear

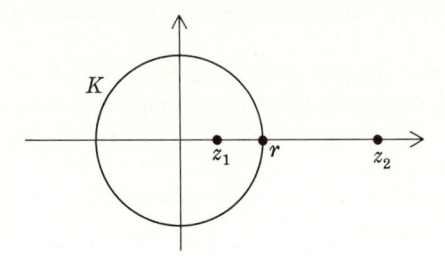

Figure 8.26

transformation f defined by

$$f(z) = \frac{z - r}{z + r}$$

maps K into the imaginary axis since $f(r) = 0, f(-r) = \infty$, and $f(ir) = i$.

Now suppose z_1 and z_2 are symmetric with respect to K. Since f preserves symmetry and since $f(z_1), f(z_2)$ are real, it follows that $f(z_1)$ and $f(z_2)$ are symmetric with respect to $f(K)$, the imaginary axis. Thus, $f(z_2) = -f(z_1)$ so that

$$\frac{z_2 - r}{z_2 + r} = -\frac{z_1 - r}{z_1 + r}. \tag{8}$$

Simplifying, we get $z_1 z_2 = r^2$. Hence, z_1 and z_2 are inverse points with respect to K.

Conversely, since $z_1 z_2 = r^2$ implies (8), we see that $f(z_2) = -f(z_1)$. Thus, $f(z_2)$ and $f(z_1)$ are symmetric with respect to $f(K)$. The symmetry preserving property of the bilinear transformation f^{-1} then shows that z_1 and z_2 must be symmetric with respect to K.

EXERCISES

8.35. Find the fixed points of the following bilinear transformations:

(a) $f(z) = \dfrac{iz}{2z + i}$.

(b) $f(z) = \dfrac{z + i}{z - i}$.

8.36. Find the bilinear transformation f satisfying $f(\infty) = 1$, $f(0) = -1$, and $f(i) = -i$.

8.37. Determine the images under the transformation $f(z) = 1/z$ of the lines and circles given by the following equations:
(a) $\text{Re}(z) = 1$.
(b) $\text{Re}(z) = 3$.
(c) $\text{Im}(z) = \frac{1}{2}$.
(d) $\text{Re}(z) = \text{Im}(z)$.
(e) $|z - 2| = 1$.
(f) $|z - i| = 1$.

8.38. Find the images of the unit circle and the real axis under the following bilinear transformations:

(a) $f(z) = \dfrac{z - 1}{z + 1}$.

(b) $g(z) = \dfrac{z}{z + 1 + i}$.

8.39. Find a bilinear transformation that maps the circle given by $|z - 1| = 2$ onto the line $\text{Im}(z) = \text{Re}(z)$.

8.40. If f is the bilinear transformation defined by

$$f(z) = \frac{iz + 1}{z - 1}$$

find the images of the following sets under f:
(a) The unit circle.
(b) The real axis.
(c) The imaginary axis.
(d) The line $\text{Re}(z) = \text{Im}(z)$.

8.41. Does there exist a bilinear transformation f such that $f(\infty) = 1$, $f(1) = i$, $f(-1) = 0$, and $f(2) = -1$? Justify your assertion.

8.42. A bilinear transformation f has the property that f maps the unit circle onto the real axis, has 1 as a fixed point, and $f(0) = i$. Find f.

8.43. Show that if z_1 and z_2 are symmetric with respect to a circle K, then there exists a constant k such that

$$K = \left\{z: \left|\frac{z - z_1}{z - z_2}\right| = k\right\}.$$

[*Hint:* Consider a bilinear transformation that maps z_1 into 0 and z_2 into ∞.]

8.44. Suppose K is a circle with center z_0 and radius r, and suppose that $f: \mathbf{C}^* \to \mathbf{C}^*$ is the transformation that leaves each point of K fixed and maps each point z not on K into the point z' such that z and z' are symmetric with respect to K. Show that

$$f = T_{z_0} \circ H_r \circ I \circ H_{1/r} \circ T_{-z_0}.$$

8.4. APPLICATIONS TO EUCLIDEAN AND NON-EUCLIDEAN GEOMETRY

In this section, we shall consider some rather diverse geometrical applications of the theory developed in the preceding sections.

Euclidean Isometries

Congruence of geometric figures is one of the most basic concepts in Euclidean plane geometry. For this reason, the class of "congruence preserving" transformations of the plane, that is, the class of transformations that map any plane figure onto a congruent plane figure, are of particular interest within the class of all mappings of the plane onto itself.

If we regard the complex plane $\mathbf{C}$ to be a model of the plane in question, then a function f from $\mathbf{C}$ onto $\mathbf{C}$ will be congruence preserving in the above sense if and only if it is distance preserving, that is, if and only if the usual Euclidean distance between any two points z_1 and z_2 in $\mathbf{C}$ is equal to the distance between $f(z_1)$ and $f(z_2)$. Such functions f are usually referred to as Euclidean isometries of the plane. More precisely, a function $f: \mathbf{C} \to \mathbf{C}$ is a *Euclidean isometry* if

$$|z_1 - z_2| = |f(z_1) - f(z_2)|$$

for all points z_1 and z_2 in $\mathbf{C}$.

It is easy to see that translations, rotations, and reflections in the real line are Euclidean isometries when such transformations are restricted to $\mathbf{C}$ from $\mathbf{C}^*$. Also note that any Euclidean isometry f of the plane is a one-to-one function. In fact, if z_1, z_2 are points in $\mathbf{C}$ and if $z_1 \neq z_2$, then $0 < |z_1 - z_2| = |f(z_1) - f(z_2)|$, so that $f(z_1) \neq f(z_2)$.

364 THE GEOMETRY OF COMPLEX NUMBERS

The following result completely characterizes Euclidean isometries of the plane.

Proposition 8.12. A function $f: \mathbf{C} \to \mathbf{C}$ is a Euclidean isometry if and only if there exist complex numbers a and b such that $|a| = 1$ and

$$f(z) = az + b, \quad z \in \mathbf{C} \tag{1}$$

or

$$f(z) = a\bar{z} + b, \quad z \in \mathbf{C}. \tag{2}$$

Isometries of type (1) are called *direct Euclidean isometries* while those of type (2) are referred to as *opposite Euclidean isometries*.

PROOF OF PROPOSITION 8.12. If f is a function of type (1) and if z_1 and z_2 are any points in $\mathbf{C}$, then

$$|f(z_1) - f(z_2)| = |az_1 + b - az_2 - b| = |z_1 - z_2|$$

since $|a| = 1$. Therefore, f is a Euclidean isometry. An entirely similar argument shows that a function of type (2) is also a Euclidean isometry.

On the other hand, suppose that b and c are any complex numbers such that $|c - b| = 1$. Then the functions h and k, defined by

$$h(z) = (c - b)z + b,$$
$$k(z) = (c - b)\bar{z} + b,$$

for all $z \in \mathbf{C}$, are both Euclidean isometries since $|c - b| = 1$. Moreover, $h(0) = k(0) = b$ and $h(1) = k(1) = c$. If g is any Euclidean isometry such that $g(0) = b$ and $g(1) = c$ and if z is any point of $\mathbf{C}$ distinct from 0 and 1, then

$$|g(z) - b| = |z - 0| = |z|,$$
$$|g(z) - c| = |z - 1|.$$

It follows that $g(z)$ must lie on the circle C_0 of radius $|z|$ centered at b and also on the circle C_1 of radius $|z - 1|$ centered at c. (See Figure 8.27.)

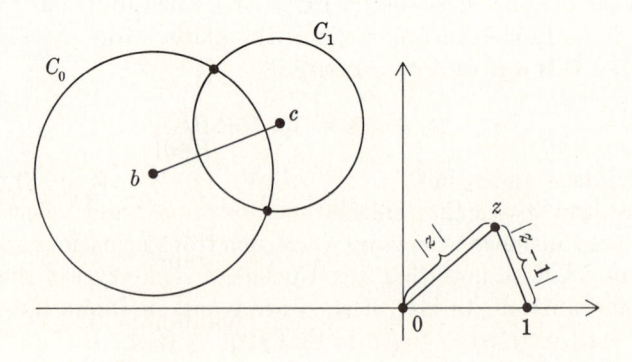

Figure 8.27

8.4. APPLICATIONS TO EUCLIDEAN AND NON-EUCLIDEAN GEOMETRY

Therefore, $g(z)$, $h(z)$, and $k(z)$ must be points of intersection of C_0 and C_1. If z is a real number, then C_0 and C_1 have only one point of intersection and this point is on the line through b and c since isometries map lines into lines (see Exercise 8.45). Therefore, $g(z) = h(z) = k(z)$ in this case. If z is not a real number, then C_0 and C_1 intersect at two points, and one of these points of intersection is $h(z)$ while the other is $k(z)$. Therefore, either $g(z) = h(z)$ or $g(z) = k(z)$ for the given z.

We shall now show that either $g = h$ or $g = k$. Suppose to the contrary that there exist z_1 and z_2 in **C** such that $g(z_1) = h(z_1) \neq k(z_1)$ and $g(z_2) = k(z_2) \neq h(z_2)$. Then, since g is an isometry, it follows that $|z_1 - z_2| = |z_1 - \bar{z}_2| = |\bar{z}_1 - z_2|$. This implies that z_1 and z_2 are real numbers, and hence $h(z_1) = k(z_1)$, $h(z_2) = k(z_2)$. Thus, we have arrived at a contradiction to the choice of z_1 and z_2, and we conclude that either $g = h$ or $g = k$.

We shall now classify direct Euclidean isometries according to their geometric properties.

Proposition 8.13. Suppose $f: \mathbf{C} \to \mathbf{C}$ is a direct Euclidean isometry and that a and b are the unique complex numbers such that $|a| = 1$ and

$$f(z) = az + b, \quad z \in \mathbf{C}.$$

Then:

(a) If $a = 1$, f is a translation by b.
(b) If $a \neq 1$, f is a rotation by $\text{ang}(a)$ around the point $z_0 = b/(1 - a)$.

PROOF. Assertion (a) is obvious. If $a \neq 1$, then Proposition 8.8 asserts that $z_0 = b/(1 - a)$ is the unique fixed point of f in **C**. Therefore, $z_0 = az_0 + b$ and so

$$f(z) = a(z - z_0) + z_0 = (T_{z_0} \circ H_a \circ T_{-z_0})(z)$$

for all $z \in \mathbf{C}$. In other words, the image $f(z)$ of a point $z \in \mathbf{C}$ is obtained by first translating by $-z_0$ (thereby translating the fixed point z_0 to the origin), then rotating by $\text{ang}(a)$, and finally translating back again by z_0. (See Figure 8.28.) Since this sequence of three geometric operations is equivalent to rotating z by $\text{ang}(a)$ around the point $z_0 = b/(1 - a)$, the proof of (b) is complete.

Our next result is the analogue to Proposition 8.13 for opposite Euclidean isometries.

Proposition 8.14. Suppose $f: \mathbf{C} \to \mathbf{C}$ is an opposite Euclidean isometry and that a and b are the unique complex numbers such that $|a| = 1$ and

$$f(z) = a\bar{z} + b, \quad z \in \mathbf{C}.$$

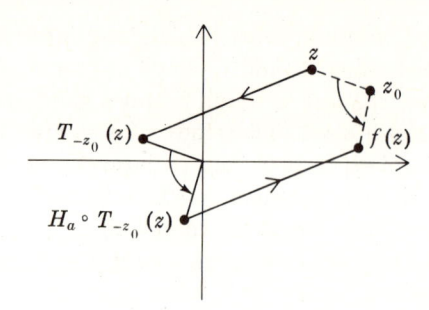

Figure 8.28

Then:

(a) If $a\bar{b} + b = 0$, f is the reflection in the line through $z_0 = b/2$ inclined at $\text{ang}(a)/2$ to the positive real axis.

(b) If $a\bar{b} + b \neq 0$, f is the reflection in the line through $z_0 = b/2$ inclined at $\text{ang}(a)/2$ to the positive real axis followed by translation by $d = (a\bar{b} + b)/2$.

Opposite isometries of type (a) are called *line reflections* while those of type (b) are referred to as *glide reflections*.

PROOF OF PROPOSITION 8.14. (a) For each $z \in \mathbf{C}$, we can write

$$a\bar{z} + b = a\overline{\left(z - \frac{b}{2}\right)} + \left(\frac{a\bar{b} + b}{2}\right) + \frac{b}{2}. \tag{9}$$

Therefore, if $a\bar{b} + b = 0$, we have

$$f(z) = (T_{b/2} \circ h \circ T_{-b/2})(z),$$

where $h: \mathbf{C} \to \mathbf{C}$ is defined by $h(z) = a\bar{z}$ for all $z \in \mathbf{C}$. Since $|a| = 1$, we see that $|h(z)| = |z|$ for all $z \in \mathbf{C}$. Also, $\text{ang}(h(z)) = \text{ang}(a) - \text{ang}(z)$ for all $z \neq 0$, that is,

$$\text{ang}(h(z)) - \frac{\text{ang}(a)}{2} = \frac{\text{ang}(a)}{2} - \text{ang}(z)$$

for all $z \neq 0$. Therefore, the image $h(z)$ of each $z \in \mathbf{C}$ is just the reflection of z in the line through the origin inclined at $\text{ang}(a)/2$ to the positive real axis. Since $f = T_{b/2} \circ h \circ T_{-b/2}$, it follows that f is simply the reflection in the parallel line through $b/2$. (See Figure 8.29.)

(b) If $a\bar{b} + b \neq 0$, then equation (9) implies that $f = T_d \circ T_{b/2} \circ h \circ T_{-b/2}$ where $d = (a\bar{b} + b)/2$. Therefore, in view of the conclusions in the preceding paragraph, f is a line reflection in the line passing through

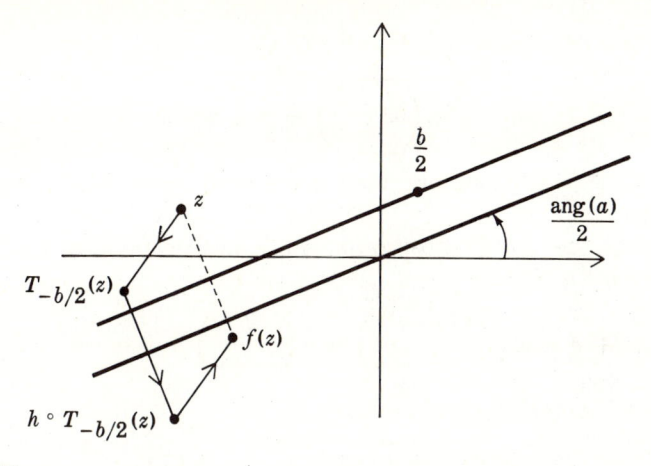

8.4. APPLICATIONS TO EUCLIDEAN AND NON-EUCLIDEAN GEOMETRY

Figure 8.29

$b/2$ inclined at $\operatorname{ang}(a)/2$ to the positive real axis, followed by a translation by $d = (a\bar{b} + b)/2$.

Combining the preceding results, we see that every Euclidean isometry of the plane is either a translation, a rotation around some point, a line reflection, or a glide reflection. The following result shows that line reflections alone can serve as "building blocks" for any Euclidean isometry.

Proposition 8.15. Every Euclidean isometry can be written as the composite of at most three line reflections.

PROOF. First of all, suppose that f is a direct Euclidean isometry and choose complex numbers a, b with $|a| = 1$ such that

$$f(z) = az + b, \quad z \in \mathbf{C}.$$

Since the identity transformation is the composite of any line reflection with itself, we shall assume that f is not the identity transformation. If $a = 1$ (that is, if f is a translation), choose a complex number c such that $c\bar{b} + b = 0$. By Proposition 8.14, the transformations g and h defined by

$$g(z) = c\overline{\left(z - \frac{b}{2}\right)} + \frac{b}{2},$$
$$h(z) = c\bar{z},$$

are both line reflections and, for all $z \in \mathbf{C}$,

$$(g \circ h)(z) = c\overline{\left(c\bar{z} - \frac{b}{2}\right)} + \frac{b}{2} = z + b = f(z)$$

since $c\bar{c} = |c|^2 = 1$. Therefore, f is the composite of two line reflections if $a = 1$.

If $a \neq 1$, then, by Proposition 8.8, f has the fixed point $z_0 = b/(1 - a)$. But for each $z \in \mathbf{C}$,

$$f(z) = az + b = a\left(z - \frac{b}{1-a}\right) + a\left(\frac{b}{1-a}\right) + b = a(z - z_0) + z_0.$$

Hence, if we define k and m on $\mathbf{C}$ by

$$k(z) = \overline{(z - z_0)} + z_0, \qquad z \in \mathbf{C}$$
$$m(z) = a\overline{(z - z_0)} + z_0, \qquad z \in \mathbf{C}$$

then k and m are line reflections and

$$(m \circ k)(z) = a(\overline{[\overline{(z - z_0)} + z_0] - z_0}) + z_0 = a(z - z_0) + z_0 = f(z)$$

for all $z \in \mathbf{C}$. Therefore, f is also the composite of two line reflections if $a \neq 1$.

Finally, if f is an opposite Euclidean isometry, there exist complex numbers a and b with $|a| = 1$ such that

$$f(z) = a\bar{z} + b.$$

If $a\bar{b} + b = 0$, then f is itself a line reflection by Proposition 8.14. If $a\bar{b} + b \neq 0$, then that same proposition implies that f is the composite of a line reflection and a translation. Since translations have already been shown to be the product of two line reflections, we conclude that k is the product of three line reflections when $a\bar{b} + b \neq 0$.

EXERCISES

8.45. Prove that each Euclidean isometry maps straight lines onto straight lines. [*Hint:* Use the fact that r is a point on the line segment joining the points s, t of the complex plane if and only if $|s - r| + |r - t| = |s - t|$.]

8.46. Find the opposite Euclidean isometry that maps i into $2 + i$ and $1 + i$ into 2.

8.47. Write the following Euclidean isometries as composites of line reflections:
   (a) $k(z) = i\bar{z} + (1 + i)$.
   (b) $h(z) = -z + (1 - i)$.

8.4. APPLICATIONS TO EUCLIDEAN AND NON-EUCLIDEAN GEOMETRY

8.48. Prove that the composite of three line reflections is a line reflection or a glide reflection.

8.49 Suppose f is a Euclidean isometry such that f composed with itself is the identity transformation. Prove that f is either the identity transformation, a line reflection, or a rotation by π around some point.

8.50. Suppose that L is a line not passing through the origin and that $z_0 \in L$ is the foot of the perpendicular to L from the origin. Show that reflection in the line L is given by

$$f(z) = -\left(\frac{z_0}{|z_0|}\right)^2 \bar{z} + 2z_0.$$

Cross Ratio

The concept of cross ratio plays a basic role in geometry, particularly in projective geometry. The *cross ratio* $[z_1,z_2,z_3,z_4]$ of four distinct points z_1, z_2, z_3, z_4 in the extended complex number system $\mathbf{C}^*$ is defined by

$$[z_1,z_2,z_3,z_4] = \frac{z_4 - z_1}{z_4 - z_3} \cdot \frac{z_2 - z_3}{z_2 - z_1}.$$

If some $z_k = \infty$, the terms involving z_k in the above definition are deleted; for example,

$$[z_1, \infty ,z_3,z_4] = \frac{z_4 - z_1}{z_4 - z_3}.$$

Since the four points are assumed distinct, the cross ratio is never 0, 1, or ∞. Note that

$$[z_1,z_2,z_3,z_4] = [z_1, \infty ,z_3,z_4] \cdot [z_1,z_2,z_3, \infty]$$
$$= [\infty ,z_2,z_3,z_4] \cdot [z_1,z_2, \infty ,z_4].$$

Also observe that

$$\operatorname{ang}[\infty ,z_2,z_3,z_4] = \operatorname{ang}(z_2 - z_3) - \operatorname{ang}(z_4 - z_3),$$

that is, the measure of the angle with vertex z_3 and initial and terminal sides passing through z_2 and z_4 respectively is equal to $\operatorname{ang}[\infty ,z_2,z_3,z_4]$. (See Figure 8.30.) Of course, a similar geometric conclusion holds for the cross ratios $[z_1, \infty ,z_3,z_4]$, $[z_1,z_2, \infty ,z_4]$, $[z_1,z_2,z_3, \infty]$.

The notion of the cross ratio of four points bears an obvious relation to bilinear transformations. In fact, the formula used to construct the unique bilinear transformation mapping three given distinct points onto three prescribed distinct points (see remarks after Proposition 8.9) yields the following result immediately.

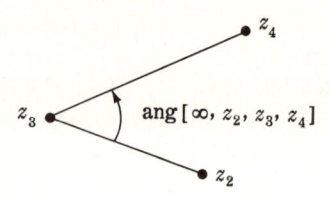

Figure 8.30

Proposition 8.16. If z_1, z_2, z_3, z_4 are distinct points of $\mathbf{C}^*$ and if f is the unique bilinear transformation such that $f(z_1) = 0$, $f(z_2) = 1$, $f(z_3) = \infty$, then

$$f(z_4) = [z_1, z_2, z_3, z_4].$$

Since there is only one bilinear transformation f such that $f(z_1) = 0$, $f(z_2) = 1$, $f(z_3) = \infty$, Proposition 8.16 implies the following important invariance property of the cross ratio.

Proposition 8.17. If z_1, z_2, z_3, z_4 are distinct points of $\mathbf{C}^*$ and if g is any bilinear transformation, then

$$[g(z_1), g(z_2), g(z_3), g(z_4)] = [z_1, z_2, z_3, z_4].$$

PROOF. Choose the unique bilinear transformations f and h such that $f(z_1) = h(g(z_1)) = 0$, $f(z_2) = h(g(z_2)) = 1$, and $f(z_3) = h(g(z_3)) = \infty$. Then $f = h \circ g$ by Proposition 8.9. Consequently,

$$[z_1, z_2, z_3, z_4] = f(z_4) = h(g(z_4)) = [g(z_1), g(z_2), g(z_3), g(z_4)].$$

By viewing cross ratio in terms of bilinear transformations, we obtain the following nice result very quickly.

Proposition 8.18. Four points z_1, z_2, z_3, z_4 in the extended complex plane lie on a line or circle if and only if their cross ratio $[z_1, z_2, z_3, z_4]$ is a real number.

PROOF. Let f denote the bilinear transformation such that $f(z_1) = 0$, $f(z_2) = 1$, $f(z_3) = \infty$. The circle preserving property of bilinear transformations implies that the image K of the real axis under f^{-1} is either a line or a circle. Also, the points z_1, z_2, z_3 belong to K. From Propositions 8.16 and 8.17, we conclude that z_4 also lies on K if and only if $f(z_4) = [z_1, z_2, z_3, z_4]$ is a real number.

Additional information concerning the relative position of the points z_1, z_2, z_3, z_4 on a line or a circle is obtained by looking at the sign of the

8.4. APPLICATIONS TO EUCLIDEAN AND NON-EUCLIDEAN GEOMETRY

real number $[z_1, z_2, z_3, z_4]$. We shall say that the points z_1, z_3 and the points z_2, z_4 are *nonseparating* if the line segment or either arc joining z_1 and z_3 contains either both z_2, z_4 or neither z_2, z_4. (In this case, the line segment or either arc joining z_2 and z_4 must contain either both z_1, z_3 or neither z_1, z_3.) (See Figure 8.31.)

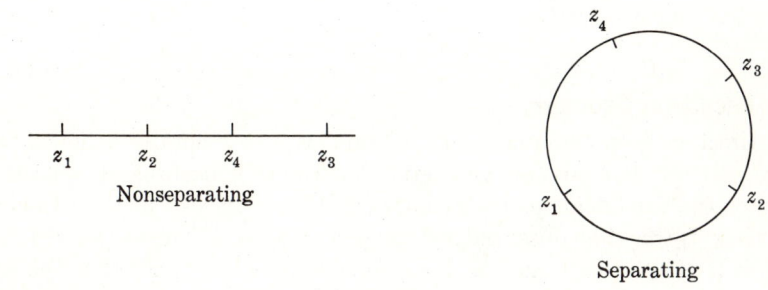

Figure 8.31

Proposition 8.19. If the complex numbers z_1, z_2, z_3, z_4 lie on a line or circle, then z_1, z_3 and z_2, z_4 are nonseparating if and only if $[z_1, z_2, z_3, z_4] > 0$.

PROOF. If the four points all lie on the real axis, then the assertion can be directly verified by checking the signs of the terms in the cross ratio. For example, if $z_1 < z_2 < z_4 < z_3$, then $z_2 - z_3$ and $z_4 - z_3$ are negative while $z_4 - z_1$ and $z_2 - z_1$ are positive, so that

$$\frac{z_4 - z_1}{z_4 - z_3} \cdot \frac{z_2 - z_3}{z_2 - z_1} > 0.$$

Similar case checking yields the fact that four real numbers are nonseparating if and only if the cross ratio is positive.

Now suppose the four points lie on a line or circle K in the complex plane. Then there exists a bilinear transformation f mapping the real axis onto K such that the four given points on K correspond to four points on the real axis. The proof is completed by noting that the bilinear transformation f preserves the nonseparation property of four points since each elementary transformation does and f is the composite of elementary transformations.

EXERCISES

8.51. Prove that the cross ratio of four distinct points is never 0, 1, or ∞.

8.52. Determine whether or not the points $2 + 2i$, $1 - i$, i, and $4 + i$ lie on a circle.

8.53. Let K be a line or circle that passes through the three points z_1, z_2, z_3. Show that z and z' are symmetric with respect to K if and only if $[z_1,z_2,z_3,z] = \overline{[z_1,z_2,z_3,z']}$.

8.54. Suppose a, b, c are three distinct points lying on a circle K. Describe the set of all points z such that:
   (a) $\text{Re}[a,b,c,z] = 0$.
   (b) $|[a,b,c,z]| = 1$.

Non-Euclidean Geometry

Until early in the nineteenth century, it was generally believed that there was only one possible geometry that could be useful and consistent, namely, the Euclidean geometry based on the axioms set forth in Euclid's *Elements*. It was then observed independently by K. F. Gauss (1777–1855), J. Bolyai (1802–1860), and N. I. Lobachevsky (1793–1856) that this system of axioms of Euclid could be altered to form the basis for new geometries which were nevertheless logically consistent mathematical systems. Such geometries are referred to as *non-Euclidean* geometries.

For centuries prior to the development of non-Euclidean geometries, mathematicians tried to prove that the *parallel postulate* of Euclid (which asserts the existence of a unique line parallel to a given line and passing through a point not on the given line) was actually a consequence of the other axioms of Euclidean geometry. The fact that no such proof can be given can be substantiated by constructing a model in such a way that all of Euclid's axioms hold for the model with the exception of the parallel postulate. We shall now indicate how one such model was constructed by H. Poincaré (1854–1912).

Henri Poincaré
David Smith Collection

8.4. APPLICATIONS TO EUCLIDEAN AND NON-EUCLIDEAN GEOMETRY

The underlying set for our model of a non-Euclidean geometry is the upper half of the extended complex plane, that is, the set H defined by

$$H = \{z \in \mathbf{C}^* : z = \infty \text{ or } \mathrm{Im}(z) \geq 0\}.$$

If $z \in H$ and if $z = \infty$ or $\mathrm{Im}(z) = 0$, then z is called an *infinite point* of H. All other points of H, that is, all $z \in \mathbf{C}$ such that $\mathrm{Im}(z) > 0$, are called *ordinary points* of H.

Now that we have described the points in our model, we proceed to specify the "lines." A *line* in H is defined to be:

(a) Any half-line in the complex plane extending upward from the real axis and parallel to the imaginary axis (including the point ∞).

(b) Any semicircle in the upper half of the complex plane that is centered on the real axis.

Figure 8.32 displays four such lines in H. Observe that each line in H contains precisely two infinite points, one on each "end," so to speak. Also, note that through each pair of distinct ordinary points of H there is one and only one line in H.

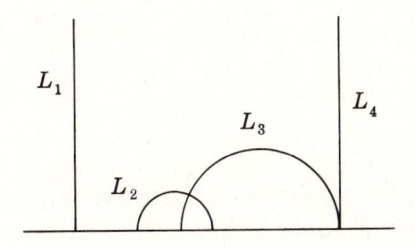

Figure 8.32

It is natural to define two lines L_1 and L_2 in H to be *parallel* in H if they intersect at an infinite point of H. For example, the lines L_1 and L_2 in Figure 8.33 are parallel. We shall now note some basic differences

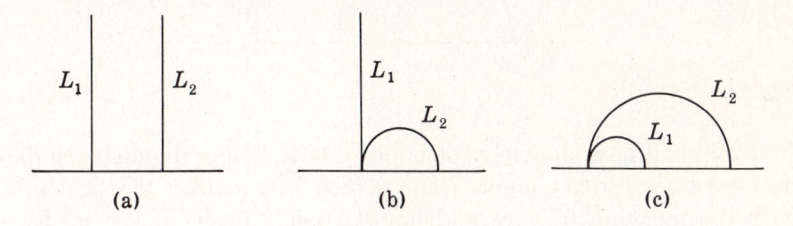

Figure 8.33

374 THE GEOMETRY OF COMPLEX NUMBERS

between parallelism in H and Euclidean parallelism. For example:

(a) If L_1, L_2, L_3 are lines in H such that L_1 is parallel to L_2 and such that L_2 is parallel to L_3, it is not necessarily true that L_1 is parallel to L_3. (See Figure 8.34.)

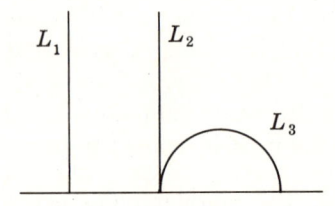

Figure 8.34

(b) If L is a line in H and if z is an ordinary point not on L, there are *two* lines L_1 and L_2 parallel to L passing through z. (See Figure 8.35.)

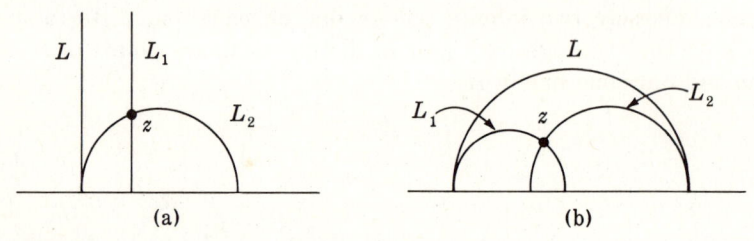

(a)   (b)

Figure 8.35

(c) If L_1 and L_2 are two lines in H that are not parallel, it is not necessarily true that L_1 and L_2 intersect. (See Figure 8.36.)

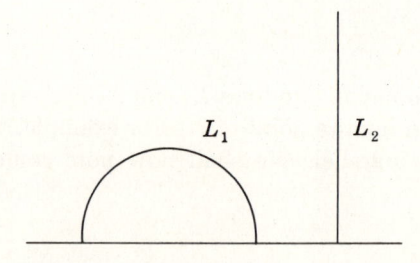

Figure 8.36

We shall now show that it is possible to define the distance $d_H(z,w)$ between two ordinary points z and w of H in a manner that is consistent with the meaning of lines and infinite points in H. If $z = w$, we shall define $d_H(z,w) = 0$. If $z \neq w$, choose the unique line in H passing through

8.4. APPLICATIONS TO EUCLIDEAN AND NON-EUCLIDEAN GEOMETRY

z and w and let r and s be the infinite points of this line. Then $[z,r,w,s]$ is a positive real number since z, r, w, s lie on a circle or a straight line in the complex plane and r, s and z, w are nonseparating (see Propositions 8.18 and 8.19). In this case, we define

$$d_H(z,w) = |\log_e[z,r,w,s]|.$$

Since

$$[z,r,w,s] = \frac{1}{[w,r,z,s]} = \frac{1}{[z,s,w,r]}$$

it follows that the definition of $d_H(z,w)$ does not depend on the order chosen for the infinite points r and s and also that $d_H(z,w) = d_H(w,z)$.

This definition of the distance $d_H(z,w)$ between two ordinary points of H may seem strange at first glance; however, the following result shows that this fits the meaning that we have assigned to lines in H.

Proposition 8.20. Suppose that L is a line in H.

(a) If z, w, and v are three ordinary points of L such that w is between z and v, then $d(z,w) + d(w,v) = d(z,v)$.

(b) If z is an ordinary point of L, then the distance $d_H(z,w)$ from z to an ordinary point w of L increases to $+\infty$ as w approaches an infinite point of L.

PROOF. If L is a half-line in the upper half of the complex plane that is parallel to the imaginary axis and if r is the infinite point of L on the real axis, then r is the real part of z, w and v. If $\text{Im}(z) = a$, $\text{Im}(w) = b$, and $\text{Im}(v) = c$, then either $a < b < c$ or $a > b > c$; consequently, either $b/a > 1$, $c/b > 1$ or $b/a < 1$, $c/b < 1$. Therefore

$$d_H(z,w) + d_H(w,v) = \left|\log_e \frac{b}{a}\right| + \left|\log_e \frac{c}{b}\right|$$

$$= \left|\log_e \frac{c}{a}\right| = d_H(z,v)$$

in either case. Also, for example, if $a > b$, then

$$\lim_{\substack{w \to r \\ w \in L}} d_H(z,w) = \lim_{b \to 0} \left|\log_e \frac{b}{a}\right| = +\infty$$

so that (b) holds.

If L is a semicircle centered on the real axis, then there is a bilinear transformation f that maps L onto a half-line L' parallel to the imaginary axis such that L' lies in the upper half of the complex plane and such that the image of one infinite point of L lies on the real axis while the image of the other is ∞. The points $f(z)$, $f(w)$, $f(v)$ are ordinary points of L' and $f(w)$ is between $f(z)$ and $f(v)$. Consequently, since the cross ratio of four points is invariant under bilinear transformations, (a) and (b) hold by virtue of the conclusions in the preceding paragraph.

The isometries of H can be identified and classified just as the Euclidean isometries were in the first part of this section. It turns out that a function f on the ordinary points of H is an isometry on H if and only if there exist *real* numbers a, b, c, d such that either

$$f(z) = \frac{az + b}{cz + d}, \qquad ad - bc = +1, z \in H \tag{1'}$$

or

$$f(z) = \frac{a\bar{z} + b}{c\bar{z} + d}, \qquad ad - bc = -1, z \in H. \tag{2'}$$

In analogy to the terminology for Euclidean isometries, isometries on H of the type (1') are said to be *direct* while those of type (2') are called *opposite*. We shall not pursue these matters further in this book. However, the reader can find additional information on this subject in books by R. Artzy and C. Carathèodory mentioned in the references at the end of this chapter.

REMARKS AND REFERENCES

The study of complex numbers can be quite valuable since it relates and unifies several topics such as number systems, vectors, trigonometry, and geometry that are considered in the secondary school mathematics program. Complex numbers are also very important in a number of areas of more advanced mathematics, some of which have significant applications in science and engineering.

Vectors in the plane are intimately related to complex numbers by virtue of the representation of complex numbers as points in the complex plane. More specifically, each complex number $z = x + iy$ corresponds to the vector V_z determined by the directed line segment in the plane that joins the origin to the point with coordinates (x,y). Moreover, addition of vectors by the parallelogram rule coincides under this correspondence to the defined meaning of addition for complex numbers as Figure 8.37 indicates. Also, multiplication of the vector V_z by a real number t corresponds to the complex number tz. The magnitude of the vector V_z is just the absolute value of z and the direction of V_z is just the angle of z.

The formula for integral powers of complex numbers in polar form, frequently associated with de Moivre, can be used in conjunction with the

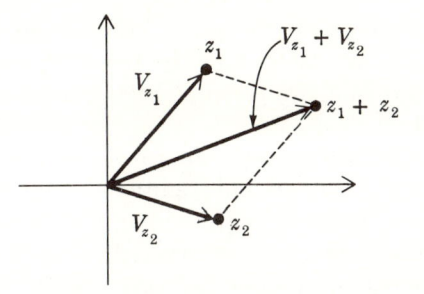

Figure 8.37

Binomial Theorem to obtain trigonometric identities for sines and cosines of multiple angles. For example, we know from this formula that

$$(\cos \theta + i \sin \theta)^4 = \cos 4\theta + i \sin 4\theta.$$

On the other hand, application of the Binomial Theorem yields

$$(\cos \theta + i \sin \theta)^4 = \cos^4 \theta + 4i \cos^3 \theta \sin \theta - 6 \cos^2 \theta \sin^2 \theta - 4i \cos \theta \sin^3 \theta + \sin^4 \theta.$$

Therefore, by equating real and imaginary parts in the two expressions for $(\cos \theta + i \sin \theta)^4$, we obtain the identities

$$\cos 4\theta = \cos^4 \theta - 6 \cos^2 \theta \sin^2 \theta + \sin^4 \theta,$$
$$\sin 4\theta = 4 \cos^3 \theta \sin \theta - 4 \cos \theta \sin^3 \theta.$$

The representation of complex numbers as points on the Riemann sphere is not ordinarily considered in textbooks written for secondary school students. This is unfortunate since this alternative geometric interpretation of the complex number system can be quite instructive. For example, it shows that the complex number system is independent of its particular geometric model and that more than one such model is possible and useful. It also clarifies the meaning of the mathematical concept of infinity in this setting. Students ordinarily become aware of the fact that this concept exists during their secondary school training; however, they are inclined to attach a certain amount of mysticism to the notion. The interpretation of the point at infinity as the North pole of the Riemann sphere should serve to place this idea in proper perspective.

The transformation defined by inversion in a circle can be used in several interesting ways in plane geometry because of its circle preserving and orthogonality preserving properties. For example, in the book by Courant and Robbins [4], inversion is used to prove the Mascheroni Theorem which asserts that all ruler and compass constructions can be performed with compass alone. (See Chapter 9 for a discussion of ruler and compass constructions and for more information on the Mascheroni Theorem.) Eves [5] also shows how the transformation of inversion in a circle can be used to simplify given geometric data and thereby facilitate the proofs of certain theorems.

The nonalgebraic, synthetic approach to geometry developed in Euclid's *Elements* and emulated in traditional secondary school geometry texts is no longer a strong force in the study of the subject. The decline of this classical approach can be attributed to two basic causes. The first was the development of analytic geometry in the seventeenth century. This permitted the powerful tools of classical algebra to be brought into the study of geometry. The second was the recognition late in the nineteenth century of the importance to geometry of the study of groups of transformations. For example, it was observed that Euclidean plane geometry may be regarded as the study of those notions and objects that remain unchanged under any transformation from the special transformation group consisting of the Euclidean isometries. Other transformation groups determine other geometries, and conversely. This transformational approach was first promoted by Felix Klein (1849–1925) in an address that he delivered at his inauguration to a professorship at the University at Erlangen in 1872. Since that time, his "Erlangen program" has had a tremendous influence on the development of geometry.

Analytic geometry is now beginning to assume its proper place in most secondary school geometry programs. The use of the transformational approach to geometry is restricted to rather small experimental programs at this time. However, the current trend is definitely in the direction of transformational rather than synthetic geometry. The study of the elementary transformations and their composites in Sections 8.2 and 8.3 as well as the study of Euclidean isometries in Section 8.4 should help the teacher to gain some insight into the transformational approach to geometry. The books by Artzy [1], Gans [6], and Klein [7] are excellent sources for further reading on this and related topics in geometry. More information concerning the geometry of complex numbers can be found in the books by Caratheodory [2], Carver [3], and Schwerdtfeger [8].

1. Artzy, R., *Linear Geometry*, Reading, Mass.: Addison-Wesley Publishing Company, Inc., 1965.
2. Carathéodory, C., *Theory of Functions*, Vol. I, Translated by F. Steinhardt. New York: Chelsea Publishing Company, 1954.
3. Carver, W. B., *The Conjugate Coordinate System for Plane Euclidean Geometry*, Slaught Memorial Paper Number 5, Washington, D.C.: The Mathematical Association of America, 1956.
4. Courant, R. and H. Robbins, *What Is Mathematics?* New York: Oxford University Press, 1941.

5. Eves, H., *A Survey of Geometry*, Vol. I, Boston: Allyn and Bacon, Inc., 1963.
6. Gans, D., *Transformations and Geometries*, New York: Appleton-Century-Crofts, Inc., 1969.
7. Klein, Felix, *Elementary Mathematics from an Advanced Standpoint—Geometry*, Translated by E. R. Hedrick and C. A. Noble. New York: Dover Publications, Inc., 1939.
8. Schwerdtfeger, Hans, *Geometry of Complex Numbers*, Mathematical Expositions No. 13. Toronto: University of Toronto Press, 1962.

Chapter 9

GEOMETRIC CONSTRUCTIONS

The art of performing geometric constructions by means of ruler and compass was first nurtured by the ancient Greeks of Pythagoras' time (ca. 550 B.C.). It has remained a source of stimulation and fascination to amateurs and scholars alike down through the centuries.

The goal of this endeavor is not skill at drafting, but rather the deductive analysis of how a prescribed geometric configuration can be obtained from given information using only these two instruments. The compass is used to draw circles and the ruler is used to draw straight lines. The term "ruler" in this context does not refer to a device for measuring distances, but rather to an unmarked instrument whose only function is to draw straight lines through pairs of given points. It is common practice to refer to it as a "straight-edge" instead of ruler in order to emphasize its limited utility. In performing geometric constructions, the ruler and compass are not to be used willy-nilly, but rather in accordance with certain restrictions that are essentially axioms in Euclidean geometry. These restrictions will be explained in the first section of this chapter.

It is assumed that the reader is already familiar with some elementary ruler and compass constructions such as bisecting an angle, constructing a line through a given point that is perpendicular to a given line, constructing a line through a given point that is parallel to a given line, and so on. This chapter is not primarily directed to developing such techniques further for more complicated constructions. Instead, we shall dwell on the more theoretical question of whether or not certain geometric constructions are possible at all. Interest in this question was first stimulated by the inability of the ancient Greeks to perform certain constructions with ruler and compass. We shall now outline several such problems briefly.

It is a rare student of mathematics who does not sooner or later encounter the statement that it is impossible to trisect angles with ruler and compass. However, unless this statement is made more precise, it can be a source of confusion. For example, it is not difficult to see that certain angles, such as 90 degrees, can easily be divided into three equal angles by a ruler and compass construction. However, the angle trisection problem asks whether or not *all* angles can be trisected through legitimate use of these two instruments. We shall show that the angle of 60 degrees cannot be trisected using ruler and compass alone.

Another construction problem posed by the ancient Greek geometers that is almost as well known as the angle trisection problem is that of doubling the cube. Legend has it that this problem arose in connection with a devastating plague that inflicted great suffering on the people of ancient Athens and caused the death of their leader Pericles. In order to alleviate their suffering, the people of Athens asked the oracle at Delos how they might end the plague and were told to double the size of Apollo's altar. As the name of the problem suggests, Apollo's altar was cubical in shape. The geometers of Plato's academy managed to accomplish this feat, but the solution was not "pure" in the sense that it lay beyond the scope of the ruler and compass. Whether the plague was actually abated or whether it simply ran its course is not known, but because of the oracle's response, the problem of doubling the cube is also referred to as the Delian problem. The Delian problem may sound like a three-dimensional problem, but it can be stated in planar terms as follows: Given a line segment that represents the edge of a cube, construct another line segment such that the cube having this line segment as edge has twice the *volume* of the original cube. As we shall show, this construction is also impossible using only ruler and compass.

Another famous ancient construction problem is that of squaring the circle. Is it possible to construct a square having the same area as that of a given circle? This problem is considerably deeper than the problems of angle trisection and cube doubling. The proof that squaring the circle by means of ruler and compass is impossible hinges on the face that π is a transcendental number; that is, π is not the root of any polynomial equation having rational coefficients. The transcendence of π was first proved in 1882 by F. Lindemann and the long and intricate proof is far beyond the scope of this book. We shall limit ourselves to accepting the fact that π is transcendental and proving the impossibility of squaring the circle with ruler and compass on that basis.

Another topic of interest to the ancient Greek geometers was that of constructing regular polygons. It was found that certain regular polygons such as the regular pentagon and hexagon, could be constructed by ruler and compass while the regular polygon of seven sides, the regular heptagon, defied their best efforts, and the regular 17-sided polygon was also beyond them. It is interesting that the ruler and compass construction of the regular heptagon is impossible, but the construction of the regular polygon of 17 sides is possible. It was not until the beginning of the nineteenth century that a youthful Karl Gauss discovered a construction of the regular polygon of 17 sides, and it is said that Gauss was so excited

Karl F. Gauss
Library of Congress

by the beauty of this work that he decided at that time to devote the rest of his life to the study of mathematics. Shortly after this, he managed to completely settle the problem of determining which regular polygons can be constructed by ruler and compass and which cannot. It is appropriate that a bronze statue of Gauss was erected in Göttingen after his death and placed on a pedestal in the shape of a regular polygon of 17 sides.

The Greek geometers were quite adept with ruler and compass, but the above problems frustrated their most vigorous efforts and continued to frustrate mathematicians for centuries. The breakthrough finally came in the nineteenth century as a result of the work of E. Galois (1811–1832). The problems were laid to rest, not with construction procedures, but with proofs that such procedures did not exist. These proofs go outside the world of geometry into the field of algebra. How does one go about proving that a certain construction is impossible to perform? The key lies in converting the geometric problems into algebraic terms and using the tools of algebra to show that the resulting equations do not possess solutions of the desired kind. The result is a beautiful interplay of geometry and algebra. This was not possible until the nineteenth century, for it was not until then that the notions of modern algebra were born and blossomed into a powerful and deep theory.

Geometers have also considered the consequences of imposing even stronger restrictions on ruler and compass constructions than those of the

Evariste Galois
David Smith Collection

ancient Greeks. These investigations resulted in several interesting theorems concerning the adequacy of the ruler and compass when used individually. Perhaps the best known and most surprising result in this direction was proved in 1794 by L. Mascheroni. His result asserts that any

Lorenzo Mascheroni
David Smith Collection

ruler and compass construction can be accomplished with compass alone! Of course, since the compass obviously cannot draw a straight line, this assertion must be interpreted properly. It is to be understood in the statement of this result that a desired line is constructed when a pair of points on that line is determined.

We shall see later that the ruler alone is not adequate for all ruler and compass constructions. However, as early as the tenth century A.D., mathematicians considered the problem of determining those ruler and compass constructions that could be performed using a ruler and a "rusty" compass, that is, a compass with a fixed radius. Even the genius Leonardo de Vinci devoted considerable effort to this cause. Finally, in 1833, Jacob Steiner (1796–1863) proved that if a single fixed circle is given in the plane,

Jacob Steiner
David Smith Collection

then any ruler and compass construction can be executed with ruler alone. Thus, if we are allowed to use a compass one time, then we may throw it away and rely on the ruler alone. We shall not prove the Mascheroni or Steiner theorems here; however, the reader can find references to proofs of these results in the Remarks and References section at the end of this chapter.

9.1. EUCLIDEAN CONSTRUCTIONS AND CONSTRUCTIBLE NUMBERS

In performing geometric constructions, the use of the ruler and compass is restricted to the following two operations:

(1) Given two distinct points A and B in the plane, the straight line passing through A and B can be constructed.

(2) Given a point P and a line segment AB of length $|AB| = r > 0$, the circle with center P and radius r can be constructed.

Of course, it is understood that the ruler and compass can be used only a finite number of times in any particular construction. We shall refer to constructions performed under these restrictions as *Euclidean constructions*.

9.1. EUCLIDEAN CONSTRUCTIONS AND CONSTRUCTIBLE NUMBERS

A familiar example of a Euclidean construction is the construction of the perpendicular bisector of a given line segment. If the line segment determined by the points A and B is given, we draw the circle with center A and radius $|AB|$ (the length of the segment AB), and also the circle centered at B of radius $|AB|$. (See Figure 9.1.) These two circles intersect

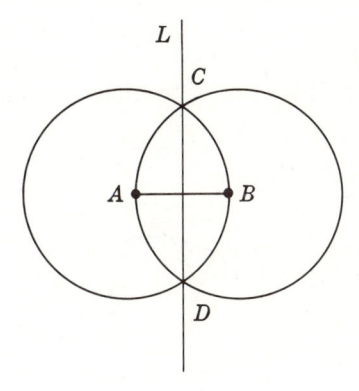

Figure 9.1

at two points, C and D, and the line L drawn through C and D is the desired perpendicular bisector of the segment AB. This may be verified by reasoning with similar triangles.

Actually, the Greek geometers of Euclid's time imposed the following restriction on ruler and compass constructions.

(2') Given a line segment AB of length $|AB| = r$, a circle centered at either A or B with radius r can be constructed.

This restriction is apparently more stringent than (2) since (2) permits "transferring" the length of the given line segment AB to a given point P which may be distinct from both A and B. In physical terms, (2) corresponds to working with a standard compass while (2') corresponds to working with a compass that collapses whenever it is raised from the paper. The use of (2') in place of (2) by the Greek geometers was motivated by their desire to exclude the use of the basic instruments as measuring devices; in particular, they felt that the use of a compass as a set of calipers to transfer distances was not legitimate.

Despite the fact that restriction (2') seems more severe than (2), we shall now show that any Euclidean construction can be performed under the restrictions (1) and (2') as well. To prove this assertion, we need to show that if we are given a line segment AB of length $|AB| = r$ and a point P other than A or B, then we can construct a circle of radius r centered at P using the operation (2'). This can be done as follows: We first draw the circle with center A that passes through P (see Figure 9.2) and the circle with center P that passes through A. These circles intersect in the points C and D. We may now draw the circle with center C that passes through B and the circle with center D that passes through B. These two circles intersect at B and also at a point Q. It is not difficult to verify that the line segment PQ has the same length as line segment AB (the line through C and D is the perpendicular bisector of both line segments AP and BQ). Thus, the circle of radius $|AB| = |PQ|$ with center P can be constructed under the restrictions (1) and (2').

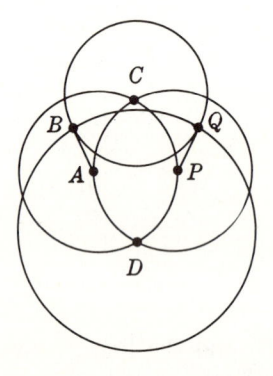

Figure 9.2

Let us now return to the study of constructions by means of ruler and compass. A typical Euclidean construction problem consists of a set of given data (points, lines, line segments, circles) and a description of the desired final configuration. The construction is to proceed from the initial data to the final configuration by using the ruler and compass a finite number of times subject to the restrictions (1) and (2). Unless stated explicitly, no particular relationship among the given elements may be assumed. For example, if we are simply given two lines, then we may not begin our construction by drawing two parallel lines or two perpendicular lines since these relations were not explicitly given.

Frequently, constructions involve an "arbitrary" line or a circle of "arbitrary" radius. For example, suppose a line L and a point P not on L are given and the line through P perpendicular to L is to be constructed. A standard approach is to first draw a circle centered at P having an arbitrary radius greater than the distance from P to L. (See Figure 9.3.) This circle then intersects L at two points, A and B, and since P is equidistant from A and B, it follows that P lies on the perpendicular bisector of the line segment AB. Thus, construction of this perpendicular bisector yields the desired line. Such arbitrary elements are legitimately introduced as long as no assumption is made concerning the relation of this

9.1. EUCLIDEAN CONSTRUCTIONS AND CONSTRUCTIBLE NUMBERS

element to the given elements. That is, we may draw an arbitrary line intersecting a given line, but we may not insist that it makes an angle of 20 degrees with the given line. However, because we are going to analyze ruler and compass constructions by means of analytic geometry and algebra, it will be necessary to have control over the introduction of auxiliary elements. This can be accomplished by the following considerations.

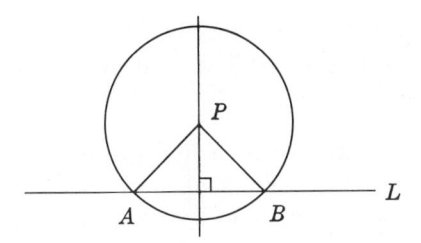

Figure 9.3

The notion of length of a line segment is dependent on a reference line segment of unit length. The statement that a line segment is 2 units in length, for example, contains implicit reference to a standard unit length against which measurements are made. Therefore, we may assume at the outset of any Euclidean construction that a reference line segment of unit length is given. This line segment may then be used to construct lines and circles instead of referring to "arbitrary" lines and circles. For example, in the previous construction of a line perpendicular to a given line and passing through a given point, we may take the radius of the auxiliary circle to be of any positive integer length greater than the distance from the point to the line. Once a segment of unit length is given, it is a simple matter to construct line segments of any positive integer length by successively marking off unit lengths on a line with the compass.

EXERCISES

9.1. Given a line L and a point P on L, construct a line perpendicular to L that passes through P.
9.2. Construct the line parallel to a given line L and passing through a given point P not on L.
9.3. Construct the angle bisector of a given angle.

The remainder of this section is devoted to an analysis of Euclidean constructions in terms of analytic geometry and algebra. The results obtained here will be used in subsequent sections to prove the impossibility of the ancient ruler and compass constructions mentioned in the introduction to this chapter. The first result shows that if we are given two line segments, we are able to construct new line segments whose lengths are the sum, difference, product, quotient, and square roots, respectively, of the lengths of the given segments. Since we refer to lengths here, it is assumed that a reference line segment of unit length is given.

Proposition 9.1. If line segments of lengths a and b are given, then line segments of lengths $a + b$, $a - b$ (if $a > b$), ab, a/b, and $\sqrt{a}$ are constructible using ruler and compass.

PROOF. The simple constructions of line segments of lengths $a + b$ and $a - b$ are left to the reader. We shall now construct a line segment of length ab. Let OA be the segment of length a and let L be the line passing through O and A. Draw line L' through O distinct from L and mark off a segment OB on L' of length b. Then mark off a line segment OU of unit length on L' as shown in Figure 9.4. A line can then be constructed passing

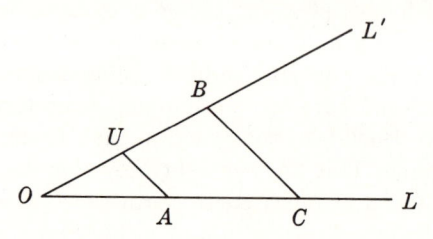

Figure 9.4

through B that is parallel to the line through U and A. This produces a point C on L, and the triangles OAU and OCB are readily seen to be similar. Hence,

$$\frac{|OA|}{|OU|} = \frac{|OC|}{|OB|}.$$

Since $|OA| = a$, $|OB| = b$, and $|OU| = 1$, we have $|OC| = ab$.

A similar construction, as indicated in Figure 9.5, yields a line segment OC of length a/b.

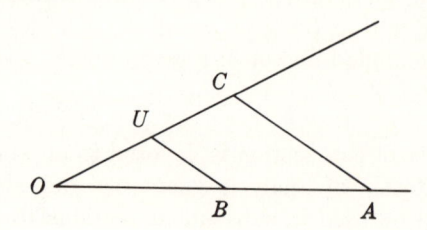

Figure 9.5

9.1. EUCLIDEAN CONSTRUCTIONS AND CONSTRUCTIBLE NUMBERS

It remains to show that we are able to take square roots with ruler and compass. Let a line segment AB of length a be given and let L be the line through A and B. Mark off a unit length on L from the point B as in Figure 9.6 and construct the circle having the resulting segment AC of length $a + 1$ as diameter. The perpendicular to L at the point B intersects this circle at a point D as in Figure 9.6. Since the triangles ABD and

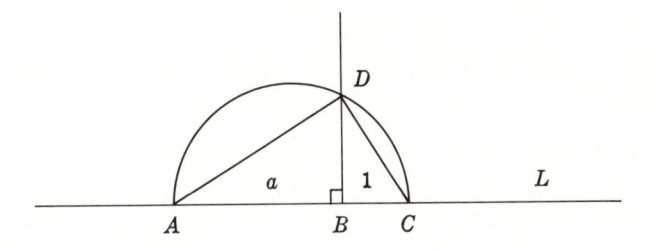

Figure 9.6

DBC are similar, we have

$$\frac{a}{|BD|} = \frac{|BD|}{1}.$$

Therefore, $|BD|^2 = a$, that is, the length of the line segment BD is $\sqrt{a}$. This completes the proof of the proposition.

Suppose we are given a collection of line segments of lengths a_1, $a_2, \cdots, a_n$ including a line segment of unit length. Then it follows from Proposition 9.1 that we may use the ruler and compass to construct line segments whose lengths are sums, differences, products, quotients, and square roots of the given numbers $a_1, a_2, \cdots, a_n$. Moreover, once a line segment has been constructed in this way, it may be used in subsequent constructions. For example, if line segments of lengths $1, \sqrt{5}, \pi$ are given, then it is possible to construct new line segments of lengths

$$\sqrt{\frac{\pi + \sqrt{5}}{3}}, \quad \sqrt{3 + 2\pi\sqrt{5}} + \sqrt{2\pi + 1},$$

and so on. Thus, the algebraic operations of addition, subtraction, multiplication, division, and taking square roots can be performed by means of ruler and compass, and a collection of real numbers can be generated from a given set of numbers by applying these operations repeatedly. In fact, any positive real number that can be expressed in terms of the given numbers $a_1, a_2, \cdots, a_n$ by using the stated arithmetic operations a finite number of times can be obtained by Euclidean construction in the sense that a line segment having the desired number as length can be constructed.

Of course, the collection of numbers generated in this way depends on the initially given set of numbers. However, as we shall see, the classical construction problems that interest us here are all based on a single given line segment, and this given line segment can be conveniently chosen to be of unit length. Therefore, we shall devote our efforts exclusively to examining the nature of those real numbers that can be obtained by ruler and compass as described above starting only with a line segment of unit length. The resulting algebraic description of this set of numbers will be the basis for the analysis of the classical constructions discussed in the introduction.

Definition. A real number a will be called a *constructible number* if there exists a line segment of length $|a|$ that can be constructed by ruler and compass starting from a given line segment of unit length. The collection of all constructible numbers will be denoted by $\mathbf{K}$.

One immediate consequence of Proposition 9.1 is the fact that the set $\mathbf{K}$ of constructible numbers is a subfield of the field $\mathbf{R}$ of real numbers; that is, the subset $\mathbf{K}$ of $\mathbf{R}$ has the following properties:

(1) If $a \in \mathbf{K}$, $b \in \mathbf{K}$, then $a + b \in \mathbf{K}$, $ab \in \mathbf{K}$, and $-a \in \mathbf{K}$.
(2) If $a \in \mathbf{K}$, $b \in \mathbf{K}$, and if $a \neq 0$, then $b/a \in \mathbf{K}$.

As a subfield of $\mathbf{R}$, $\mathbf{K}$ inherits all of the other arithmetic properties of a number field such as commutativity, associativity, and distributivity of addition and multiplication. In short, $(\mathbf{K}, +, \cdot)$ is a field in its own right.

Property (1) of the subfield $\mathbf{K}$ together with the fact that $1 \in \mathbf{K}$ imply that any integer is in $\mathbf{K}$. Therefore, Property (2) of $\mathbf{K}$ implies that any rational number is an element of $\mathbf{K}$. The following result summarizes these conclusions for later reference.

Proposition 9.2. The set $\mathbf{K}$ of constructible numbers is a subfield of the field $\mathbf{R}$ of real numbers and $\mathbf{K}$ contains the field $\mathbf{Q}$ of rational numbers.

Thus, we have $\mathbf{Q} \subset \mathbf{K} \subset \mathbf{R}$, and furthermore, the inclusions are proper. $\mathbf{Q} \neq \mathbf{K}$ since $\sqrt{2}$ is an irrational constructible number. The fact that $\mathbf{K} \neq \mathbf{R}$ will follow from later work. The field of constructible numbers therefore lies between the field of rational numbers and the field of real numbers. In order to analyze the structure of $\mathbf{K}$ more carefully, we now bring the tools of analytic geometry into the picture.

The given line segment of unit length can be used to introduce a coordinate system into the plane. If OU is the given line segment, then the line through O and U will be the x-axis of our coordinate system with O as the origin and with U located on the positive ray of the x-axis. The

9.1. EUCLIDEAN CONSTRUCTIONS AND CONSTRUCTIBLE NUMBERS

line perpendicular to the line through O and U that passes through O will be the y-axis. (See Figure 9.7.) We may now introduce a Cartesian coordinate system on the plane based on these coordinate axes and the given unit distance in the usual way. Each point P in the plane thus corresponds to an ordered pair (a,b) of real numbers, and conversely.

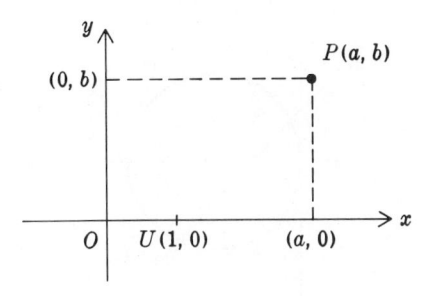

Figure 9.7

It is natural to call a point in the plane a *constructible point* if it is attainable by Euclidean construction starting from the unit line segment OU. That is, a point is regarded as constructible if it can be realized as a point of intersection of two lines, a line and a circle, or two circles constructed in a finite number of steps starting from the given line segment OU. These points are simply described in terms of the above coordinate system as follows.

Proposition 9.3. A point P is a constructible point if and only if both of its coordinates are constructible numbers.

PROOF. If P can be attained in a construction involving a finite number of operations with ruler and compass, then the construction of the perpendiculars through P to the coordinate axes shows that each of the coordinates of P is in fact a constructible number. Conversely, if P has coordinates (a,b) and both a and b are constructible numbers, then the lines perpendicular to the coordinate axes at the points $(a,0)$ and $(0,b)$, respectively, can be constructed with ruler and compass. Since the point of intersection of these two lines is just P, we see that P is a constructible point.

If we call a point with rational coordinates a *rational point*, then it follows from the preceding propositions that each rational point is a constructible point. However, the set of rational points does not exhaust all of the constructible points.

The following example indicates a pattern that will evolve shortly.

392 GEOMETRIC CONSTRUCTIONS

Example 9.1. Let L be the line passing through $(1,0)$ and $(0,1)$ and let C be the circle with center $(0,2)$ and radius 2. (See Figure 9.8.) Since L and C are determined by rational points, they are constructible with ruler and compass. Therefore, the points of intersection of L and C are constructible points. We shall now determine these points of intersection.

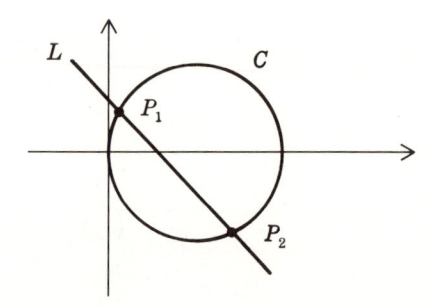

Figure 9.8

The equations of L and C are easily found to be

$$L: \quad x + y = 1$$
$$C: \quad x^2 - 4x + y^2 = 0.$$

If we solve these two equations simultaneously, we see that the points of intersection of L and C are

$$P_1(\tfrac{3}{2} - \tfrac{1}{2}\sqrt{7}, -\tfrac{1}{2} + \tfrac{1}{2}\sqrt{7}),$$
$$P_2(\tfrac{3}{2} + \tfrac{1}{2}\sqrt{7}, -\tfrac{1}{2} - \tfrac{1}{2}\sqrt{7}).$$

Neither P_1 nor P_2 are rational points.

We observe that in the preceding example, L and C were determined by rational points and numbers and their equations involved only rational coefficients. While the points P_1 and P_2 of intersection were not rational points, their coordinates all had a common form, namely, the form $a + b\sqrt{7}$ where a and b are rational numbers. It is convenient to express this phenomenon in terms of the algebraic notion of a field extension.

Let $\mathbf{Q}(\sqrt{7})$ denote the set of all numbers $a + b\sqrt{7}$ where a and b are in the field $\mathbf{Q}$ of rational numbers. If $a_1 + b_1\sqrt{7}$ and $a_2 + b_2\sqrt{7}$ are elements of $\mathbf{Q}(\sqrt{7})$, then their sum and product are also in $\mathbf{Q}(\sqrt{7})$ since

$$(a_1 + b_1\sqrt{7}) + (a_2 + b_2\sqrt{7}) = (a_1 + a_2) + (b_1 + b_2)\sqrt{7}$$

and

$$(a_1 + b_1\sqrt{7})(a_2 + b_2\sqrt{7}) = (a_1a_2 + 7b_1b_2) + (a_1b_2 + a_2b_1)\sqrt{7}.$$

9.1. EUCLIDEAN CONSTRUCTIONS AND CONSTRUCTIBLE NUMBERS

In order to conclude that $\mathbf{Q}(\sqrt{7})$ is a subfield of $\mathbf{R}$, we must also know that the reciprocal of any nonzero element $a + b\sqrt{7}$ of $\mathbf{Q}(\sqrt{7})$ is also in $\mathbf{Q}(\sqrt{7})$. If $a + b\sqrt{7} \neq 0$, then $a - b\sqrt{7} \neq 0$ since $\sqrt{7}$ is irrational, and thus,

$$\frac{1}{a + b\sqrt{7}} = \frac{1}{a + b\sqrt{7}} \cdot \frac{a - b\sqrt{7}}{a - b\sqrt{7}} = \frac{a}{a^2 - 7b^2} - \frac{b}{a^2 - 7b^2}\sqrt{7}$$

is an element of $\mathbf{Q}(\sqrt{7})$. Hence, $\mathbf{Q}(\sqrt{7})$ is a field, and since we may have $b = 0$, we see that $\mathbf{Q}(\sqrt{7})$ contains the rational field $\mathbf{Q}$ as a subfield.

We may think of $\mathbf{Q}(\sqrt{7})$ as being obtained from $\mathbf{Q}$ by adjoining a certain collection of numbers to $\mathbf{Q}$, and therefore we call $\mathbf{Q}(\sqrt{7})$ an *extension field* of $\mathbf{Q}$. Because the numbers adjoined to $\mathbf{Q}$ involve the square root of a particular number, $\mathbf{Q}(\sqrt{7})$ is called a *quadratic extension field* of $\mathbf{Q}$.

In the preceding example, we saw that the construction of a line and circle based on rational data led to points whose coordinates belong to $\mathbf{Q}(\sqrt{7})$. Also, since the rational numbers and $\sqrt{7}$ are constructible numbers, it follows that any point whose coordinates lie in $\mathbf{Q}(\sqrt{7})$ can be constructed with ruler and compass.

We shall now turn to the general situation and perform an analysis along the lines of the particular case just discussed. Instead of the field $\mathbf{Q}$ of rational numbers, suppose we deal with an arbitrary subfield F of the field $\mathbf{K}$ of constructible numbers. Note that since $1 \in F$, such a field F must contain $\mathbf{Q}$ as a subfield. We wish to examine how this given field F must be extended when constructions with ruler and compass are executed. As we shall see, the situation is analogous to the example just discussed in that intersections of lines and circles determined by F can lead to nothing more complicated than a quadratic extension field of F.

Definition. If F is a subfield of $\mathbf{K}$ and if k is a positive number in F such that $\sqrt{k} \notin F$, then

$$F(\sqrt{k}) = \{a + b\sqrt{k} : a, b \in F\}$$

is called a *quadratic extension field* of F.

An argument similar to that given for $\mathbf{Q}(\sqrt{7})$ shows that $F(\sqrt{k})$ is in fact a field that contains F as a subfield. We leave the verification of this as an exercise.

In order to examine the algebraic nature of the points of intersection of lines and circles determined by points whose coordinates belong to F, we must first examine the equations of lines and circles. This is done in the next proposition. For convenience, we shall refer to a point in the plane whose coordinates belong to F as an *F-point*. For example, a rational point would be a **Q**-point, while a constructible point would be a **K**-point.

Proposition 9.4. Let F be a given subfield of **K**.
 (a) If L is a line containing two F-points, then L is given by an equation $ax + by = c$ where a, b, c belong to F.
 (b) If C is a circle whose center is an F-point and whose radius is a number in F, then C is given by an equation $x^2 + y^2 + ax + by + c = 0$ where a, b, c belong to F.

PROOF. (a) Suppose L passes through the distinct F-points (x_1, y_1) and (x_2, y_2). If L is a vertical line, then $x_1 = x_2$ and L has equation $x - x_1 = 0$, which has the desired form. If L is not vertical, then it has slope

$$m = \frac{y_2 - y_1}{x_2 - x_1}.$$

Since F is a field, we see that $m \in F$. The point-slope form of the equation for a straight line yields the following equation for L:

$$y - y_1 = m(x - x_1).$$

We can rewrite this equation as

$$mx - y = mx_1 - y_1,$$

which is an equation of the desired form.
 (b) Suppose the circle C has center (x_0, y_0) and radius r where x_0, y_0, and r belong to F. Then C is given by the equation

$$(x - x_0)^2 + (y - y_0)^2 = r^2,$$

or equivalently,

$$x^2 + y^2 - 2x_0 x - 2y_0 y + (x_0^2 + y_0^2 - r^2) = 0.$$

Thus, we have obtained an equation for C of the desired form.

EXERCISES

9.4. Find the points of intersection of the line through $(1,0)$ and $(0,-1)$ and the circle with center $(0,0)$ and radius 3. In what extension field of **Q** do the coordinates of these points belong?

9.5. Show that the converse assertions of Proposition 9.4 are valid. That is, if the line L has an equation of the stated form, then L contains at least two F-points. Also, if the circle C is given by an equation of the stated form, then its center is an F-point and its radius is a number in F.

9.1. EUCLIDEAN CONSTRUCTIONS AND CONSTRUCTIBLE NUMBERS

9.6. Give an example of a straight line that does not contain more than one rational point.

9.7. Let F be a subfield of $\mathbf{K}$ and let $F(\sqrt{k})$ be a quadratic extension field of F.
   (a) Show that if $a + b\sqrt{k} \in F(\sqrt{k})$, then $a + b\sqrt{k} = 0$ if and only if $a = 0$ and $b = 0$.
   (b) Show that $F(\sqrt{k})$ is a field that contains F as a subfield.

9.8. Prove that $\sqrt{1 + \sqrt{2}}$ does not belong to $\mathbf{Q}(\sqrt{2})$.

In a Euclidean construction, points in the plane are determined as points of intersection in three possible ways, namely, as intersections of:

(1) Two lines.
(2) A line and a circle.
(3) Two circles.

We now show that if the lines or circles are determined by F-points, where F is a given subfield of $\mathbf{K}$, then the coordinates of the points of intersection obtained in these three ways belong to either F itself or a quadratic extension field of F. We first show that if only the ruler is used, giving us case (1) above, then the points of intersection are again F-points. Thus, many Euclidean constructions require the use of the compass as well as the ruler.

Proposition 9.5. Let F be a subfield of $\mathbf{K}$ and let L_1 and L_2 be distinct intersecting lines, each line determined by two F-points. Then the point of intersection of L_1 and L_2 is an F-point.

PROOF. By Proposition 9.4, the lines are given by equations

$$L_1: \quad a_1 x + b_1 y = c_1,$$
$$L_2: \quad a_2 x + b_2 y = c_2,$$

where the coefficients belong to F. If we solve these equations simultaneously, we find that the point (x_0, y_0) of intersection of L_1 and L_2 has coordinates given by

$$x_0 = \frac{b_2 c_1 - b_1 c_2}{a_1 b_2 - a_2 b_1},$$

$$y_0 = \frac{a_1 c_2 - a_2 c_1}{a_1 b_2 - a_2 b_1}.$$

Note that $a_1 b_2 - a_2 b_1 \neq 0$ since the lines are not parallel. Therefore, since F is a field, x_0 and y_0 belong to F and hence (x_0, y_0) is an F-point.

We now examine case (2), the intersection of a line and a circle.

Proposition 9.6. Let F be a subfield of $\mathbf{K}$ and let L be a line determined by two F-points that intersects a circle C whose center is an F-point and whose radius is a number in F. Then the points of intersection of L and C are either F-points or $F(\sqrt{k})$-points where $F(\sqrt{k})$ is a quadratic extension field of F.

PROOF. By Proposition 9.4, L and C have equations

$$L: \quad ax + by = c,$$
$$C: \quad x^2 + y^2 + dx + ey + f = 0,$$

where a, b, c, d, e, f belong to F. Either a or b must be nonzero, and for definiteness we assume $b \neq 0$. (If $b = 0$, then $a \neq 0$ and the argument is similar.) The coordinates of the points of intersection are found by solving the equations for L and C simultaneously. If we solve the equation of L for y in terms of x and substitute into the equation of C, we obtain a quadratic equation in x

$$\alpha x^2 + \beta x + \gamma = 0,$$

where α, β, γ will again belong to F. Explicitly,

$$\alpha = a^2 + b^2, \qquad \beta = b^2 d - 2ac - abe, \qquad \gamma = b^2 f + c^2 + bce.$$

The solutions of this quadratic equation are

$$x = \frac{-\beta \pm \sqrt{\beta^2 - 4\alpha\gamma}}{2\alpha}.$$

If we set $k = \beta^2 - 4\alpha\gamma$, $A = -\beta/2\alpha$, $B = 1/2\alpha$, then these solutions can be written

$$x = A \pm B \sqrt{k}$$

and A, B, and k are numbers in F. Note that $k \geq 0$ since L and C intersect by hypothesis. If $k = 0$, then L and C have one point of intersection, namely, $(A, (c - aA)/b)$, which is an F-point. If $k > 0$, then there will be two points of intersection, namely:

$$\left(A + B \sqrt{k}, \quad \frac{c - aA}{b} - \frac{aB}{b} \sqrt{k} \right),$$
$$\left(A - B \sqrt{k}, \quad \frac{c - aA}{b} + \frac{aB}{b} \sqrt{k} \right).$$

If k is such that $\sqrt{k} \in F$, then these are again F-points. However, if $\sqrt{k} \notin F$, then the coordinates of these points will not lie in F, but in the quadratic extension field $F(\sqrt{k})$ of F.

9.1. EUCLIDEAN CONSTRUCTIONS AND CONSTRUCTIBLE NUMBERS

We now come to the last case (3) in which two circles intersect. It may appear at first glance that the situation could be more complicated than the previous case, but actually it turns out to be the same as case (2).

Proposition 9.7. Let F be a subfield of $\mathbf{K}$, and let C_1 and C_2 be two intersecting circles whose centers are F-points and whose radii are numbers in F. Then the points of intersection are either F-points or $F(\sqrt{k})$-points where $F(\sqrt{k})$ is a quadratic extension field of F.

PROOF. The equations of the circles have the form

$$C_1: \quad x^2 + y^2 + a_1 x + b_1 y + c_1 = 0,$$
$$C_2: \quad x^2 + y^2 + a_2 x + b_2 y + c_2 = 0,$$

where the coefficients lie in F. If we subtract the equation of C_2 from the equation for C_1, we get

$$(a_1 - a_2)x + (b_1 - b_2)y = c_2 - c_1,$$

which is the equation of a line L that must pass through the points of intersection of C_1 and C_2. (See Figure 9.9.) Since the coefficients in the

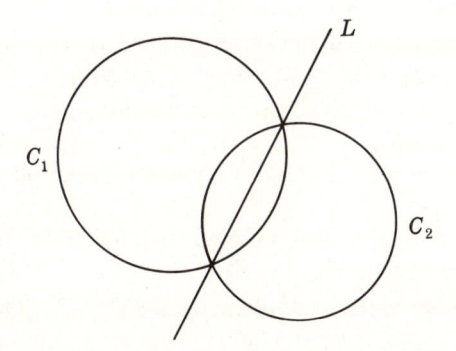

Figure 9.9

equation for L belong to F, the line L contains at least two F-points. (See Exercise 9.5.) Hence, $C_1 \cap C_2 = L \cap C_2$ and the result now follows from Proposition 9.6.

We are now in a position to algebraically characterize constructible numbers in terms of quadratic extension fields. The idea is that as each step in a Euclidean construction is executed, the resulting points of intersection of the lines and/or circles have coordinates which can involve only square roots and arithmetic combinations of the coordinates obtained in the previous step of the construction. That is, algebraically speaking, each step of the construction progresses from one field of numbers to at worst a quadratic extension of that field. This leads to the idea of a quadratic extension chain.

Definition. *A quadratic extension chain of* $\mathbf{Q}$ is a finite sequence of fields $F_0, F_1, \cdots, F_n$ such that

$$F_0 = \mathbf{Q}, \qquad F_1 = F_0(\sqrt{k_1}), \qquad \cdots, \qquad F_n = F_{n-1}(\sqrt{k_n}),$$

where for each $i = 1, 2, \cdots, n$, k_i is a positive number in F_{i-1} such that $\sqrt{k_i}$ does not belong to F_{i-1}. That is, each field F_i in the chain is a quadratic extension field of the previous field F_{i-1} for $i = 1, 2, \cdots, n$.

Suppose x is a given constructible number. Then there is a ruler and compass construction starting from the given unit line segment such that the final step in the construction determines x. Each stage in the construction involves the intersection of either two lines, a line and a circle, or two circles. The construction begins in the field $\mathbf{Q}$ of rational numbers and will either stay inside this field or be forced outside of $\mathbf{Q}$ at some stage with the introduction of an irrational square root $\sqrt{k_1}$. In the latter case, we will move from $\mathbf{Q}$ to a quadratic extension $F_1 = \mathbf{Q}(\sqrt{k_1})$ of $\mathbf{Q}$. If a subsequent step of the construction introduces the square root of a number k_2 in F_1 such that $\sqrt{k_2} \notin F_1$, then we go from F_1 to $F_2 = F_1(\sqrt{k_2})$. Thus, as we follow the steps of the construction, we move along a quadratic extension chain $F_0, F_1, F_2, \cdots, F_k, \cdots$ and since the ruler and compass construction for x involves only a finite number of steps, we have $x \in F_n$ for some n.

Conversely, if x is a real number belonging to the field F_n in a quadratic extension chain $F_0, F_1, F_2, \cdots, F_n$ of $\mathbf{Q}$, then x must be a constructible number since each element of $F_n = F_{n-1}(\sqrt{k_n})$ can be constructed from elements of F_{n-1}, each element of $F_{n-1} = F_{n-2}(\sqrt{k_{n-1}})$ can be constructed from the elements of F_{n-2}, and so on back to $\mathbf{Q}$.

Summarizing these observations, we have the following algebraic criterion for a real number to be constructible.

Proposition 9.8. *A real number x is constructible if and only if there exists a quadratic extension chain $F_0, F_1, F_2, \cdots, F_n$ of $\mathbf{Q}$ such that $x \in F_n$.*

In general, there are many different ruler and compass constructions that will lead to a desired final configuration. Consequently, there will be different quadratic extension chains that can be used to reach a given constructible number x. No assertion of uniqueness can be made in

9.1. EUCLIDEAN CONSTRUCTIONS AND CONSTRUCTIBLE NUMBERS

Proposition 9.8 concerning the quadratic extension fields $F_1, \cdots, F_n$. In particular, the number n of links in the chain leading to x is not unique. However, it does follow from the Well-Ordering Property for the natural number system that among all possible quadratic extension chains of $\mathbf{Q}$ leading to x, there will be one containing a least number of links. This important observation will be useful later.

The following example indicates how the above characterization of constructible numbers can be applied to decide whether a desired Euclidean construction is possible.

Example 9.2. Show that it is possible to construct with ruler and compass the legs of a right triangle having a hypotenuse of length 9 units and an area of 20 square units.

If x and y denote the lengths of the legs of the desired triangle, then we are given that
$$\tfrac{1}{2}xy = 20,$$
$$x^2 + y^2 = 81.$$

Since $y = 40/x$, we see that x must satisfy the equation
$$x^2 + \frac{1600}{x^2} = 81.$$

Solving for x^2, we get
$$x^2 = \frac{81 \pm \sqrt{161}}{2}$$

and thus
$$x = \sqrt{\frac{81 \pm \sqrt{161}}{2}}.$$

These values of x are constructible numbers, and hence we see that two such triangles may be constructed. However, it turns out that these two triangles are congruent. The corresponding values of y are easily found by using the relation $xy = 40$.

It was mentioned earlier that the set of constructible numbers falls short of exhausting the set of all real numbers. This assertion will follow from our next proposition. Recall that each real number is one of two types, algebraic or transcendental. A real number is called *algebraic* if it is the root of some polynomial equation having rational coefficients. A real number that is not algebraic is called *transcendental*. The term transcendental is used for these numbers since they "transcend" algebra, in the sense that they cannot be obtained from the integers by the algebraic operations of addition, subtraction, multiplication, division, or extraction of roots of any order. We now use the criterion of Proposition 9.8 to show that constructible numbers are all algebraic.

Proposition 9.9. Let $F_0, F_1, \cdots, F_n$ be a quadratic extension chain of **Q**. If $x_0 \in F_n$, then for each m, $1 \leq m \leq n$, x_0 is the root of a polynomial equation of degree 2^m having coefficients in F_{n-m}.

PROOF. Since x_0 belongs to $F_n = F_{n-1}(\sqrt{k_n})$, there exist a and b in F_{n-1} such that $x_0 = a + b\sqrt{k_n}$. Then $(x_0 - a)^2 = b^2 k_n$, and hence x_0 is a root of the polynomial equation

$$x^2 - 2ax + (a^2 - b^2 k_n) = 0,$$

which has coefficients in F_{n-1}. Thus, the assertion is true for $m = 1$. Proceeding by induction, we assume the statement is true for a given m, where $1 < m \leq n$. This means that x_0 satisfies

$$x_0^{2^m} + \sum_{i=0}^{2^m-1} a_i x_0^i = 0, \tag{*}$$

where each coefficient a_i belongs to $F_{n-m} = F_{n-m-1}(\sqrt{k_{n-m}})$. That is, $a_i = b_i + c_i \sqrt{k_{n-m}}$ where b_i and c_i lie in F_{n-m-1} for each i. Substituting these expressions for a_i into (*), we obtain

$$x_0^{2^m} + \sum_{i=0}^{2^m-1} b_i x_0^i = -\sqrt{k_{n-m}} \sum_{i=0}^{2^m-1} c_i x_0^i.$$

Squaring both sides shows that x_0 is a root of the polynomial equation

$$\left(x^{2^m} + \sum_{i=0}^{2^m-1} b_i x^i\right)^2 - k_{n-m} \left(\sum_{i=0}^{2^m-1} c_i x^i\right)^2 = 0,$$

which has degree 2^{m+1} and coefficients in F_{n-m-1}. The proposition now follows.

Corollary. Each constructible number is algebraic.

PROOF. If x is constructible, then there exists a quadratic extension chain $F_0, F_1, \cdots, F_n$ of **Q** such that $x \in F_n$. Taking $m = n$ in Proposition 9.9, we see that x is a root of the polynomial equation of degree 2^n with rational coefficients. Hence, x is an algebraic number.

EXERCISES

9.9. Show that $\sqrt{1 + \sqrt{2} + \sqrt{3}}$ is a constructible number by exhibiting an appropriate quadratic extension chain of **Q**.

9.10. Prove that $\sqrt{2} + \sqrt{3}$ is a constructible number, but that it cannot belong to $Q(\sqrt{k})$ for any positive rational number k. [*Note:* $\sqrt{\frac{3}{2}}$ is irrational.]

9.11. Show that if x_0 lies in a quadratic extension field $F(\sqrt{k})$ of a field F, then there is a polynomial P of degree two with coefficients in F such that x_0 is a root of $P(x) = 0$.

9.12. Show that if P is the polynomial $P(x) = ax^2 + bx + c$, where a, b, c lie in a field F, then the real roots of $P(x) = 0$ lie in a quadratic extension field $F(\sqrt{k})$ of F.

9.13. Show that the roots of $x^4 - 7x^2 + 5 = 0$ are constructible numbers.

9.14. Find a polynomial equation with rational coefficients that has the number $1 + \sqrt{5} + \sqrt{2}$ as a root.

9.15. Show that it is possible to construct with ruler and compass the base of an isosceles triangle having sides of unit length and area $\frac{1}{8}$.

9.16. Show that if A is a positive constructible number, then it is possible to construct an equilateral triangle of area A with ruler and compass.

9.2. THE IMPOSSIBILITY OF THE CLASSICAL GREEK CONSTRUCTIONS

In this section, we will apply the criterion for a real number to be a constructible number to prove that it is impossible to find Euclidean constructions for doubling a cube, trisecting all angles, and squaring a circle. The problem of squaring a circle is basically different from the other two problems and we shall dispose of it first.

Given a circle of specified radius, we are asked to construct a square with ruler and compass such that this square has area equal to the area of the given circle. Once the circle has been given, we may take the line segment specifying its radius as the reference segment of unit length for our coordinate system. That is, we may regard the circle as having unit radius. Since this circle has area π, the problem is to construct a line segment of length x such that $x^2 = \pi$. Thus, the impossibility of constructing such a line segment with the Euclidean tools would be proved if we could show that $\sqrt{\pi}$ is not a constructible number. For this, it would suffice to know that π is not a constructible number since a positive real number a is a constructible number if and only if $\sqrt{a}$ is a constructible number. However, π is known to be transcendental (see the introduction to this chapter), and the corollary to Proposition 9.9 asserts that every constructible number is algebraic, that is, *not* transcendental. Therefore, $\sqrt{\pi}$ is not a constructible number and we conclude that the circle cannot be squared by means of ruler and compass.

We now turn to proving that cubes cannot be doubled nor all angles trisected by Euclidean constructions. The proofs proceed by relating a cubic polynomial equation to each of these construction problems in such a way that the construction problem is solvable if and only if the associated cubic equation has a root that is a constructible number. It is then shown that these cubic equations do not have any roots that are constructible numbers, and hence, the desired constructions are impossible to perform.

Let us first derive the cubic equation associated with the problem of doubling the cube. Since this problem only involves the relative lengths of two line segments, we may take the line segment representing the edge of the given cube to be the reference segment of unit length for our coordinate system. We are then asked to construct a cube whose edges have length x_0 such that the volume of the constructed cube is twice that of the given cube of unit edge (see Figure 9.10). Since the given cube has

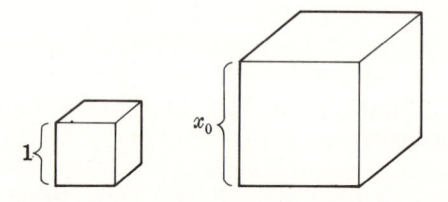

Figure 9.10

volume 1 and the cube to be constructed has volume x_0^3, the problem can be restated as follows: Given a line segment of unit length, construct with ruler and compass a line segment of length x_0 such that $x_0^3 = 2$. Thus, the desired number x_0 must be a root of the cubic equation

$$x^3 - 2 = 0.$$

The construction of doubling the cube of unit volume is possible if and only if $\sqrt[3]{2}$ is a constructible number. We shall later prove that the above cubic equation has no roots that are constructible numbers, and hence, that it is not possible to double the cube with ruler and compass.

We now turn to the problem of trisecting angles with ruler and compass. For convenience, we may assume that a given angle has its vertex at the origin and its initial side along the positive x-axis of the coordinate system. The problem of angle trisection is approached by using the cosine of the angles involved.

The relationship between the angle θ, that is, the angle whose measure is θ, and the number $\cos \theta$ is exhibited in Figure 9.11. We see that if the angle is regarded as given, then it can be used to construct with ruler and compass a line segment of length $\cos \theta$ (or $- \cos \theta$ if $\cos \theta$ is negative).

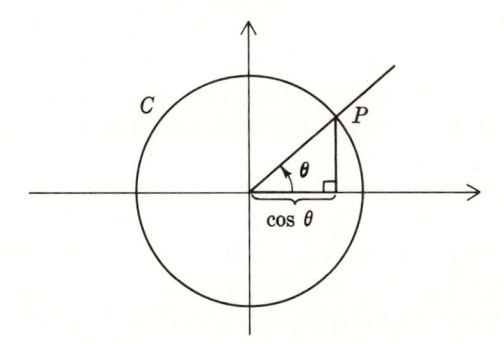

9.2. THE IMPOSSIBILITY OF THE CLASSICAL GREEK CONSTRUCTIONS 403

Figure 9.11

This is done by constructing the perpendicular to the x-axis which passes through the point P of intersection of the terminal side of the given angle and the circle C of unit radius centered at the origin. [*Note:* This does not necessarily imply that $\cos\theta$ is a constructible number since the given terminal side of the angle might not be determined by two constructible points.]

Conversely, if the number $\cos\theta$ is given, then it is possible to construct with ruler and compass the angle θ by erecting a perpendicular to the x-axis at the point $(\cos\theta, 0)$. This perpendicular intersects C at a point P which determines the terminal side of the desired angle.

Now suppose we are given the angle 3θ and we wish to construct the angle θ with ruler and compass. The preceding observations tell us that this is possible if and only if the point $(\cos\theta, 0)$ can be constructed by starting from the segment of length $|\cos 3\theta|$ and the segment of unit length. We want to apply the theory of constructible numbers developed in the previous section, and in order to do this, we will restrict our attention to the case that $\cos 3\theta$ is itself a constructible number. In this case, we can make the following assertion.

Proposition 9.10. If $\cos 3\theta$ is a constructible number, then the angle 3θ can be trisected with ruler and compass if and only if $\cos\theta$ is a constructible number.

PROOF. If $\cos 3\theta$ is a constructible number, then the angle 3θ can be obtained by Euclidean construction. Therefore, if this angle can be trisected with ruler and compass, it follows that the angle θ can also be constructed with ruler and compass, and hence, $\cos\theta$ is a constructible number. Conversely, if $\cos\theta$ is a constructible number, then the angle θ can be constructed from the segment of unit length alone. The angle of measure 3θ is therefore certainly trisectable with ruler and compass.

In the case that $\cos \theta$ is a constructible number, then $\cos 3\theta$ is also a constructible number since it is an easy matter to triple an angle by Euclidean construction. Thus, we see that *if $\cos \theta$ is a constructible number, then the angle 3θ can be trisected with ruler and compass*. This observation assures us that certain angles are certainly trisectable with ruler and compass. For example, $\cos 30° = \sqrt{3}/2$ and $\cos 45° = \sqrt{2}/2$ are constructible numbers since they involve only square roots and rational numbers, and therefore the angles 90 degrees and 135 degrees can be trisected by means of ruler and compass. What we wish to prove is that it is impossible to trisect *every* angle by a Euclidean construction. This will be accomplished by showing that the angle of 60 degrees can resist all efforts to be trisected with the Euclidean instruments. Since $\cos 60° = \frac{1}{2}$ is a constructible number, this will be established by proving that $\cos 20°$ is *not* a constructible number.

A cubic equation can be associated with the problem of angle trisection by using trigonometric identities as follows. The identity for the cosine of the sum of two angles yields

$$\cos 3\theta = \cos(\theta + 2\theta) = \cos \theta \cos 2\theta - \sin \theta \sin 2\theta.$$

Now using the identities $\cos 2\theta = 2\cos^2 \theta - 1$, $\sin 2\theta = 2 \cos \theta \sin \theta$, and $\sin^2 \theta = 1 - \cos^2 \theta$, and collecting terms, we get

$$\cos 3\theta = 4 \cos^3 \theta - 3 \cos \theta.$$

If we set $x_0 = 2 \cos \theta$, this becomes

$$x_0^3 - 3x_0 - 2 \cos 3\theta = 0.$$

Therefore, if an angle 3θ is given, then the number $x_0 = 2 \cos \theta$ must be a root of the cubic equation

$$x^3 - 3x - 2 \cos 3\theta = 0.$$

For the case $3\theta = 60$ degrees, the last equation becomes

$$x^3 - 3x - 1 = 0.$$

If we can prove that this cubic equation has no roots that are constructible numbers, then it will follow that the root $x_0 = 2 \cos 20°$ is not a constructible number, and hence 60 degrees cannot be trisected with ruler and compass.

We now come to the basic result concerning the constructibility of roots of cubic equations. In the proof, we shall make use of the following simple fact.

Lemma. If r_1, r_2, r_3 are the roots of the equation $x^3 + ax^2 + bx + c = 0$, then $r_1 + r_2 + r_3 = -a$.

PROOF. The given polynomial equation may be written in terms of the roots as $(x - r_1)(x - r_2)(x - r_3) = 0$. Carrying out the indicated multiplication and comparing the resulting coefficient of x^2 with that of the originally given equation, we see that $-r_1 - r_2 - r_3 = a$, or equivalently, $r_1 + r_2 + r_3 = -a$.

Proposition 9.11. If the equation $x^3 + ax^2 + bx + c = 0$, where a, b, c are rational numbers, has a constructible root (that is, a root that is a constructible number), then it has a rational root.

REMARK. This result may be stated in the following equivalent form: If $x^3 + ax^2 + bx + c = 0$ has no rational roots, then it has no constructible roots. Thus, in order to determine whether or not a particular cubic equation with rational coefficients has any constructible roots, it suffices to determine whether or not the equation has any rational roots, and this is frequently quite easy to do.

PROOF OF PROPOSITION 9.11. Suppose that a constructible root of the equation exists, but that no rational roots exist. This will eventually lead to a contradiction. Since the equation has a constructible root, there exists a quadratic extension chain

$$F_0 = \mathbf{Q}, \qquad F_1 = F_0(\sqrt{k_1}), \quad \cdots, \quad F_n = F_{n-1}(\sqrt{k_n})$$

such that some root x_1 of the equation lies in F_n and such that n is the least natural number with this property. The Well-Ordering Principle for the natural number system assures us that such an n exists. Note that $n \geq 1$ since no root is rational. Thus, the root x_1 can be written in the form

$$x_1 = p + q\sqrt{k_n},$$

where p, q, k_n are in F_{n-1} and $\sqrt{k_n} \notin F_{n-1}$. We now show that

$$x_2 = p - q\sqrt{k_n}$$

is also a root of the equation.

If we substitute $x_1 = p + q\sqrt{k_n}$ into the equation $x^3 + ax^2 + bx + c = 0$ and rearrange the terms, we get

$$(p^3 + 3pq^2k_n + ap^2 + aq^2k_n + bp + c) \\ + (3p^2q + q^3k_n + 2apq + bq)\sqrt{k_n} = 0.$$

Let us write this expression simply as

$$A + B\sqrt{k_n} = 0.$$

This latter equation holds if and only if $A = 0$ and $B = 0$; for otherwise, we could conclude that $\sqrt{k_n} \in F_{n-1}$ which is not the case. A look at the expression above shows us that A involves only even powers of q while B contains only odd powers of q. Consequently, replacement of q by $-q$ leaves A unchanged and changes B to $-B$. Therefore, if we substitute $x_2 = p - q\sqrt{k_n}$ into the cubic equation, we obtain

$$x_2^3 + ax_2^2 + bx_2 + c = A - B\sqrt{k_n} = 0,$$

since both $A = 0$ and $B = 0$. Hence, x_2 is also a root of the equation.

The third root x_3 is now easily obtained by using the lemma preceding the proposition. We have

$$x_3 = -a + x_1 + x_2 = -a + 2p.$$

Therefore, we see that the root x_3 is a member of F_{n-1} since a is rational and $p \in F_{n-1}$. However, this contradicts the choice of n. Therefore, the cubic equation has no constructible roots if it has no rational roots.

In order to apply Proposition 9.11, we shall use the Rational Root Test, which is stated as follows: If $P(x) = a_n x^n + a_{n-1} x^{n-1} + \cdots + a_1 x + a_0$ is a polynomial with integer coefficients where $a_n \neq 0$, and if $x_0 = c/d$ is a root of $P(x) = 0$ where c and d are integers with no common factors, then c divides a_0 and d divides a_n. For a proof of this result, see Proposition 3.10.

We are finally in a position to prove the impossibility of doubling the cube and trisecting all angles by ruler and compass.

Proposition 9.12. It is impossible to double the cube of unit edge by Euclidean construction.

PROOF. We observed earlier that if the construction were possible, then the equation $x^3 - 2 = 0$ would have a constructible root. It would then follow from Proposition 9.11 that this equation would have a rational root. However, the Rational Root Test shows us that the only possible rational roots of $x^3 - 2 = 0$ are ± 1, ± 2, and a direct check shows us that none of these numbers are in fact roots. Hence, the equation has no rational roots and therefore no constructible roots. We conclude that doubling the cube by Euclidean construction is impossible.

The problem of angle trisection is settled in a similar manner.

Proposition 9.13. It is impossible to trisect the angle 60 degrees by Euclidean construction.

9.2. THE IMPOSSIBILITY OF THE CLASSICAL GREEK CONSTRUCTIONS

PROOF. We saw earlier that if the angle of 60 degrees were trisectable with ruler and compass, then the equation $x^3 - 3x - 1 = 0$ would have a constructible root. Proposition 9.11 would then imply the existence of a rational root for this equation. But the only possible rational roots are ± 1 by the Rational Root Test, and direct substitution shows that these numbers do not satisfy the equation. Hence, the root $x_0 = 2 \cos 20°$ is not a constructible number which implies that $\cos 20°$ is not a constructible number. Since $\cos 60° = \frac{1}{2}$ is a constructible number, it follows from Proposition 9.10 that 60 degrees cannot be trisected by Euclidean construction.

The above techniques can also be applied to other construction problems.

Example 9.3. Is it possible to construct an isosceles triangle of perimeter 8 and area 1 with ruler and compass?

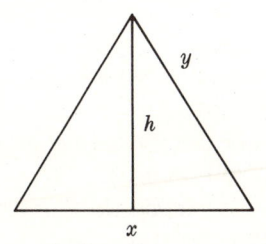

Figure 9.12

Denote the base length of the triangle by x and the length of a side by y as shown in Figure 9.12. The altitude h is then given by

$$h = \sqrt{y^2 - (\tfrac{1}{2}x)^2}.$$

We are given that the perimeter is $2y + x = 8$ and the area is

$$\tfrac{1}{2}xh = \tfrac{1}{4}x\sqrt{4y^2 - x^2} = 1.$$

Substituting $2y = 8 - x$ into the latter equation, we obtain the equation $x^3 - 4x^2 + 1 = 0$. Since this equation has no rational roots, it has no constructible roots, and hence the base length of the triangle is not a constructible number. Therefore, the triangle cannot be constructed by ruler and compass.

EXERCISES

9.17. Is it possible to construct with ruler and compass an isosceles triangle of area 1 inscribed in a circle of radius 1? What if the altitude of the triangle is required to be strictly greater than 1? (See Figure 9.13.)

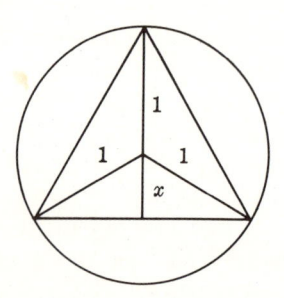

Figure 9.13

9.18. Show that it is impossible to construct with ruler and compass the radius of a sphere whose volume is twice the volume of a given sphere.

9.19. Show that 45 degrees can be trisected with ruler and compass. [*Hint:* cos 15° = cos (60° − 45°).]

9.20. Prove that none of the angles 75 degrees, 30 degrees, 15 degrees can be trisected with ruler and compass.

9.21. Is it possible to trisect 3θ if $\cos 3\theta = \frac{1}{3}$? What if $\cos 3\theta = \frac{11}{16}$?

9.22. Suppose the coefficients in the equation $x^3 + ax^2 + bx + c = 0$ are all constructible numbers. Prove that if this equation has at least one constructible root, then all roots which are real numbers must be constructible numbers.

9.23. Consider a rectangular box with square end of edge x and length y whose volume is 3 and whose surface area is 7 (Figure 9.14). Is it possible to construct the edges of such a box with ruler and compass?

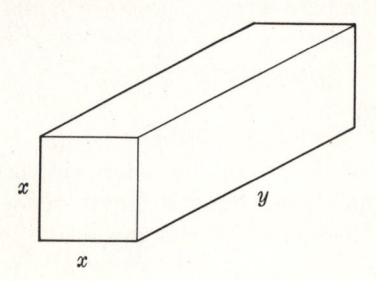

Figure 9.14

9.3. EUCLIDEAN CONSTRUCTION OF REGULAR POLYGONS

Besides the problems already discussed, the ancient Greek geometers were interested in many other construction problems. In particular, they studied the problem of constructing regular polygons with ruler and compass. They managed to obtain constructions for several of the regular polygons, but others frustrated their every effort. For centuries it was not known if these were impossible to construct or if they were only very difficult to obtain. It was not until 1801 that the complete solution to the problem was presented in Gauss' masterpiece *Disquisitiones arithmeticae*.

We shall call a regular polygon with n sides a *regular n-gon*. For the purposes of our analysis, we can always assume that we are given only a circle of unit radius and that we are to inscribe a regular n-gon in this circle using ruler and compass. The n vertices of the regular n-gon are then n equally spaced points on the circle and each side of the n-gon is a chord of the given circle that subtends a central angle of $2\pi/n$ radians or $360/n$ degrees.

One of the regular polygons that is quite easy to construct is the regular 6-gon or hexagon. Observe that since the angle subtended by a side of the regular hexagon is 60 degrees and since the triangle AOB in Figure 9.15 formed by a side and two radii is isosceles, the triangle AOB must in

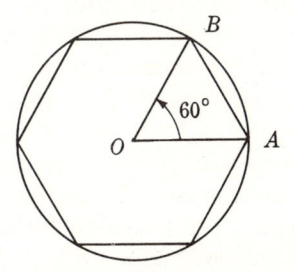

Figure 9.15

fact be equilateral. Hence, each side of the regular hexagon is equal in length to the radius of the circle in which the hexagon is inscribed. Thus, once the circle is drawn, we may use the compass to successively mark off six chords of unit length and so construct the regular hexagon.

The regular 3-gon, that is, equilateral triangle, can be obtained from the regular hexagon by simply joining every second vertex of the regular hexagon. (See Figure 9.16.) Also, the regular 12-gon can be obtained from the regular hexagon by constructing the perpendicular bisector of each side of the regular hexagon. (See Figure 9.17.) Similarly, successive bisections yield the regular 24-gon, 48-gon, and in general, the regular $3 \cdot 2^k$-gon for any natural number k.

410   GEOMETRIC CONSTRUCTIONS

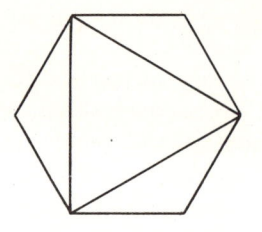

Figure 9.16

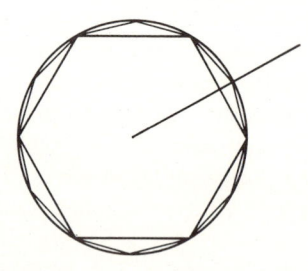

Figure 9.17

Another regular polygon that is quite easy to construct is the regular 4-gon, or square, since we can simply draw the four line segments joining the points of intersection of the unit circle with the coordinate axes. Bisection of the sides then gives us the regular 8-gon, or octagon, and additional bisection can be performed to construct any regular 2^n-gon for $n = 2, 3, \cdots$.

The first regular n-gons to present any challenge in so far as ruler and compass construction is concerned are the pentagon, $n = 5$, and the decagon, $n = 10$. Constructions for these regular polygons were known to the Greek geometers, but before describing the classical constructions we will use the methods of the previous sections to prove that they are in fact constructible with ruler and compass.

Consider a regular decagon inscribed in a circle of unit radius and let x denote the length of one side of the decagon. Consider the triangle AOB formed by a side of the decagon and two radii of the circle as in Figure 9.18. Since the central angle subtended by a side of a regular decagon is 36 degrees, the isoceles triangle AOB has equal base angles of 72 degrees. The bisection of the angle at the vertex B of this triangle produces another triangle ABC. (See Figure 9.19.) The triangle ABC must also be isoceles since $\angle ABC = 36°$ and $\angle BAC = 72°$ so that $\angle ACB = 72°$ again. Thus, $|BC| = x$. Furthermore, since the triangle BOC is also isoceles, we have $|OC| = x$, and therefore, $|AC| = 1 - x$.

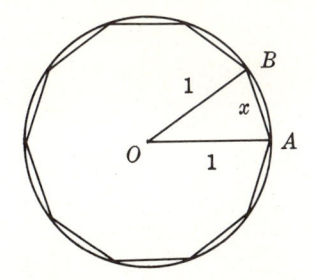

9.3. EUCLIDEAN CONSTRUCTION OF REGULAR POLYGONS

Figure 9.18

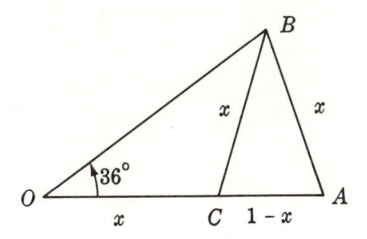

Figure 9.19

We now use the fact that the triangles AOB and ABC are similar to obtain

$$\frac{1}{x} = \frac{x}{1-x}.$$

Hence,

$$x^2 + x - 1 = 0.$$

Thus, we see that each side of the regular decagon has length

$$x = \frac{-1 + \sqrt{5}}{2}.$$

Since x is clearly a constructible number, we conclude that the regular decagon can be constructed with ruler and compass. By joining every second vertex of the regular decagon, we obtain the regular pentagon. Hence, the regular pentagon is also possible to construct with ruler and compass.

The following relationship between the lengths of the sides of the regular n-gon and the sides of the regular $2n$-gon will enable us to use the above formula for the length of the sides of the regular decagon to compute the length of the sides of the regular pentagon.

Proposition 9.14. If the length of each side of the regular n-gon inscribed in a circle of unit radius is denoted by s_n for each integer $n \geq 3$, then

$$s_n^2 = s_{2n}^2(4 - s_{2n}^2).$$

PROOF. In Figure 9.20, we have $s_{2n} = |AB|$, $s_n = 2|BC|$, and $|AD| = 2$.

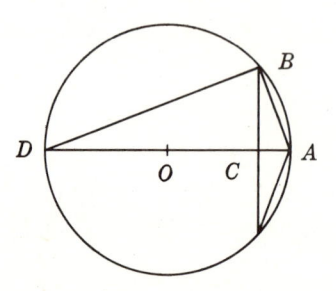

Figure 9.20

These quantities are related by the fact that the area of the right triangle ABD is given by both $\frac{1}{2}|AD| \cdot |BC|$ and $\frac{1}{2}|AB| \cdot |BD|$. Also, the Pythagorean Theorem gives us

$$|BD| = \sqrt{|AD|^2 - |AB|^2} = \sqrt{4 - s_{2n}^2}.$$

Then, equating the two expressions for the area of triangle ABD and squaring both sides, we obtain

$$s_n^2 = s_{2n}^2(4 - s_{2n}^2).$$

This completes the proof.

Since we have previously found that $s_{10} = (-1 + \sqrt{5})/2$, it follows from Proposition 9.14 that the length of each side of the regular pentagon inscribed in a circle of unit radius is

$$s_5 = \frac{\sqrt{10 - 2\sqrt{5}}}{2}.$$

It is interesting that line segments representing the side of the regular decagon and regular pentagon can be obtained simultaneously by a single simple construction. This classical construction discovered by the ancient Greek geometers is performed as follows: Given a circle of unit radius, choose two mutually perpendicular radii and draw the circle centered at the midpoint of one of the radii and that passes through the endpoint of the other radius as shown in Figure 9.21. In terms of the coordinate system, the drawn circle is centered at the point A with coordinates

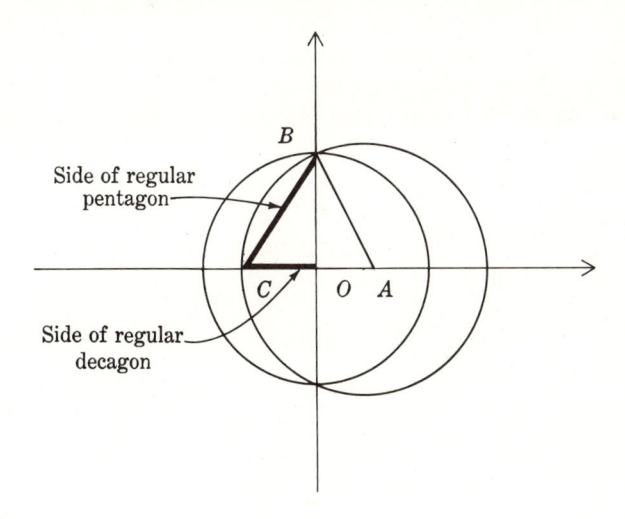

9.3. EUCLIDEAN CONSTRUCTION OF REGULAR POLYGONS

Figure 9.21

($\frac{1}{2}$,0) and passes through the point B with coordinates (0,1). By applying the Pythagorean Theorem, one can show that the line segment BC in Figure 9.21 has length s_5 and the line segment OC has length s_{10}. Thus, we obtain the sides of both the regular pentagon and the regular decagon.

Up to this point, we have shown that the regular n-gon can be constructed with ruler and compass for $n = 3, 4, 5, 6$ and the bisection process produces constructions if these numbers are multiplied by a power of 2. The following proposition shows that known constructions can be used to construct additional regular polygons.

Proposition 9.15. If it is possible to construct a regular n-gon and a regular m-gon with ruler and compass and if m and n are integers that are relatively prime, then it is possible to construct a regular mn-gon with ruler and compass.

PROOF. Since m and n are relatively prime, there exist positive integers a and b such that $an - bm = 1$. (See Proposition 3.8.) Then

$$a\frac{2\pi}{m} - b\frac{2\pi}{n} = \frac{2\pi}{mn}.$$

This states that the angle $2\pi/mn$ for a regular mn-gon is just the difference of positive integer multiples of the angles for the regular m-gon and n-gon. Thus, if the angles $2\pi/m$ and $2\pi/n$ are constructible, then the angle $2\pi/mn$ is also constructible, and the proposition follows.

Example 9.4. The regular 15-gon can be constructed from the regular 3-gon and 5-gon as follows. Since $1 = 2 \cdot 3 - 1 \cdot 5$, we have $2\pi/15 = 2 \cdot (2\pi/5) - 2\pi/3$. We now construct the regular 3-gon and 5-gon such that they have a common vertex A. (See Figure 9.22.) Then, moving counter-clockwise from A, we connect the second vertex of the regular 5-gon (since $2\pi/5$ is multiplied by 2) and the first vertex of the regular 3-gon. This yields the side of the regular 15-gon.

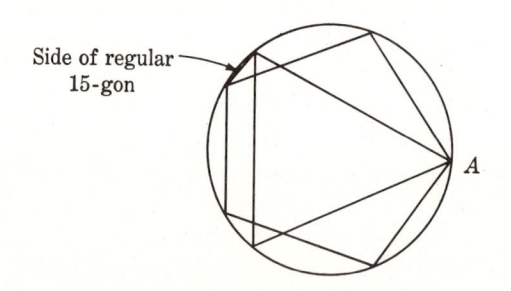

Figure 9.22

We now turn from the possible to the impossible. We first note that the regular 9-gon cannot be constructed with ruler and compass. This assertion follows immediately from the fact that the angle 20 degrees cannot be constructed, that is, the angle 60 degrees cannot be trisected with ruler and compass. For if the regular 9-gon could be constructed, then the angle $40° = 360°/9$ could be constructed, and bisection of this angle would result in a construction of the angle 20 degrees. Hence, it is impossible to construct the regular 9-gon with ruler and compass.

We now prove that the regular 7-gon, or heptagon, is impossible to construct with ruler and compass. This can be approached through trigonometric identities as follows: The central angle subtended by a side of the regular heptagon is $\theta = 2\pi/7$. Since $3\theta + 4\theta = 2\pi$, it follows that $\cos 3\theta = \cos 4\theta$. Utilizing trigonometric identities similar to those used in discussing angle trisection, we can express this equality as

$$4 \cos^3 \theta - 3 \cos \theta = 2(2 \cos^2 \theta - 1)^2 - 1.$$

If we set $x_0 = 2 \cos \theta$, we see that x_0 is a solution of the equation

$$x^4 - x^3 - 4x^2 + 3x + 2 = 0,$$

which can be written as

$$(x - 2)(x^3 + x^2 - 2x - 1) = 0.$$

The number x_0 is a root of this equation if and only if either $x_0 = 2$ or x_0 is a root of the cubic equation $x^3 + x^2 - 2x - 1 = 0$. Now $x_0 = 2$ would mean that $\cos \theta = 1$, that is, $\theta = 0$, which is not the case. Thus, x_0 satisfies $x^3 + x^2 - 2x - 1 = 0$. Application of the Rational Root Test and Proposition 9.11 shows us that this equation has no constructible roots. Thus, $\cos \theta$ is not a constructible number and it follows that the regular heptagon is impossible to construct with ruler and compass.

We conclude our discussion of regular polygons by stating the theorem of Gauss that provides a necessary and sufficient condition for a regular n-gon to be constructible with ruler and compass. Unfortunately, the proof of this beautiful theorem is far too complicated to present here.

Proposition 9.16. *(Gauss)* A regular n-gon can be constructed with ruler and compass if and only if either
    (1) n is a prime number of the form $2^{2^k} + 1$ or a product of distinct primes of this form.
    (2) n is a power of 2.
    (3) n is a product of numbers satisfying (1) and (2).

The significant portion of this theorem is the assertion that a regular polygon having a prime number of sides can be constructed with ruler and compass if and only if the prime is of the form $2^{2^k} + 1$ for some nonnegative integer k. Prime numbers of this form are known as Fermat primes. Although considerable effort has been expended in the search for Fermat primes, only five have been discovered to date, namely, those corresponding to $k = 0, 1, 2, 3, 4$. That is, the only five known Fermat primes are 3, 5, 17, 257, and 65,537, and by Gauss' theorem, the regular n-gons for these values of n can be constructed with ruler and compass. Actual constructions for the last two values mentioned are enormously complicated, but they have been performed. In the mid-nineteenth century, a construction for the regular 257-gon was given that filled a book of 194 pages. Professor Hermes of Lingen worked 10 years to obtain a construction for the regular 65,537-gon.

EXERCISES

9.24. Show that a regular polygon of 18 or 21 sides cannot be constructed with ruler and compass. (Do not use Proposition 9.16.)
9.25. Show how to construct a regular polygon of 12 sides.
9.26. Is it possible to construct a regular polygon of 150 sides? 630 sides?
9.27. Use Gauss' theorem to find all values of $n \leq 100$ such that the regular n-gon can be constructed with ruler and compass.
9.28. Verify that the construction illustrated in Figure 9.21 yields the sides of the regular pentagon and the regular decagon.

9.29. Use Proposition 9.14 to show that the length of each side of the regular octogon inscribed in a circle of unit radius is

$$s_8 = \sqrt{2 - \sqrt{2}}$$

and the length of each side of the regular 16-gon is

$$s_{16} = \sqrt{2 - \sqrt{2 + \sqrt{2}}}.$$

In general, show that if $n = 2^k$ for $k \geq 2$, then

$$s_n = \sqrt{2 - \sqrt{2 + \sqrt{2 + \cdots + \sqrt{2}}}},$$

where there are $k - 1$ square roots.

9.30. Prove that the regular pentagon can be constructed with ruler and compass by proving as follows that the angle 72° can be constructed.
  (a) Show that if $\theta = 72°$, then $\cos 2\theta = \cos 3\theta$.
  (b) Use (a) to derive a cubic equation for which $x_0 = 2 \cos \theta$ is a root.
  (c) Show that the roots of the cubic equation in (b) are all constructible numbers.

APPENDIX: ANGLE TRISECTION BY OTHER MEANS

If the restrictions on ruler and compass constructions are relaxed, there are many ways in which angles can be trisected. The ancient Greek geometers invented several devices and curves for the purpose of solving the problems of doubling the cube, trisecting angles, and squaring circles. Although these solutions were not as "pure" as ruler and compass constructions would be, since they assumed curves other than lines and circles could be drawn, they nevertheless did provide methods that produced solutions. In this appendix we shall describe a few of these "impure" methods of trisecting angles.

The simplest way to trisect a given angle is to allow ourselves the liberty of marking the ruler. The following technique of inserting a line segment of prescribed length is attributed to Archimedes. Let us mark our ruler at two points a distance r apart where r is any positive number. Given an angle 3θ, we first draw a circle of radius r centered at the vertex of the angle. Let A and B be the points on the circle such that $\angle AOB = 3\theta$. We now slide the ruler into a position such that the resulting line passes through B and such that the line segment CD has length r as in Figure 9.23. Then, $\angle OCD = \theta$. To see this, we note that the exterior angle

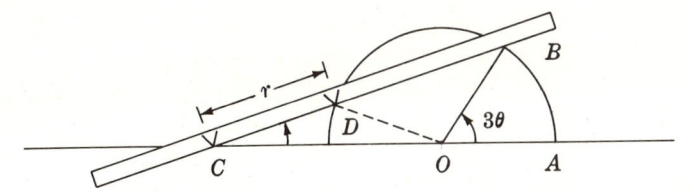

Figure 9.23

∢ODB equals 2∢OCD since triangle OCD is isosceles. Also, ∢AOB = ∢OCD + ∢OBD for a similar reason. Hence,

$$3\theta = ∢AOB = ∢OCD + ∢OBD$$
$$= ∢OCD + 2∢OCD$$
$$= 3∢OCD.$$

Many of the curves with exotic names found in some geometry books were invented for the sole purpose of trisecting angles. We now discuss one of these curves, the *conchoid of Nicomedes*. Nicomedes, who lived around 150 B.C., was very proud of his discovery and even manufactured instruments so that the conchoid could be easily drawn. The locus definition of the conchoid is as follows: Fix a point O and a line L not passing through O, and let b be a given length. Then the conchoid determined by O, L, and b consists of all points P lying on lines drawn through O such that P is at a distance b from L, where the distance between P and L is measured along the line through O. For example, the point C in Figure 9.24 lies on the conchoid since the distance from C to L along the line L' is b. The point A on L' also lies on the conchoid for the same reason.

The conchoid is shown in Figure 9.24 where coordinate axes are drawn with origin at O. Let a denote the distance from 0 to L. (In Figure 9.24 we have $a < b$. If $b \leq a$, then there is no loop on the conchoid.) A point with coordinates (x,y) lies on the conchoid if and only if

$$\frac{\sqrt{x^2 + y^2}}{|x|} = \frac{b}{|x - a|}.$$

See Figure 9.25 for the appropriate similar triangles from which this equality is obtained. A similar picture is valid for points on the other side of L. Thus, we see that the equation of the conchoid with respect to the indicated coordinate system is

$$b^2 x^2 = (a - x)^2 (x^2 + y^2).$$

A conchoid may be used to trisect a given angle as follows. Place the vertex of a given angle ∢AOB at O with points A and B on L as

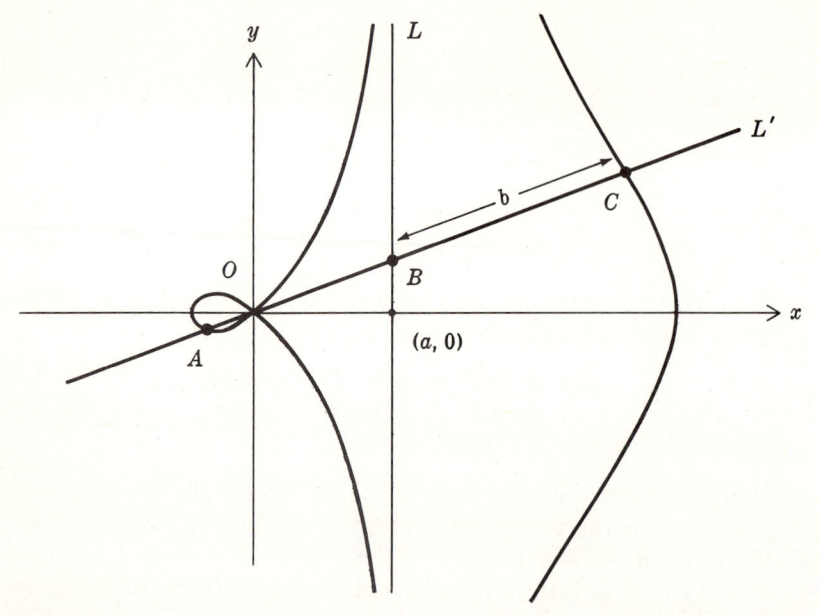

Figure 9.24

shown in Figure 9.26. Now a conchoid is drawn with distance $b = 2|OB|$. The perpendicular to L at B intersects the conchoid at D as in Figure 9.26. Let M be the midpoint of the line segment OD. It can now be shown by using the isosceles triangles BDM and BOM, and exterior angles that $\measuredangle AOD = \frac{1}{3} \measuredangle AOB$. We leave the details to the reader as an exercise.

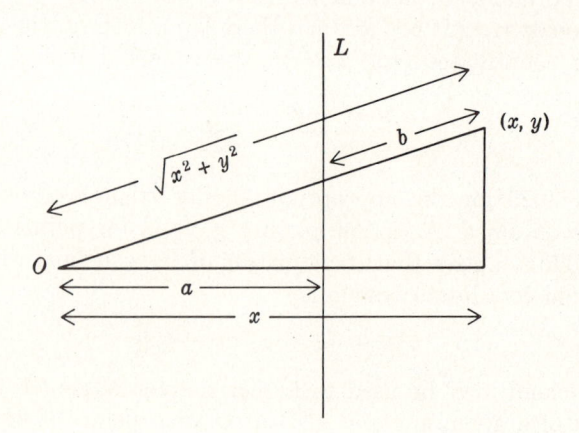

Figure 9.25

APPENDIX: ANGLE TRISECTION BY OTHER MEANS 419

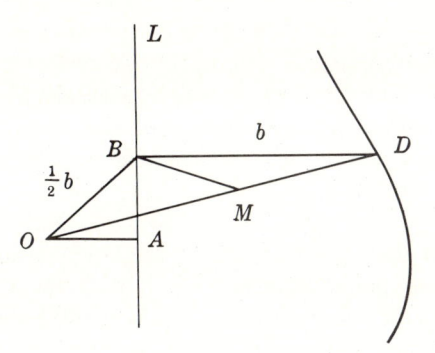

Figure 9.26

It is also possible to use the conic sections for the purpose of trisecting angles. As an example, we show how the parabola $y = x^2$ can be used to this end. Let 3θ be the given angle and let $a = \cos \theta$. Draw the circle C with center $(a, 2)$ that passes through the origin, and let P be the intersection of C and the parabola $y = x^2$. (See Figure 9.27.) Drop a perpendicular from P to the x-axis to obtain the point Q lying on the circle $x^2 + y^2 = 4$. Then the line through Q and the origin determines the angle θ. This can be seen analytically by using the equation $x^2 - 2ax + y^2 - 4y = 0$ for

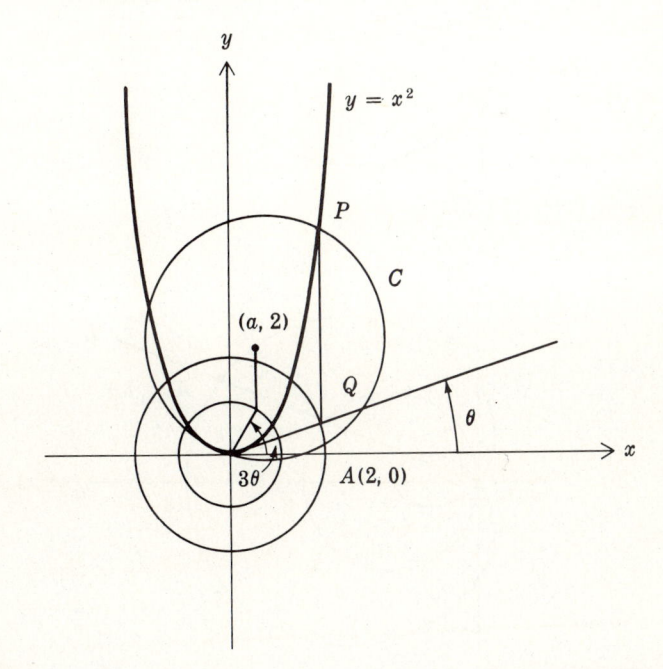

Figure 9.27

420 GEOMETRIC CONSTRUCTIONS

C. Substituting $y = x^2$ into this equation, we obtain $x(x^3 - 3x + 2a) = 0$. Since $x = 0$ corresponds to the origin, the x-coordinate of Q must satisfy $x^3 - 3x + 2a = 0$. Since $a = \cos \theta$, this is precisely the equation derived previously in Section 9.2 in the discussion of angle trisection. The positive solution, which must be the x-coordinate of Q, is just $x = 2 \cos \theta$. Hence, $\measuredangle AOQ = \theta$.

We next show the trisection of any angle is possible if we are given the limacon having polar equation $r = 1 + 2 \cos \theta$. This curve was invented by Pascal for other purposes, but was later shown to permit the trisection of arbitrary angles. Its graph is given in Figure 9.28. This limacon is related to the circle K of unit radius centered at $(1, 0°)$ as follows. Draw the circle K and the limacon and a line L as shown below in Figure 9.29. Then L intersects the limacon at the point A with the polar

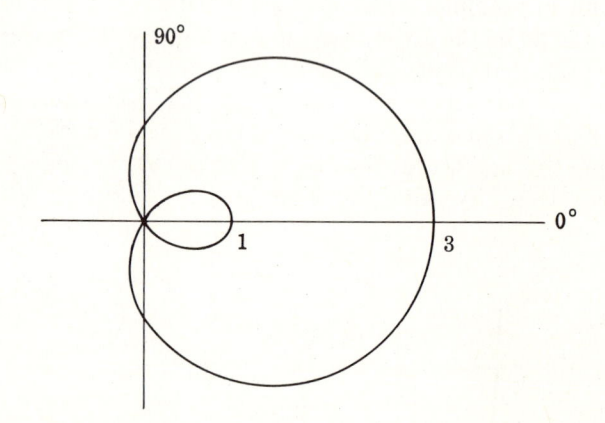

Figure 9.28

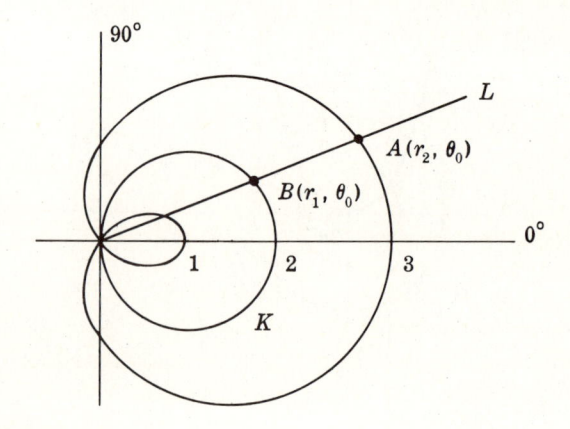

Figure 9.29

coordinates (r_2, θ_0) and the circle at the point B with polar coordinates (r_1, θ_0). The distance between A and B is $r_2 - r_1$. But since the polar equation of the circle is $r = 2 \cos \theta$, this distance is just $r_2 - r_1 = 1 + 2 \cos \theta_0 - 2 \cos \theta_0 = 1$. Thus, the limaçon may be described as the set of points whose distance from the circle K measured along a line through the origin is constantly 1. [*Note:* The same property also holds for points on the inner loop of the limaçon.]

Now suppose an angle $\sphericalangle CBA$ is given where the vertex B is the center of the circle K and the points A, C lie on the limaçon as shown in Figure 9.30. Draw the line L through the origin O and point A. Since

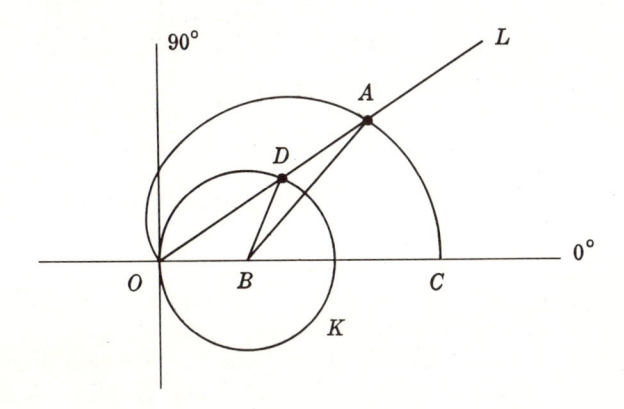

Figure 9.30

$|AD| = |BD| = |BO| = 1$ (see Figure 9.30), the triangles BOD and BDA are isosceles. Therefore, $\sphericalangle DBA = \sphericalangle DAB = \frac{1}{2} \sphericalangle BOD$. Also, $\sphericalangle CBD = 2 \sphericalangle BOD$ since $\sphericalangle CBD$ is an exterior angle of triangle BOD. Hence, $\sphericalangle CBA = \sphericalangle CBD - \sphericalangle CBA = \frac{3}{2} \sphericalangle BOD$ from which it follows that $\sphericalangle BAD = \frac{1}{3} \sphericalangle CBA$.

EXERCISES

9.31. (The cone trisector.) Suppose a right circular cone is constructed with slant height three times as large as the base radius. Given an angle θ, place the cone so that the vertex of the angle is at the center of the base. Now wrap a piece of paper around the cone and mark the points A and B determined by the angle and the vertex V of the cone on the paper. (See Figure 9.31.) Show that when the paper is removed and flattened, the angle $\sphericalangle AVB$ is $\frac{1}{3}\theta$.

9.32. Verify the validity of the construction for trisecting an angle by means of the conchoid.

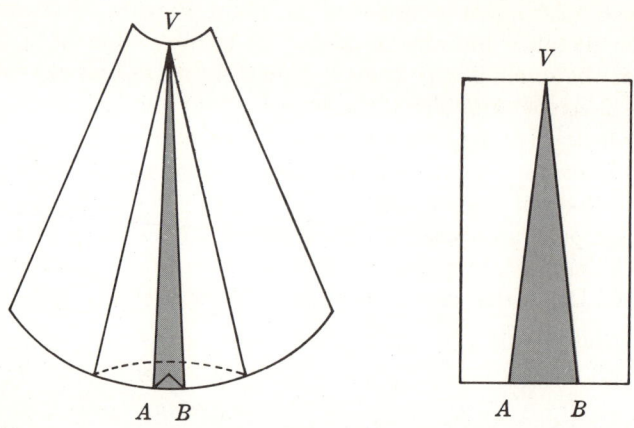

Figure 9.31

9.33. (The hyperbolic trisection.) Let the angle $\measuredangle AOB$ be given where A and B are points on a circle of unit radius and O is the center of the circle, and let the line L bisect the given angle. (See Figure 9.32.) Now draw the curve consisting of those points whose distance from A is twice their distance from L and those points whose distance from B is twice their distance from L. The resulting curve is a hyperbola with directrix L and foci A and B. Show that the angle $\measuredangle COD$ determined by the points C and D of intersection of the circle and the hyperbola is one third as large as $\measuredangle AOB$.

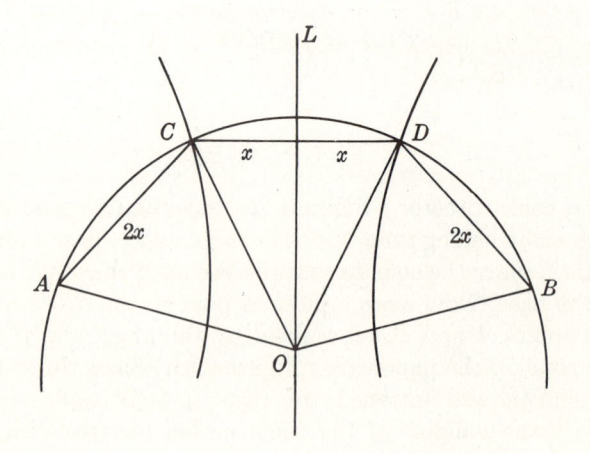

Figure 9.32

9.34. The parabola $y = x^2$ can also be used to find cube roots, and thus will solve the problem of doubling the cube. Let $a > 0$ be given. Draw the circle C with center $(a/2, \frac{1}{2})$ that passes through the origin. Show that the x-coordinate of the intersection of C and the parabola $y = x^2$ is $\sqrt[3]{a}$. (See Figure 9.33.)

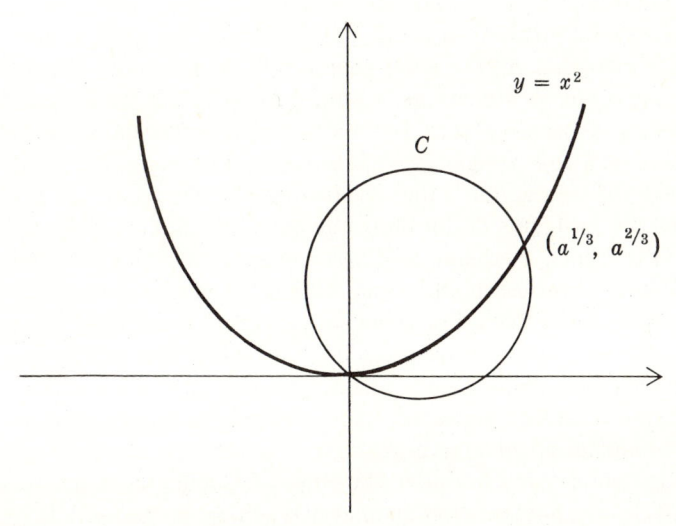

Figure 9.33

REMARKS AND REFERENCES

The study of ruler and compass constructions is one of the most fascinating topics in elementary plane geometry. It provides the student with the technical skill to draw the accurate figures required to discover, understand, and prove geometrical theorems. However, the objectives of the study of ruler and compass constructions go well beyond this descriptive function. The analysis and verification of particular geometric constructions constitute challenging and instructive exercises in the application of the theorems of plane geometry. Moreover, they provide the opportunity to discuss useful geometric ideas and results in a recreational context that is both entertaining and stimulating.

The objective of this chapter has been to develop the algebraic theory of ruler and compass constructions. Every teacher of geometry should be familiar with this aspect of the subject even though the theory of constructible numbers and the impossibility proofs for angle trisection, cube doubling, circle squaring, and so on, are beyond the scope of the standard secondary school geometry program. A teacher who understands these ideas can help his students gain some appreciation of the beauty of this theory without pursuing all of the details.

Frequently, the constructions that we have proved to be impossible are inaccurately described as "unsolved" problems. Also, due to their lack of mathematical experience, students often interpret the term "impossible" to mean "very difficult," akin to putting a man on the moon, for instance. As a result, people still spend many fruitless hours attempting these constructions, thinking of the fame that would follow from solving a problem that has baffled mathematicians for centuries. Still others seek to turn ignorance into profit. For example, a booklet was advertised not long ago in which "solutions" to the trisection problem as well as to the problems of doubling the cube and squaring the circle were offered to teachers, students, and other interested parties for only $1.95. This "scientific breakthrough" that "shook the mathematical world" was even praised in the U.S. Congress and excerpts from the Congressional Record were used in promoting the booklet. Alleged constructions of this sort usually result from a basic misunderstanding of the problem. Those that are not complete nonsense either provide an approximation to the desired construction or else make use of additional data or instrument capabilities not allowable under the rules of Euclidean constructions. In any event, it is important that geometry teachers understand how matters actually stand on this subject.

We shall now list some references for further reading. The SMSG text by Kutuzov [5] considers a variety of questions related to geometric constructions including: (a) the useful construction technique known as the method of two loci, (b) construction techniques with ruler and compass, with compass alone, with ruler alone, and with double-edged ruler, (c) the impossibility of trisecting all angles, doubling cubes, squaring the circle, and constructing the regular heptagon with ruler and compass. Courant and Robbins [2] contains a readable account of geometric constructions including the algebraic theory of constructions, the Mascheroni theorem, linkages, and so on. The discussion of ruler and compass constructions in Fehr [3] includes a simple construction for the regular 17-gon. A proof of Gauss' theorem on the constructibility of regular polygons can be found in the monograph by Klein [4]. The book by Ball and Coxeter [1] is an excellent source of supplementary material and historical information concerning ruler and compass constructions. The short book by Yates [6] contains an extremely nice account of angle trisection, including material on mechanical trisectors, approximations, and several examples of false solutions to the problem.

REMARKS AND REFERENCES

1. Ball, W. W. R. and H. S. M. Coxeter, *Mathematical Recreations and Essays*, New York: The Macmillan Company, 1962.
2. Courant, R. and H. Robbins, *What Is Mathematics?*, New York: Oxford University Press, 1941.
3. Fehr, H., *Secondary Mathematics, A Functional Approach for Teachers*, Boston: D. C. Heath and Company, 1951.
4. Klein, F., W. F. Sheppard, P. A. Macmahon, and L. J. Mordell, *Famous Problems and Other Monographs*, New York: Chelsea Publishing Company, 1962.
5. Kutuzov, B., *Studies in Mathematics, Volume IV: Geometry,* Translated by L. I. Gordon and E. S. Shater. Stanford, California: School Mathemathics Study Group.
6. Yates, R. C., *The Trisection Problem*, Baton Rouge, La.: The Franklin Press, 1942.

INDEX

Absolute value, of a complex number, 91, 325
   in an ordered field, 76
Absorption properties, 245
Addition, Boolean, 248
   modulo m, 32
   principle for counting, 173
Algebraic, number, 42, 399
   system, 31
Anchor-ring surface, 304, 312
AND-network, 256
Angle of a complex number, 326
   trisection, 380, 401, 406, 416
Antireflexive relation, 18
Antisymmetric relation, 13
a posteriori probabilities, 225
a priori probabilities, 225
Archimedes, 416
Archimedean ordered field, 85
Argand, J., 323
Armstrong, J. W., 226
Associative property for, addition and multiplication, 66
   binary operations, 34
   Boolean algebras, 239
   union and intersection, 8
Axiom of, choice, 48
   the empty set, 45
   extensionality, 44
   infinity, 47
   the power set, 46
   specification, 47
   union, 45
   unordered pairs, 46
Axis, imaginary, 325
   real, 325

Banach's matchbox problem, 235
Bartle, R. G., 60
Basic parallel network, 256
   series network, 256
   solutions of a congruence, 144
Bayes' Theorem, 224
Bernoulli, Jakob, 189, 233
Bernoulli trials, 233
Bilinear transformation, 352
Binary, adder, 263, 268
   digit, 264
   operation, 29
   representation, 263
Binomial coefficients, 164
Binomial Theorem, 168
Birthday problem, 217
Bolyai, J., 372
Boole, G., 238, 239
Boolean, addition, 248
   algebra, 11, 239
   function, 243
   sum, 69, 247
Boolean formula, 243
   dual, 244
   equivalent, 243

Bound, greatest lower, 86
　least upper, 86
　lower, 86
　upper, 86
Bounded, above, 86
　below, 86
Bridge hand, 215

Cancellation property, 66
Canonical prime factorization, 117
Cantor, G., 35, 37, 42
Cantor's hotel, 36
Cap operation, 240
Cardano, G., 189
Carry, outputs, 265
　switches, 265
Cartesian, plane, 12
　product of sets, 12
Casting out nines, 138
Cauchy, A., 94
Cayley, A., 306, 307
Certain event, 194
Charged state, 265
Chinese remainder theorem, 152
Choice function, 48
Circuit, horizontal, 303
　minimal, 294
　simple, 294
　vertical, 302
Clock arithmetic, 108
Closed, operation, 30
　state, 256
Coconut problem, 148
Cohen, P. J., 48
Colored cubes problem, 316
Combination, 174
Commutative ring with unit, 72
Commutative property for, addition, 66
　binary operations, 34
　Boolean algebras, 239
　multiplication, 66
　union and intersection, 8
Compatability property for, order and addition, 66
　order and multiplication, 66
Complement, 5, 240

Complementary event, 195
Complete, graph, 292
　ordered field, 87
Completeness property, 87
Complex, conjugate, 91, 325
　numbers, 89, 94
　plane, 325
Composite, function, 26
　number, 115
Computer, 268
　human, 269
Conchoid of Nicomedes, 417
Conditional probability, 220
Cone trisector, 421
Congruence, linear, 144
　preserving transformation, 363
　reduced, 146
Congruent integers, 21, 137
Connected graph, 285
Connectivity, 279
Constant, 2
Constructible, number, 390
　point, 391
Convex polyhedron, 296
Countable additivity, 208
Countably infinite set, 36
Cross ratio, 369
Cup operation, 240

Decagon, 410, 412
Decimal, 95
　fraction, 95
　periodic, 101
　representation, 95, 263
　terminating, 101
Degree of a vertex, 283
Delian problem, 381
De Moivre, A., 189
De Moivre's formula, 332
De Morgan, A., 306
De Morgan Identities, 10, 245
Denumerable set, 36
Dependent events, 228
Derangement, 185
Descartes, R., 276
Determinant of a transformation, 352
Dictionary order, 17

Diophantine equations, 130
Diophantus, 130
Disjoint sets, 9
Distributive property for, Boolean
  algebras, 240
 integers, 66
 least common multiples and greatest
   common divisors, 129
 unions and intersections, 8
Divisibility tests, 138
Division, algorithm, 78
Divisor, 114
 in a Boolean algebra, 251
 greatest common, 122
Domain, of a function, 24
 integral, 72
 ordered, 72
 of a variable, 2
Doubling the cube, 381, 401, 406, 416
Duality principle, 244

Edge, 282
 chain, 285
 condition, 166
Element, 2
 integral, 77
 prime, 251
 rational, 84
Elementary transformation, 344, 349
Empty set, 5, 45
Engelbart, D. C., 269
Equally likely probability measure, 202
Equivalence, class, 20
 relation, 19
Equivalent, Boolean formulas, 243
 networks, 259
Erlangen Programm, 323, 378
Euclid, 113, 118, 121, 372
Euclid's lemma, 126
 for Boolean algebras, 252
Euclidean, algorithm, 123
 construction, 384
 isometry, 363
Euler, L., 113, 121, 275, 276
Euler's formula, 276, 295, 303
Even vertex, 283

Events, 194, 209
 certain, 194
 complementary, 195
 dependent, 228
 impossible, 194
 independent, 228
 mutually exclusive, 195
Extended, complex number system, 339
 complex plane, 339
Extension field, 393
 quadratic, 393
Exterior face, 295
Extraordinary set, 43

Face, 295
 exterior, 295
 interior, 294
Factor, 114
False theorem, 62
Ferguson, W. A., 136
Fermat, Pierre de, 113, 121, 160, 189
Fermat primes, 121, 415
Fermat's, last theorem, 113, 160
 theorem, 142
Field, 80
 Archimedean ordered, 85
 ordered, 83
 σ-, 209
Finite, additivity property, 200
 set, 36
Five color theorem, 311
Fixed point, 356
Flush, 177
Four color problem, 276, 306
Four letter word, 206
Four-of-a-kind, 177
F-point, 394
Fractions, 20
Fraenkel, A., 49
Frege, G., 42
Full house, 176
Function, 24
 Boolean, 243
 composite, 26
 inverse, 28
 left inverse, 27

Function (*continued*)
  one-to-one, 27
  onto, 27
  right inverse, 27
Fundamental theorem of arithmetic, 116

Galileo, 189
Galois, E., 382
Gauss, K. F., 59, 113, 120, 323, 372, 382, 409, 415, 424
Gillies, D., 120
Glide reflection, 366
Gödel, K., 49
Golden needle problem, 60
Graph, 282
  complete, 292
  connected, 285
  isomorphic, 284
  minimal complete, 305
  planar, 293
  utilities, 299
Greatest, common divisor, 122, 124, 241
  integer function, 25
  lower bound, 86

Hamilton, W. R., 94, 289
Hamiltonian lines, 289
Hamilton's game, 276, 289
Heawood, P. J., 306
Heptagon, 381, 414
Highway inspection problem, 290
Homology, 346
Huygens, C., 189
Hyperbolic trisection, 422

Idempotent properties for, Boolean algebras, 244
  union and intersection, 10
Identity function, 27
ILLIAC II, 120
Image of a set, 25
Imaginary, axis, 325
  part, 91
Impossible event, 194
Inclusion relation, 17

Independent events, 228
Infimum, 86
Infinite, point, 373
  probability spaces, 207
  set, 36
Instant Insanity, 321
Integral, domain, 72
  element, 77
Interior face, 294
Intersection of sets, 5
Inverse, element, 80
  function, 28
  left, 27
  right, 27
Inversion, in an arbitrary circle, 360
  in the unit circle, 346
Irrationality of $\sqrt{2}$, 109
Isometry, direct, 364, 376
  Euclidean, 363
  non-Euclidean, 376
  opposite, 364, 376
Isomorphism, for algebraic systems, 33
  for graphs, 284

Join operation, 240

Kempe, A. B., 306
Klein, F., 323, 378
Kolmogorov, A. N., 191
Königsberg bridge problem, 275, 276
k-permutation, 173
Kreisverwandtschaften, 354
k-subset, 174

Laplace, Pierre de, 189
Least, common multiple, 128, 241
  upper bound, 86
Lebesgue, Henri, 209
Lebesgue, measurable sets, 209
  measure, 209
Leibnitz, G. W., 189
Leonardo da Vinci, 384
Lexicographic order, 17
Limacon, 420
Lindemann, F., 381
Line, non-Euclidean, 373
  reflection, 366

Linear congruence, 144
Loaded die, 201
Logical connectives, 7
Long division, 99
Loop, 282
Lower bound, 86

Map, 307
   regular, 308
Mapping, 24
Mascheroni, L., 383
Mascheroni Theorem, 377, 383
Match, 185, 205
Mathematical Induction, Principle of, 55
Meet operations, 240
Méré, Chevalier de, 189, 234
Mersenne, Marin, 120
Mersenne prime, 120
Method of infinite descent, 158
Minimal, circuit, 294
   complete graph, 305
Mistake, 114, 382
Möbius, A. F., 306, 323, 354
Möbius, strip, 313
   transformation, 352
Modular identities for sets, 10
Modulus of a complex number, 91
Morse code, 180
M-prime, 118
Multinomial coefficients, 179
Multiple, 114
   least common, 128
Multiplication, modulo m, 32
   principle for counting, 173
Mutually exclusive events, 195

Natural number system, 54
n-colorable, 307
Negative elements, 66, 73
Network, AND-, 256
   basic series, 256
   basic parallel, 256
   equivalent, 259
   OR-, 256
   series-parallel, 257
Neumann, J. von, 49

Nicomedes, 417
Non-Euclidean, geometry, 372
   isometries, 376
Nonseparating points, 371
North pole of the Riemann sphere, 336
nth roots of unity, 333
Number, algebraic, 42, 399
   constructible, 390
   prime, 115
   transcendental, 42, 381, 399

Occupancy problems, 217
Odd vertex, 283
One-to-one, correspondence, 28
   function, 27
Onto function, 27
Open state for switches, 256
Open statement, 3
Operation, binary, 29
   closed, 30
   unary, 29
Order relation, 16
   strict, 18
   usual, 16
Ordered, domain, 72
   field, 83
   pair, 12
   sample, 212
   sample space, 212
Ordinary, point, 373
   set, 43
Ore and Stemple, 306, 311
Orientable surface, 313
OR-network, 256
Orthogonal circles and lines, 342
Orthogonality preserving, 342

Palindrome, 141
Parallel, lines (non-Euclidean), 373
   network, 256
   postulate, 372
Partition of a set, 22
Pascal, Blaise, 163, 166, 189
Pascal's Triangle, 165
Path, 285
Peano axioms, 54
Peano, G., 54

Pentagon, 410, 412
Perfect number, 120
Pericles, 381
Periodic decimal, 101
Permutation, 173
Planar graph, 293
Platonic bodies, 297
Poincaré, H., 372
Point, infinite, 373
   ordinary, 373
Point at infinity, 338
Point probability, 200
Poker hand, 176, 203
Polar form of a complex number, 326
Pollack, Barry, 160
Polyhedron, 296
   convex, 296
   regular, 296
Positive cone, 79
Power set, 5, 46
Preimage, 25
Prime, element in a Boolean algebra, 251, 255
   Fermat, 121, 415
   M-, 118
   Mersenne, 120
   number, 115
   pairs, 121
Prime factorization, 117
   in a Boolean algebra, 252, 255
Prime Number Theorem, 120
Primitive Pythagorean triple, 156
Principle, of Duality, 244
   of Inclusion-Exclusion, 182, 184
   of Mathematical Induction, 55
Probability, *a posteriori*, 225
   a priori, 225
   conditional, 220
Probability measure, 199
   equally likely, 202
   product, 232
Probability space, 199
   infinite, 207
Problem, Banach's matchbox, 235
   birthday, 217
   coconut, 148
   colored cubes, 316

four color, 276, 306
golden needle, 60
highway inspection, 290
Königsberg bridge, 275
matching, 205
occupancy, 217
shortest path, 169
traveling salesman, 290
utilities, 299
waiting, 210
Product measure, 232
Proper subset, 18
Properly colored maps, 305
Property, Archimedean, 85
   completeness, 87
   nested interval, 88
   well-ordering, 62
Pythagoras, 112, 380
Pythagorean, equation, 156
   triple, 156

Quine, W. V., 44

Range of a function, 24
Rational, element, 84
   point, 391
   root test, 127, 406
Real axis, 325
Real part of a complex number, 91
Recursion formula for binomial coefficients, 166
Reflection, 346
   glide, 366
   line, 366
Regular, map, 308
   n-gon, 409
   polygon, 381, 409
   polyhedron, 296
Relation, 13
   antisymmetric, 13
   equivalence, 19
   inclusion, 17
   order, 16
   reflexive, 13
   symmetric, 13
   transitive, 13
Relative frequency, 190

INDEX

Repeated trials, 230
Representation theorem for finite Boolean algebras, 253
Representative of equivalence class, 20
Residue class, 22, 31
Riemann, G. F. B., 323
Riemann sphere, 324, 337
Roots, of complex numbers, 331
   of unity, 333
Rotation, 346
Rotation-dilation, 345
Ruler, 380
Russell, Bertrand, 42
Russell's paradox, 43, 47

Sample, ordered, 212
   unordered, 212
Sample space, 192
Sampling, 196, 212
   with replacement, 196, 215
   without replacement, 196, 212
Selfridge, John, 160
Separating points, 371
Series network, 256
Series-parallel network, 257
Set, 2
   countably infinite, 36
   denumerable, 36
   empty, 5
   extraordinary, 43
   finite, 36
   infinite, 36
   ordinary, 43
Shadow isomorphism, 296, 301, 312
Shannon, C., 239
Shortest paths, 169
Sieve of Eratosthenes, 115
σ-field of sets, 209
Simple circuit, 294
Solution set, 3
South pole of the Riemann sphere, 336
Square-free positive integer, 242
Squaring the circle, 381, 401
State, charged, 265
   uncharged, 265
Statement, 3
   open, 3

Steiner, J., 384
Stereographic projection, 337
Stone, M. H., 254
Straight-edge, 380
Strict order relation, 18
Subset, 4
   proper, 18
Supremum, 86
Switches, two-state, 255
Symmetric difference of sets, 69
Symmetric, points, 359
   relation, 13

Terminating decimal, 101
Terms of a decimal representation, 95
Theaetetus, 297
Theorem, Bayes', 224
   binomial, 168
   Chinese remainder, 150, 152
   Fermat's, 142
   Fermat's last, 113, 160
   five color, 311
   Gauss', 415
   prime number, 120
Torus, 302, 312
Total order property, 66
Trace, 285
Transcendence of π, 381
Transcendental number, 42, 381, 399
Transformation, bilinear, 352
   elementary, 344, 349
   Möbius, 352
Transitive relation, 13
Translation, 344
Traveling salesman, 290
Trials, Bernoulli, 233
   repeated, 230
Trisection of angles, 380, 401, 406, 416
Truth set, 3
Truth table, 7
Tuckerman, B., 121
Twin primes, 121
Two-state switches, 255

Unary operation, 29
Uncharged state, 265
Union of sets, 5

Unit element, 66, 73
  of a Boolean algebra, 240
University of Illinois, 2, 120
Unordered sample, 212
Upper bound, 86
  least, 86
Usual order relation, 16
Utilities graph, 299
Utilities problem, 299

Values, of a function, 24
  of a variable, 3
Variable, 2
Vector, 21, 376
Venn diagram, 5
Vertical circuit, 302

Vertices, 282
  even, 283
  odd, 283

Well-ordering property, 62, 66
Wessel, C., 323
Whitehead, Alfred North, 44
With replacement, 196, 215
Without replacement, 196, 212
Word, 181

Zermelo, Ernst, 44, 49
Zermelo-Fraenkel axioms, 49
Zero divisor, 74
Zero element, 66, 73
  of a Boolean algebra, 240